AUTOMOBILE ELECTRICAL AND ELECTRONIC EQUIPMENT

Theory and Practice for Students, Designers, Automobile Electricians and Motorists

A. P. YOUNG
OBE, MIEE, MIMechE, FIWM

and

L. GRIFFITHS
MIMechE, AMIEE

Revised by

G. E. FARDON
CEng, MIMechE, MIEE

Butterworths
London - Boston - Durban - Singapore - Sydney - Toronto - Wellington

First published 1933
Eighth edition 1970
   reprinted 1977, 1978
Ninth edition 1980
   reprinted 1981
   reprinted (with corrections and revisions) 1986

   © A. P. Young and L. Griffiths, 1980

British Library Cataloguing in Publication Data

Young, Arthur Primrose
   Automobile electrical and electronic equipment.
   – 9th ed.
   1. Automobiles – Electrical equipment
   I. Title   II. Griffiths, Leonard
   III. Fardon, George Edward
   IV. Automobile electrical equipment
   629.2'54   TL272

   ISBN 0-408-00463-0

Typeset by Butterworths Litho Preparation Department
Printed in England by Butler & Tanner Ltd, Frome, Somerset

PREFACE

In this ninth edition extensive revision has been made and additional chapters have been included to greatly extend the scope of the book.

Attention has been given to the use of automobile equipment in world-wide environments and particularly in European Common Market countries as well as the UK. The increasing use of electronic devices has been given special attention, particularly those dealing with such items as alternator control, automatic transmissions, instrumentation, ignition and fuel injection systems.

To assist the student, the number of illustrations, line drawings and diagrams have been greatly increased and the subject matter dealt with from a theoretical angle to meet the requirements of students taking HNC and the City and Guilds courses, Motor Vehicle Craft Studies 380/381 and Motor Vehicle Technicians Certificate 390. At the same time additional practical and constructional features have been included.

G. E. Fardon

ACKNOWLEDGEMENTS

The authors and revisor wish to acknowledge the valuable assistance given by the following for the provision of technical information and illustrative material.

Mr. G. H. Townsend, Coventry College of Further Education
Alpinair Ltd.
Associated Engineering Developments, Ltd.
AC-Delco, Ltd.
Robert Bosch, Ltd. (Mr. A. Roger)
British Leyland
Brown Bros. Ltd.
Butec Electrics Ltd.
CAV Ltd.
Chrysler UK Ltd.
Champion Sparking Plug Co. Ltd.
City & Guilds of London Institute
Crypton-Triangle Ltd.
e.d. Veglia UK, (Mr. J. Wood)
Ford Motor Co. Ltd.
Inertia Switches Ltd.
Javelin Electronics Ltd.
Kenlowe Accessories & Co. Ltd.
Kistler Instruments Ltd.
Klamix Marketing Services
Lucas Electrical Ltd. (Messrs. A. Cox, L. J. Nevett, & W. A. Clarke)
Lucas Kienzle Instruments Ltd.
Lumenition Ltd.
S.E.V. Marchal Ltd.
Mobilec Ltd.
Otter Controls Ltd.
Renault Ltd.
Ripaults Ltd.
Rists Wires & Cables Ltd.
Rolls Royce Motors Ltd.
Siba Electric Ltd.
Smiths Industries Ltd.
Stellar Components Ltd.
Telma Retarders (Mr. Leslie F. Hall)
Veglia Borletti Ltd.
Watchdog Products Ltd.

CONTENTS

1. Fundamental principles   1
2. The complete electrical equipment, including instrumentation   36
3. The alternator and d.c. generator   88
4. The starting motor   148
5. The battery   183
6. Automobile lighting and signalling   206
7. Electrical ignition systems   251
8. Magnetos   288
9. Sparking plugs   292
10. Ancillary equipment   307
11. Switches and switchgear   332
12. Wiring harnesses   353
13. Heating, ventilating and air conditioning   365
14. Fuel injection equipment   379
15. Electrical control in automatic transmission systems   409
16. Radio interference suppression   428
17. Electro-magnetic eddy current brakes   439
Appendix   451
Index   454

CHAPTER 1

FUNDAMENTAL PRINCIPLES

Although the effects of static electricity were known to the Greeks 2500 years ago, it is only within the last century that the science of electricity has been applied to any appreciable extent in the service of mankind. This now has a tremendous effect on the daily lives of the world's population.

Following Faraday's discovery in 1831 of electro-magnetic induction, the generation and widespread use of electricity became commercially possible and has also resulted in the development of a vast electrical engineering industry supplying light and power to every kind of industry from the oldest forms of agriculture to space travel and communications.

The development of electrical equipment for all forms of transport has also followed this discovery of electro-magnetic induction which forms the basic principle of automobile ignition equipment, the starting motor, the lighting generator and numerous electro-mechanical and electro-magnetic devices incorporated in the automobile.

Nature of electricity

Scientists throughout this century have contributed in outstanding measure to our present knowledge of electricity. In fact, they have completely revolutionised our ideas, not only in regard to the nature of electricity, but also in regard to matter.

Hitherto, the atom was regarded as the ultimate sub-division of matter. Today the atom is regarded as a purely electrical system, in which a nucleus of positively charged particles called *protons* is surrounded by still smaller negatively charged units or *electrons*. In other words, matter is now considered to be made up of electricity, the smallest constituent being the electron, the size of which is so infinitesimal that many millions of electrons would be required to fill the space occupied by the finest speck of dust.

1

Conductors of electricity

According to the electronic theory, the electron is responsible for a flow or current of electricity. Good conductors are considered to be those substances in which there are present free electrons in constant but indiscriminate motion between the atoms. Under the action of an electromotive force, these free electrons are caused to move in some definite direction, resulting in a constant stream of electrons flowing at a phenomenally rapid rate in the conductor. In the case of a uni-directional or direct current (d.c.), this electronic stream is always in one direction, whilst with an alternating current (a.c.) the electronic stream reverses its direction of motion with regular frequency.

All pure metals are good conductors of electricity, silver being the best since it offers the least resistance to a flow or current of electricity. Copper is very nearly as good a conductor as silver and being very much cheaper is extensively used for electrical apparatus of all kinds. Until recently, copper was universally employed in all automobile electrical equipment, but quite extensive use is now made of anodised aluminium foil for heavy current windings.

The electric system and Ohm's Law

The three fundamental characteristics of the electric system are pressure, current and resistance, and there is a definite relationship between them, as expressed by Ohm's Law for a direct current circuit. This states that the current is directly proportional to the pressure and inversely pro-portional to the resistance. We may express this law as an equation, thus:

$$I = E/R$$

where I = current in amperes (A)
         E = pressure or electromotive force in volts (V)
         R = resistance in ohms (Ω).

If the values of any two of the factors involved are known, it is quite evident from the equation that the value of the third can be readily determined. If, for example, we know the amperes and volts, we have only to re-write the equation as:

$$R = E/I$$

to determine the resistance in ohms. Further, should we desire to determine the electromotive force (e.m.f.) in volts, the equation is then written as:

$$E = IR$$

Practical units of electricity

It is advisable at this stage to define the three fundamental characteristics of the electric circuit and also other practical units with which the reader will become acquainted in dealing with the subject of automobile electrical equipment:

(*a*) *Pressure or electromotive force*. The practical unit is the *volt*, which from Ohm's Law may be defined as the value of e.m.f. necessary to produce a current of 1 A through a resistance of 1 ohm. It is also defined as 0.6974 of the voltage of a standard Clark cell at 15°C.

(*b*) *Current*. The *ampere* is the unit of current and is defined as that value of unvarying current which will deposit silver at a rate of 0.001118 gramme per sec, when passed through a specified solution of silver nitrate.

(*c*) *Resistance*. The *ohm* is the unit of resistance and is equal to the resistance of a column of mercury of uniform cross-section, 106.3 cm in length, having a mass of 14.452 grammes at 0° C.

(*d*) *Quantity*. Unit quantity of electricity passes when a current of 1 A flows for 1 sec, this unit being termed a *coulomb*. The product of the current in amperes and the time in seconds therefore gives the quantity in coulombs.

(*e*) *Power*. Power is the rate of doing work and the electrical unit of power is the *watt*. Electrical power in watts is for direct current the product of the e.m.f. in volts and the current in amperes.

Thus $W = EI$

where W = power in watts
E = e.m.f. in volts
I = current in amperes.

Where the power is great, as in the case of public supply undertakings, the unit adopted is the kilowatt, which is equal to 1000 W.

The mechanical unit of power is the horse-power and is equal to 746 W and from this we get:

$$1 \text{ kW} = 1000/746 = 1.34 \text{ hp}$$

The use of this mechanical unit of power (h.p. is becoming obsolete with the introduction of the SI (international) system and the Kilowatt is now universally used.

(*f*) *Energy*. Energy is the product of power and time, the electrical unit being the *joule* or *watt-second*. This is a very small unit and is only employed where the energy is quite small. A more convenient unit is the *watt-hour*, which is equal to 3600 joules.

The kilowatt-hour is the unit of electrical energy standardised through-out the world for the measurement of electrical energy supplied for power and lighting.

Resistance of conductors

Even the best conductors offer some resistance to the flow of electricity, in consequence of which there is a drop of electrical pressure in the conductor. Where heavy currents are concerned this pressure or voltage drop may be appreciable, as the following example will show.

A starting motor is connected to a 12-V battery by cables having a total resistance of 0.005 ohm. On closing the starter switch, an initial current of 200 A flows in the cables, as a result of which there is a voltage drop in the cables of 1 V, since $E = IR = 200 \times 0.005 = 1V$. Therefore, instead of 12 V being applied at the motor terminals, we only have 11 V, the other volt being expended in urging the current of 200 A through a cable resistance of 0.005 ohm.

The resistance of any conductor is directly proportional to its length and specific resistance, and is inversely proportional to the cross-sectional area of the conductor. The specific resistance ρ is a constant depending upon the nature of the material employed as the conductor, and is defined as *the resistance between the two opposite faces of a cube maintained at $0°$ C and having sides each one centimetre long.* For all metals, the value of the specific resistance is extremely small, and it is usual to express specific resistance in *microhms*, or millionths of an ohm. For copper, the specific resistance value is 1.59 microhms; for pure nickel 10.5; whilst with alloys specially manufactured for use as resistances the value varies from 30 to 100 microhms.

Therefore, in order to keep the resistance of any conductor at a minimum, it is necessary to use the minimum length of a material having the lowest specific resistance and the largest cross-sectional area com-mensurate with cost. In the case of starting motor circuits, every consideration should be given to keeping the length of the cables as short as possible.

Formula for conductor resistance

The resistance of a conductor may be computed from the formula:

$$R = \rho \times \frac{l}{A}$$

where R = resistance in ohms
        l = length in cm
        A = cross-sectional area in cm^2
        ρ = specific resistance in ohms.

Let us illustrate the foregoing by a numerical example. Find the resistance of a No. 33 SWG copper conductor 1000 cm long:

$$\text{Diameter of No. 33 SWG} = 0.001 \text{ in or } 0.0254 \text{ cm}$$

$$\text{Cross-sectional area,} \quad A = \frac{\pi D^2}{4} = \frac{3.14 \times 0.0254^2}{4}$$

$$= 0.000506 \text{ cm}^2$$
$$l = 1000 \text{ cm}$$
$$\rho = 1.59 \text{ microhms at } 0° \text{ C.}$$

Therefore:

$$R = \rho \frac{l}{A} = \frac{1.59}{1\,000\,000} \times \frac{1000}{0.000506} = 3.15 \text{ ohms at } 0°\text{C.}$$

Effect of temperature on conductor resistance

With but few exceptions, conductors of electricity increase their resistance with increase in temperature. Metals such as silver, copper and aluminium increase their resistance by approximately 0.4% for each degree Centigrade rise in temperature. In the case of certain alloys, such as Eureka and Manganin, the resistance is, for all practical purposes, unaffected by change in temperature.

Table 1.1 SPECIFIC RESISTANCES AND TEMPERATURE
COEFFICIENTS OF VARIOUS METALS

Material	*Specific resistance (ρ): microhms per cm^3 at $0°$ C*	*Temperature coefficient (a)*
Silver	1.54	+ 0.004
Copper	1.59	+ 0.00428
Aluminium	3.05	+ 0.00423
Iron	8.96	+ 0.00625
Nickel	12.12	+ 0.00623
German Silver (Cu + Zn + Ni)	32.3	+ 0.00027
Manganin (Cu + Mn + Ni)	47.6	+ 0.00002
Eureka (Cu + Ni)	49.5	Zero
Carbon	3000 to 40 000	− 0.0005

Such alloys are specially used for electrical measuring instruments, in order that the accuracy of the instrument shall not be affected by temperature variation. Carbon is noteworthy for its decrease in resistance with increase in temperature.

The resistance of a conductor at any temperature can be calculated from the formula:

$$R_T = R_0(1 + aT)$$

where R_T = resistance at $T°$ C
R_0 = resistance at $0°$ C
a = temperature coefficient of the material.

The temperature coefficient is the change in the resistance per $1°$ C change in temperature occurring when the resistance at $0°$ C is 1 ohm. In Table 1.1 above are given the specific resistances and temperature coefficients of a number of well-known materials.

Current-carrying conductors or cables generally operate at a temperature which is higher than that of the surrounding air — normally $15°$ C — whilst the windings of automobile electrical apparatus are frequently operated at temperatures as high as $80°$ C.

When the apparatus is cold the winding temperature is equal to that of the surrounding air, say $15°$ C. The resistances as then measured relate to this temperature. When operating under normal conditions, the windings are raised to some higher temperature and, by again measuring the resistances under the steady hot condition, it is possible to calculate the temperature rise in the windings as follows:

$$R_{T2} = R_0(1 + aT_2)$$
$$R_{T1} = R_0(1 + aT_1)$$

where R_{T2} = resistance under hot condition
T_2 = temperature corresponding to R_{T2}
R_{T1} = resistance under cold condition
T_1 = temperature corresponding to R_{T1}.

We get from these two fundamental equations the following:

$$R_{T2} - R_{T1} = R_0a(T_2 - T_1)$$

or

$$T_2 - T_1 = \text{temperature rise} = \frac{R_{T2} - R_{T1}}{R_0a}$$

Substituting for R_0 the expression $\dfrac{R_{T1}}{1 + aT_1}$ we finally get:

$$T_2 - T_1 = \left(\frac{1 + aT_1}{a}\right)\left(\frac{R_{T2} - R_{T1}}{R_{T1}}\right)$$

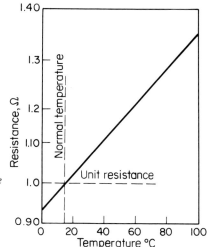

Figure 1.1 Resistance-temperature curve for copper

For copper windings $a = 0.0043$, and if we assume that T_1 is normally $15°$ C, this equation can be simplified to:

$$\text{Temperature rise } °C = 250\left(\frac{R_{T2} - R_{T1}}{R_{T1}}\right) \text{ approximately}$$

In other words, *the temperature rise is approximately the percentage increase in resistance of the winding multiplied by 2.5.*

For example, if a winding when cold (at air temperature of $15°$ C) has a resistance of 1 ohm, and the resistance when hot is found to be 1.28 ohms, then:

$$\text{Increase in resistance} = 28\%$$
$$\text{Approximate temperature rise} = 2.5 \times 28 = 70° \text{ C.}$$

This means that the actual temperature of the winding when hot is $85°$ C.

The straight-line curve shown in Figure 1.1 for all practical purposes provides a simple method of determining the change in resistance of any conductor or winding for a given rise in temperature. Suppose, for example, we know that a winding has a resistance of 3.56 ohms at normal temperature ($15°$ C) and we desire to know what the resistance will be at $55°$ C. By taking the resistance value from the curve for this temperature and multiplying it by the normal temperature resistance of the winding, we get the required resistance value, thus:

$$1.16 \times 3.56 = 4.13 \text{ ohms.}$$

Temperature rise of conductors and windings

Whenever electricity flows in a conductor, heat is generated resulting in a rise in temperature of the conductor. The heat so generated is directly proportional to the resistance value and to the square of the current value. In order to avoid high temperatures likely to damage the insulation around the conductor, it is therefore necessary to keep the current density within prescribed limits.

For small sized conductors of relatively short length, a safe working current density is 4000 A/sq in. In the case of electrical machinery, it is usual to limit the temperature rise of the windings when operating continuously at full load to 50° C. In determining the temperature rise, the resistance of the windings is measured at normal air temperature and also after the machine has been operating for some hours at full load. The rise in temperature in ° C is then computed from these values by multiplying the percentage increase by 2.5.

Resistances in series and in parallel

When resistances are connected in series as shown in Figure 1.2, the combined resistance is the sum of the individual resistances. Thus, if R_1 and R_2 have values of 3 and 6 ohms respectively, then the combined resistance is $3 + 6 = 9$ ohms.

In Figure 1.3, the resistances are shown connected in parallel, which is the general method of connecting up the lamps to the battery in a car. In this case, the combined resistance is determined from the formula:

$$R = \frac{1}{\frac{1}{R_1} + \frac{1}{R_2}}$$

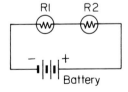

Figure 1.2   Resistances connected in series

Figure 1.3   Resistances connected in parallel

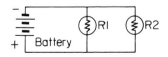

and if we again assume the resistances R_1 and R_2 to be 3 and 6 ohms respectively, then the combined resistance will be:

$$R = \frac{1}{\frac{1}{3} + \frac{1}{6}} = 2 \text{ ohms.}$$

Magnetism

If we were to take a steel bar about 200 mm long and 6 mm diameter and magnetise it, we should find that the bar (which is now a magnet) has had communicated to it properties which were not previously possessed by the steel. In the first place, the ends of the bar are able to attract to themselves pieces of iron, and the force of attraction is so appreciable that quite an effort has to be exerted to remove from these ends any pieces of iron that may have adhered. Secondly, if suspended at its centre by a fine silk thread, the bar will swing to and fro, and finally come to rest in a definite position.

The ends of a magnet are called the *poles*. In the second experiment just described, that end or pole which points to the north geographical pole of the earth is called the *north* pole of the magnet, and the other end or pole the *south* pole of the magnet. Now suppose we take another bar magnet similar to the one that has been suspended by a silk thread, and bring the north pole of this second magnet slowly towards the north pole of the suspended magnet, holding the south pole well away. The suspended magnet would begin to revolve as if there were some invisible force *repelling* the two north poles. If the experiment be repeated by bringing the south pole of the second magnet towards the north pole of the suspended magnet, the latter would try to revolve in the opposite direction, as if there existed an invisible *attractive* force between the north and south poles.

Summarising, we can state the following facts in regard to a magnet:

1. The ends of a magnet are termed the poles. There are two poles, known as North (N) and South (S). The imaginary line running through the body of the magnet from one pole to the other is called the magnetic axis.

2. A magnet will attract to itself particles of iron and steel. These metals, because of this, are termed magnetic. The force of attraction is always greatest at the poles.

3. A bar magnet if suspended on a silk thread, or supported at its centre on a jewelled bearing, so that the friction of the support is negligible, will always point in a definite direction. That pole which points to the geographical North is known as the *N pole*.

4. The N pole of one magnet will always *repel* a N pole of another magnet; whilst a N pole of one magnet will always *attract* a S pole of another magnet. From this we deduce the law that:

 'Like poles repel and unlike poles attract.'

Magnetic lines of force

We must now explain the law of attraction and repulsion. That is, we must consider that curious state of the medium surrounding a magnet, which is associated with the presence of what are called *magnetic lines of force*. These invisible magnetic lines did not exist in the space surrounding the steel bar before it was magnetised and they were therefore brought into being by the process of magnetising. Every magnet possesses a magnetic field which is made up of a large number of invisible magnetic lines extending from the body of a magnet right out into infinite space. It is important to note that each magnetic line forms in itself a complete circuit, a portion of which actually passes through the body of the magnet in a direction that is parallel to the magnetic axis.

Distribution of magnetic lines

In Figure 1.4 we get the magnetic field associated with a bar magnet, as determined by an iron-filings experiment, whilst Figure 1.5 shows

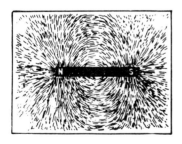

Figure 1.4   A magnetic field associated with a bar magnet

the distribution of magnetic lines, first between two S poles placed opposite one another; and, second, between a N and a S pole similarly disposed. In studying these diagrams, we must bear in mind that these magnetic lines, although quite invisible, always behave as if they had a material existence and possessed certain physical properties.

A correct understanding of the true nature of a magnetic field is essential before one can hope properly to understand the action of any piece of electrical apparatus, and for this reason some space is devoted to this important, though very elementary point. Our present-day conception of a magnetic field is due to Faraday, and it is both interesting and significant to record that the advances made since he gave his ideas to the world over one hundred years ago have not produced

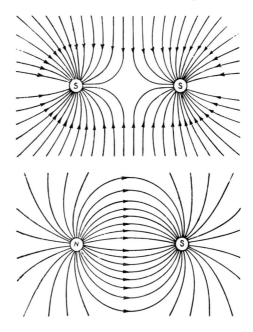

Figure 1.5   Lines of force

any radical modification of the views he then formulated regarding the matter. He conceived for the invisible and intangible magnetic line two properties which it is important, and indeed necessary, that we should remember when explaining magnetic actions. These are:

1. A magnetic line always tends to shorten itself from end to end.
2. Magnetic lines tend to repel one another.

In mechanical language, this conception can be expressed by saying that there is always tension along the lines and pressure across them. It will be found convenient, when explaining any magnetic action, to assume that the magnetic lines have a physical reality and that they take the form of stretched elastic bands.

If we again look at Figure 1.5, remembering that the magnetic lines in either diagram behave as elastic bands, it is easy to understand that the lines flowing between the N and S poles are trying to unstretch or shorten themselves, thus pulling together the two poles to which their ends are securely anchored; whilst in the other diagram the lines crowded into the space immediately above and below the poles are repelling one another, and thus forcing the two poles apart. In the first case we get *attraction*, and in the second case *repulsion*, as shown by the experiment already described.

PERMANENT MAGNETS

A permanent magnet is defined by IEC standards as a magnet which requires no power to maintain its field and is generally made from iron or steel containing ferro-magnetic material which has become magnetised by being placed in a magnetic field of a permanent magnet or electro-magnet.

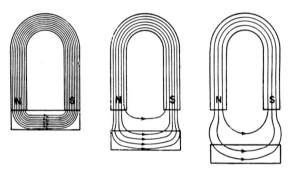

Figure 1.6   Lines of force with a horseshoe magnet

The earliest form of magnet used in ignition and allied equipment was the 'Horseshoe' type shown in Figure 1.6. This gives three diagrams of the magnetic fields for a horseshoe magnet similar to that fitted to early magnetos. It is assumed that a soft-iron keeper is brought near the poles and the diagrams illustrate the change in the magnetic field that occurs as soon as the keeper is allowed to move towards the poles due to the attraction they exert upon it. Here again the magnetic lines (which pass readily through the keeper, because iron offers to the flow of magnetic lines anything from 1/1000 to 1/3000 of the resistance offered by air having the same volume) are behaving as elastic bands and in the act of shortening themselves, are pulling the keeper towards the magnet poles.

As the air space between poles and keeper is reduced, the total number of lines flowing (called the flux) increases and at a greater rate the pull on the keeper. In the case of an ordinary magneto magnet, when the keeper is very near the poles, the pull reaches many pounds, as can be demonstrated by trying to separate the keeper from the poles. Actually a force of something like 50 lb has to be exerted to pull the keeper away.

Development of permanent magnets

The manufacture of permanent magnets from special forms of steel alloys has developed over a period of some fifty years resulting in the production of vastly smaller and lighter weight magnets. This in turn has enabled much smaller and more compact designs of electro-magnetic devices to be produced. Evidence of this was very clear in the aircraft magneto and motor cycle fields and more recently in the wide range of small permanent magnet motors in use on present day automobiles.

Prior to 1920 magnet steels most extensively used contained 6% tungsten and 2% chromium. In 1920 a steel containing 35% cobalt and much smaller percentages of either molybdenum, chromium or tungsten was developed which enabled magnets to be reduced to $\frac{1}{3}$ of their earlier volume and weight. Around 1930 aluminium-nickel alloy magnet steels were developed which gave an energy content twice that of the 35% cobalt steel. These magnet steels were designated *Alni, Alnico* and in 1938 further advances were made in both the alloys and forms of heat treatment which produced *Alcomax* and *Columax* magnet steels having directional properties and great increases in energy content and referred to as anisotropic alloys.

In the utilisation of the anisotropic alloys, it is essential in order to obtain the specified magnetic characteristics, to magnetise the magnets along the same axis as that employed in the process of manufacture of the alloy. When magnetised in a direction at right angles to the pre-determined or treated axis, the characteristics are inferior to those of the Alni alloy. For this reason, the magnet manufacturer must be advised of the intended magnetic axis in the case of all magnets specified in anisotropic alloys.

Difficulties associated with magnetising during cooling in a strong magnetic field, necessitate that magnets of anisotropic alloys shall be of simple design.

Two important characteristics of a permanent magnet are its coercive force and remanence. The remanence is the magnetic flux density in the magnet steel under closed circuit conditions, after the steel has been fully magnetised. By applying a demagnetising force, after the steel has been fully magnetised, the flux density can again be reduced to zero and the value of the demagnetising force necessary to do this is termed the coercive force.

Quality of magnet steels

For many years, these two characteristics were regarded as the chief criteria of magnetic quality. It is established that the true criterion of magnetic quality is the maximum product of the demagnetising force and flux density, now referred to as the BH_{max} value.

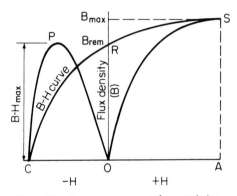

Figure 1.7   Permanent magnet characteristics

A clearer understanding of the characteristics of a permanent magnet will be obtained by reference to Figure 1.7. If we assume that a magnet is wound over its whole length with a magnetising coil and has its poles bridged by a substantial soft-iron keeper making good contact with them, the flux density B in the magnet on passing current through the coil will increase in accordance with the curve OS. As the current in the magnetising coil is increased, so will the flux density in the magnet increase until a point S is reached corresponding to a magnetising force (+H) equal to OA, when the maximum flux density B_{max} is attained. Further increase in current results in no appreciable increase in flux density.

Coercive force and remanence

Upon withdrawing the magnetising force, the flux density will not fall to zero again but will remain at some value B_{rem}, this being the remanence.

In order to reduce the flux density to zero, it is necessary to reverse the current in the magnetising coil and thereby apply a demagnetising force. As the reversed current is gradually increased the flux density will fall along the B–H curve RC. The value of the demagnetising force

−H necessary to reduce the flux density to zero is the coercive force H_c of the magnet. If we take corresponding values of B and −H and plot their products, we obtain a curve OPC, the maximum point of which gives the maximum product of B and −H (BH_{max}). It is this value that is now considered to be the true criterion of magnetic quality.

The values of the magnetising and demagnetising forces H applied to the magnet can be computed for various currents flowing in the coil around the magnet, from the fundamental formula:

$$H = \frac{4\pi}{10} \frac{I S}{L} \quad \text{C.G.S. units}$$

where I = current in amperes
        S = number of turns on the coil
        L = mean length of magnet in cm.

The determination of the characteristics of a permanent magnet by the fundamental ballistic galvanometer method is a somewhat lengthy process and is essentially a laboratory method. For the routine testing of magnets a much quicker method is required and the testing apparatus must necessarily be more robust than the usual laboratory equipment.

Other noteworthy advantages

Apart from their remarkable magnetic properties, aluminium-nickel-iron alloys have other noteworthy advantages. They are unaffected by temperatures up to 700° C, which enables magnets to be ground in a heat-treated state without fear of impairing the magnetic characteristics; it also permits the casting of the magnet in aluminium and thereby opens up new possibilities in the design of electrical apparatus utilising permanent magnets. These alloys are also much less susceptible to the demagnetising effects of mechanical shock or vibration and have a slightly lower specific gravity as compared with the tungsten and cobalt steels.

Some limitations

Owing to their large crystalline structure, magnets of aluminium-nickel-iron alloys are mechanically weaker than other alloy magnets and therefore should not be subjected to shear and tensile stresses if fracture is to be avoided. The magnet design must be simple in form as it is generally only possible to cast magnets in these alloys. Machining is limited to grinding, which precludes the use of small holes for fixing screws or other purposes.

In recent years some progress has been made in the direction of machineable alloys, though generally by sacrificing some of the magnetic properties.

Sintered magnetic alloys

Another interesting development in recent years is the manufacture of aluminium-nickel-iron alloy magnets by the sintering process. In this process the basic ingredients, reduced to metallic powder, are heated to a temperature approaching their melting point and then moulded by great pressure to the required shape.

The particular advantage of the sintering process of manufacture is that magnets can be formed in complex shapes to very close limits and the only machining necessary is the grinding of the working faces. Magnets produced in this way are also mechanically stronger than cast magnets and a further advantage is that soft-iron pole faces can be formed integrally with the magnet during the process of sintering. The relative costs of cast and sintered magnets will depend largely upon the size and shape of the magnet concerned.

The high cost of the powdered alloy generally precludes the use of sintered alloys for other than small magnets, such as employed for electrical instruments, and even then the qualities need to be large to justify the necessarily explosive dies.

In magnetic properties the sintered magnets are slightly inferior to those of cast magnets of the same alloy, notably in respect to the remanence and other induction values throughout the B–H curve, which result in a 10% lower energy content.

Ceramic permanent magnets

A noteworthy new addition to permanent magnet materials, likely to find a future application for small generators and small flywheel magnetos, is the ceramic class, the basic ingredients of which are iron oxide (Fe_2O_3) and barium carbonate ($BaCO_3$). Correctly mixed and heated in air at about $1000°$ C, these ingredients form a barium ferrite compound ($BaFe_{12}O_{19}$). After reduction to a very small particle size, the powdered compound, together with a suitable binder is pressed to the required shape by normal powder metallurgy techniques and subsequently fired at $1200°$ C in an oxidising atmosphere.

Typical of materials commercially available under a variety of trade names is 'Feroba' which is available in two grades — isotropic and anisotropic. In the latter, the powder particles are magnetically aligned in the required direction during the pressing process, with consequent improvement in the magnetic characteristics.

An outstanding feature of this material is its exceptionally high coercive force — nearly three times that of Alcomax. The remanence,

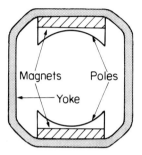

Figure 1.8 Application of ceramic permanent magnets

however, is low and even for the anisotropic grade, the remanence is less than one-third of the value for Alcomax. Nevertheless, the BH_{max} value for the anisotropic grade is relatively high as the following comparison with the characteristics of Alcomax indicates.

Material	H_c	B_R	BH_{max}
	oersteds	*gauss*	*m.g.o.*
Feroba I	1700	2100	0.9
Feroba II	1750	3700	2.0
Alcomax III	650	12600	5.4

In consequence of the high coercivity characteristic of the ceramic magnet, the length is greatly reduced as compared with even an Alcomax magnet for a given purpose. A thin magnet of large sectional area, as needed owing to the low remanence characteristic, can be quite conveniently housed in a generator or motor, especially if two magnets, each of half the required length, are used in series as illustrated in Figure 1.8.

The manner of producing ceramic magnets necessitates simplicity in form, which should be cylindrical, ring or rectangular blocks. A magnetising force of not less than 7,500 ampere-turns per centimetre length is necessary for efficient magnetisation. Once fully magnetised, the only simple way to completely demagnetise ceramic magnets is to heat them slowly in a furnace to a temperature of 450–500° C and allow to cool to room temperature in the furnace. These magnets have a density of 5 gm/c.c. as compared with 7.35 gm/c.c. for Alcomax and compare quite favourably for cost providing the magnet design is efficient. In this respect the advice of the suppliers should always be solicited in new applications.

ELECTROMAGNETISM

More than a hundred years ago a discovery was made which had the effect of interlinking the two sciences of electricity and magnetism. Prof. Oersted of Copenhagen demonstrated, for the first time, that an electric current creates, in the space surrounding the conducting wire, a magnetic field which has exactly the same nature and properties as those already assigned to the magnetic field of a magnet, or to a piece of natural lodestone.

The simple apparatus required for performing Oersted's classical experiment, which first established the magnetic nature of an electric current, is shown in Figure 1.9. The conducting wire is fixed in a N–S direction, so that the compass needle placed immediately below it will point in the same direction. As soon as the circuit is closed and

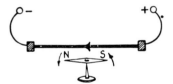

Figure 1.9   Oersted's experiment

current flows through the wire, the compass needle moves and tends to set itself at right angles to the conductor. When the current is reversed the compass needle moves in the opposite sense, still striving to place itself at right angles to the wire.

The explanation of these movements is that the electric current creates a magnetic field, and if we determine the distribution of the magnetic field, we shall find that it comprises a large number of magnetic lines which take the form of concentric circles having their common centre on the axis of the wire.

The compass needle, in endeavouring to set itself along the magnetic lines, must place itself at right angles to the conductor, so that it lies tangentially to the direction of the field. In Figure 1.10 we see the actual field distribution for a fundamental case of this kind; whilst in Figure 1.11 we have two diagrams showing the direction in which the magnetic lines flow around two parallel conductors carrying currents in two directions.

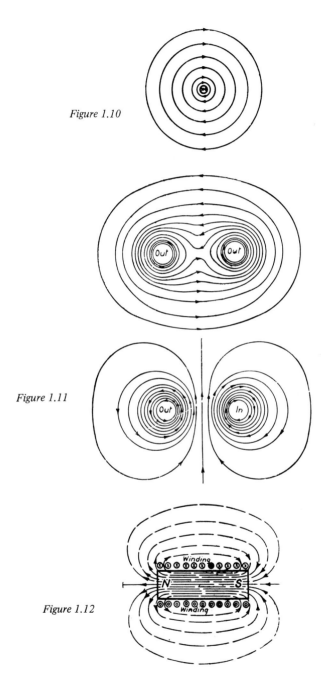

Figure 1.10

Figure 1.11

Figure 1.12

If instead of a simple straight conductor we take a coil of insulated wire and pass current through the coil, it is obvious from what has already been said that a large number of magnetic lines will be generated (*see* Figure 1.12) and it is important to note that each of these will *pass through* the coil and complete its circuit in the surrounding space. That is, *each magnetic line will be truly linked with each turn in the coil.*

Let us go a step further and analyse what happens when current is passed through two coils which are threaded on the respective limbs of a U-shaped piece of soft iron. The iron core becomes, for the time being, a very powerful magnet, and it is able to exert an enormous tractive force on a soft-iron keeper in close proximity to the ends of the U-shaped core. Such an arrangement is called an electromagnet, and the magnetic field distributions will be similar to those already given for a horseshoe magnet and keeper (Figure 1.6). Magnetically, the two cases are identical.

If we imagine the current to increase, a corresponding change will occur in the magnetic field, and more and more magnetic lines will be crowded into the coils through the iron core. Should the current decrease, the total number of magnetic lines (or the flux) will be reduced, whilst if the direction of the current be suddenly reversed the flux will first die down to zero and then grow again in the opposite direction.

This alternate increase and decrease of the magnetic field corresponds to a displacement of energy exactly as in the case of a flywheel which is made to increase and decrease its speed of rotation. When run up to speed, energy is applied to and absorbed by the rotating mass, and the energy so stored is given back again when the wheel is allowed to come to rest. Similarly with a coil of wire carrying current. When the circuit is broken and the magnetic field thus destroyed, the energy stored in the field reappears as a spark at the two points in the circuit where the current is ruptured. It is evident, then, that a change in the magnetic lines linked with a winding can induce an electric current and this will now be explained.

Electromagnet induction

Dealing with the work of Faraday, which established principles that lie at the very foundation of the electrical industry. Up to the year 1831 the only available means for producing electric currents was the chemical battery, the forerunner of the modern accumulator. In the autumn of 1831, Faraday showed an entirely new method of production, in which use was made of a magnetic field. Briefly, he reversed Oersted's experiment. Oersted passed current through a conducting circuit and obtained a magnetic field. *Faraday moved a magnetic field across a conducting circuit and created in that circuit an electric current.* This basic principle, known as electromagnetic induction, is utilised in the construction of all forms of motors and generators.

Figure 1.13 shows four diagrams which represent the four most famous experiments of the wonderful series made by Faraday in establishing this principle. We shall concern ourselves with the top

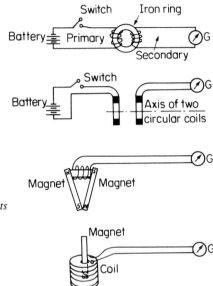

Figure 1.13   Faraday's experiments

diagram, which relates to his most classical experiment. The apparatus comprised an iron wire ring, upon which we wound two coils of insulated copper wire. The left-hand coil (called the primary) was connected through a switch to a battery, whilst the right-hand coil (called the secondary) was connected directly to a galvanometer, which is a sensitive instrument that will detect the presence of a minute current.

Faraday's discoveries

Faraday found by experiment that:

1. On closing the switch, there was a disturbance of the galvanometer needle, indicating a current in the secondary circuit. The current was transitory; that is, it only occurred at *the instant* when the switch was closed and current was forced through the primary winding.

2. On opening the switch, much the same thing happened, except that the current in the secondary circuit now flowed in the opposite direction.

Let us analyse what happens in this kind of experiment. When the switch is first closed, the current in the primary circuit will *start from zero and grow until the final* value is reached. It will take a certain time (which is a small fraction of a second) to do this. Correspondingly, this current will create a magnetic field, and during the period when the current is growing, this magnetic field (which will be constrained to circulate around the iron core) will also grow. We therefore have the condition that for a small fraction of a second *immediately* following the closing of the primary circuit, *magnetic lines are virtually being thrust into the secondary coil.* It is this thrusting of magnetic lines into the secondary coil that generates an e.m.f. (electromotive force) or voltage, which in turn gives rise to a current in the secondary circuit.

When the switch is opened, *the primary current dies away to zero and with it the magnetic field*. Correspondingly, we get the condition that *immediately* following the opening of the primary circuit and the rupturing of the primary current, the magnetic field will rapidly fall to zero, so that *magnetic lines are virtually being withdrawn from the secondary coil.* This withdrawal gives rise to an e.m.f. (of opposite sign to that occurring in the previous case) and a reverse current in the secondary circuit.

Law of e.m.f. production

It is clear from this that a current is induced in the secondary circuit whenever the flux linked with that circuit is changed, and *the current only persists for the period during which a change in the flux occurs*. In general, if we have any conducting circuit, and by some means or another we produce a change in the number of magnetic lines linked with the circuit — either by moving the magnetic field and keeping the circuit stationary, or vice versa — a voltage will be induced in the circuit which will give rise to a current if the circuit be closed. The law underlying this method of producing e.m.f. is simple, and can be stated as follows:

1. The induced e.m.f. is directly proportional to the rate at which the flux linkage changes.
2. When this rate of change is 100 000 000 (=10^8) line-turns per second, the induced e.m.f. is exactly 1 V. Thus, a change of one million lines per second in a 100 turn coil would induce an e.m.f. of 1 V.

By 'flux linkage' we mean the product of the flux threading the winding and the number of turns in the winding. A better conception of what is meant by linkage can perhaps be obtained by referring to Figure 1.14, which diagrammatically depicts the fundamental case of

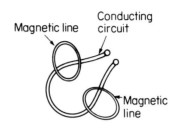

Figure 1.14   Concept of 'flux linkage'

a coil of one turn truly linked with two magnetic lines. We finally arrive at the basic formula:

$$E = \frac{N \times S}{t} \times 10^{-8}$$

where N = change in flux linked with winding in t seconds
S = number of turns in winding
E = induced e.m.f. in volts.

For example, if we thrust the N pole of a magnet from which 30 000 lines radiate into a circular coil having 1000 turns, performing this operation in $\frac{1}{10}$ second, there will be a momentary voltage induced between the ends of the coil equal to

$$\frac{30\ 000 \times 1000}{0.1} \times 10^{-8} = 3\ \text{V}$$

When the induced voltage is steady, as in a dynamo running at a constant speed, the current flowing in the closed circuit can be easily calculated from Ohm's Law, already enumerated.

ELECTRICAL INSULATIONS

The subject of insulating materials is of interest in a book of this kind and information is quoted from the author's experience in this field.

Insulation coverings for wire

Copper wire is used almost universally in the manufacture of electrical apparatus for automobiles and is provided with some form of protective covering which may be of cotton, silk, plastic compound or enamel. Though silk is now usually replaced by Terylene.

Enamel-covered wire is used extensively particularly where the size of wire is very small, since this form of covering provides a maximum insulation for a minimum thickness of material. It is generally necessary to obtain the maximum copper section for a given winding space and with coverings such as cotton or silk the space occupied by the insulation is relatively great for small sizes of wire. A notable illustration of enamel covered wire is to be found in ignition coil or magneto secondary windings, consisting of several thousand turns of about No. 44 SWG wire, which is very little thicker than a human hair.

Table 1.2 gives the relative thicknesses of covering for different classes of wire insulation and from this it will be seen that the thickness of cotton covering is about three times that of silk and two to eight times that of enamel. On the smaller sized wires the thickness of covering is very important and may easily occupy 50% of the winding space. In many applications it is therefore no longer necessary to use cotton or silk or some forms of plastic to give the requisite mechanical protection during winding operations, particularly if tough enamels such as Thermex and Bicolon are used. This also means that such windings take up less space resulting in a more compact and smaller overall size of unit. Such coverings are substantially non-hygroscopic and no not necessarily require to be varnish treated in order to render them resistant to moisture.

Table 1.2 WIRE INSULATIONS

Covering	Abbreviation	Thickness of covering in mils
Enamel	En.	0.4 – 4.0
(Ordinary) single cotton	S.C.C.	4.0 – 8.0
Specially fine single cotton	S.F.S.C.C.	3.5 – 6.5
(Ordinary) double cotton	D.C.C.	8.0 – 13.0
Specially fine double cotton	S.F.D.C.C.	5.5 – 10.0
Single silk	S.S.C.	1.2 – 3.0
Double silk	D.S.C.	2.2 – 4.0
Enamel and single cotton	En. & S.C.C.	4.5 – 10.0
Enamel and single silk	En. & S.S.C.	1.6 – 6.0

Anodised aluminium foil

As a result of recent advancements in the manufacture of anodised aluminium foil, increasing use is now being made of this material for electrical windings. Anodised aluminium foil is now readily obtainable with an oxide film of 0.2 mil, providing adequate between-turn insulation for many applications. Tests on several tightly wound coils on $1\frac{3}{8}$ in diameter formers have indicated d.c. breakdown voltages in excess of 240 V, which is more than adequate for automobile electrical devices. This, together with the fact that the windings are laminar, as distinct from round wire layers, can effect an appreciable increase in the volume of conductor material relative to the winding volume, i.e., in the space factor.

For example, a coil designed for 2000 ampere-turns with a current consumption of 10 A on 12 V, had a space factor of 87% with anodised aluminium foil as compared with 52% for copper wire. The increase of 68% in space factor in this case just compensates for the higher resistivity of aluminium as compared with that of copper, as borne out by the comparable consumption current. Since the specific gravity of aluminium is one-third that of copper, the weight of the winding was halved by the use of anodised aluminium foil. This weight advantage can be of great importance in some applications, especially where large windings are concerned.

The overall advantage of using aluminium foil will depend upon the design of the winding, but for certain heavy current types, quite appreciable economy in current consumption and cost can be effected. Moreover, by using foil in single or parallel strips where the required single strip thickness becomes difficult to manipulate, the winding operation is generally facilitated and the need for interlayer or barrier insulations is eliminated. Another advantage is that anodised aluminium foil windings can operate at a very much higher temperature than comparable copper windings, since the insulating properties of the oxide film are retained up to a temperature of 600° C, which is just below the melting point of the material.

Some difficulty may be presented in making connections to external leads but this is largely solved by the nature of the foil windings which lends itself to some form of strip or bar connector to which the winding foil is secured by clamping or screws. For joining foil ends, pressure welding using specially designed tools, can be employed.

Varnished cloths and papers

Varnish cloths and papers are used to a large extent for interlayer insulations on ignition equipment together with varnished and plastic impregnated papers and cotton fabrics. Synthetic and bituminous varnishes are used in the treatment of windings where resistance to warm humid conditions is required.

In addition to strip insulation, tubings made from extruded plastics such as polyvinyl chloride have generally replaced varnished cotton and silk tubing; they have the advantage of being extremely flexible, non-hygroscopic, resistant to oil, with good ageing properties, but have a rather low softening point.

It is important to store insulating materials in a dry atmosphere maintained at a temperature of 20–25° C to ensure their electric strength and flexibility will be maintained for long periods.

Sheet insulations

Sheet insulating materials such as vulcanised fibres, presspahn, synthetic resin bonded paper and nylon, varying in thickness down to 0.005 in and various forms of plastic film are used for a variety of purposes on automobile electrical apparatus. With the older vulcanised fibres and presspahn it is necessary to provide some form of varnish treatment to retard moisture absorption for such purposes as armature slot insulation, slot wedges, small insulating washers and packing strips.

Synthetic bonded fabric material is used extensively for insulating base boards and supporting plates and owing to its high tensile strength and excellent wearing characteristics is used for gear wheels. Moulded nylon gears are also being used increasingly, however, where the mechanical loading is relatively low and silence of operation is essential.

Moulded insulations

Moulded insulations are used extensively on ignition and allied apparatus and the needs of the automobile industry has led to extensive development of these materials, mainly in two groups, rubber compound

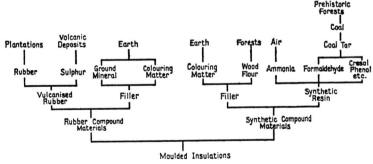

Figure 1.15   Applications of rubber and synthetic compound materials

materials and synthetic compound materials. Figure 1.15 shows the origin of these two compositions.

Rubber compound materials

The chief members of this group are more familiarly known as ebonite and stabalite. Ebonite usually refers to the produce made from pure rubber vulcanised hard with sulphur, whilst stabalite consists of rubber and sulphur to which is added a percentage of filling matter, giving body and rigidity to the vulcanised rubber.

Synthetic compound materials

The first stage in the manufacture of synthetic compound materials is the production of the synthetic resin or gum, this is formed by the hardening action of formaldehyde on the coal-tar products cresol and carbolic acid. The hardening is accelerated by the use of certain acid or alkaline catalysts.

The resin is substantially dehydrated by distillation under reduced pressure and after cooling down is then ready for the second stage of manufacture which is the incorporation of certain fibrous fillers. The filler imparts mechanical strength, facilitates moulding and reduces the shrinkage of material during moulding to the required shape. The moulding time is very short, and apart from removing flashes the moulding is in a finished state requiring no further machining.

Where resistance to shock or vibration is of importance in such components as ignition distributor caps, and rotor arms, fabrio-filled powders are extensively used and have replaced mouldings made from rubber based compound.

Polytetrafluoroethylene

More generally referred to as PTFE and produced in this country by ICI Ltd. under the trade name of Fluon, this plastic material is a resin-like substance with good electrical and thermal properties and is a derivitive of ethylene and closely allied to polythene.

The material is noteworthy for its chemical inertness, solvent resistance and good mechanical properties over a wide temperature range. It has no known solvents, is attacked only by molten alkali, metals and fluorine and has zero water absorption. The serviceable temperature range is 100° to 250° C, it is non-inflammable and non-tracking in that arcing over the surface does not cause carbonisation.

With a dielectric strength of about 1500 V/mil for thicknesses of the order of 0.005 in, very low dielectric losses over an extreme frequency range, PTFE is particularly suitable for high voltage and high frequency applications.

Mouldings of simple form may be sintered under comparatively low pressure — complicated shapes under higher pressure. The material may also be extruded in rod or tube form using a temperature approaching 400° C.

Mica, porcelain and Steatite

Mica is probably the best known of electrical insulators having a very high electric strength of the order of 3000 V/mil and exceptional heat resisting properties and also remarkable cleavage properties which enables it to be split into sheets of thicknesses ranging from 1 to 2 mils.

This material was used for many years for sparking plug insulators and also insulation strips between the segments of commutators and for ignition capacitors but owing to cost is now only used for some heater plugs for compression ignition engines and laboratory and aircraft instruments.

Porcelain and Steatite are in widespread use for sparking plug insulators and the essential features of the material for this use are heat, electrical and weather resistance, durability and hardness. To meet these requirements the important constituents are china clay, felspar and quartz which contribute respectively to the heat resisting, electrical, and mechanical strength of the finished product.

Steatite insulators as now used are generally produced by first pulverising the natural rock (or speckstein as it is called) and then compressing and moulding to shape and firing or baking at a temperature of 1200–1400° C, in a similar manner to porcelain. Although steatite has a higher expansion coefficient than porcelain, it is mechanically stronger. It can also be moulded to size with accuracy. Compositions of steatite and porcelains combine the physical and electrical properties of these two materials.

Modern ceramic insulations

The term 'ceramic' is applicable to a wide range of products and includes materials made in fused alumina. Examples of this specially developed material in widespread use for sparking insulators are Corundite and

Sintox which are both sintered materials with fused aluminium oxide as their base and are much less susceptible to surface erosion than earlier materials.

SOLID STATE DEVICES

Since the first commercial application of a solid state electronic or semi-conductor device – the radio set crystal detector used extensively in the early days of sound broadcasting – considerable progress has been made in the science of solid state physics. Physicists today have a very comprehensive knowledge of the highly mathematical Band Theory of solids underlying the properties of semi-conductors. These devices have greatly influenced techniques in automobile electrical engineering since the latter 1940's and are now in widespread use on all types of equipment.

For this reason, some reference to the fundamental theory underlying these devices is advisable, even though it can only be brief. Nevertheless, it aims to give the uninitiated a useful, if very superficial, understanding of these principles. For a more comprehensive treatment, readers are referred to available publications on the subject.

Earlier in this chapter a flow or current of electricity has been attributed to a stream of electrons passing between atoms. A feature of good conductors of electricity is that they have very many electrons free to act as carriers of current. In insulators, and to a lesser extent in semi-conductors, the electrons are closely bound to their associated atoms and only a very small percentage of the electrons are free to conduct current.

The band theory of solids

Each atom of a solid is normally electrically neutral, the negative charge represented by electrons balancing the positive charge of the nucleus. The electrons rotate around the nucleus in fixed orbits or energy levels. Since the number of electrons in each orbit or level is restricted, the greater the number of electrons, the more will be the orbits occupied. The electrons in the outer orbits are referred to as valence electrons and are at higher energy levels than those in orbits nearer to the nucleus. According to the Band Theory, the manner in which these separate at distinct energy levels and, due to the interaction of the atoms, form themselves into bands, determines the conductivity characteristic of solids.

The band diagrams illustrating this theory (Figure 1.16) have some semblance to a solution in a test tube in which there is a heavy sediment-filled band at the bottom; a clear band free from particles at the top; and, conceivably, an intermediate area only partly filled with particles in suspension, due in the electronic case to overlapping bands. Movement of the particles within the filled band is precluded and to raise particles from the filled to the empty band requires some expenditure of energy.

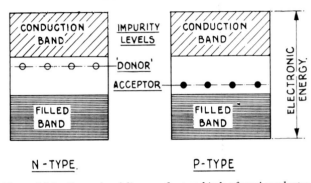

Figure 1.16   Energy band diagrams for two kinds of semi-conductor materials

In like manner, to move electrons from a filled to an empty or conduction band necessitates the application of energy from some external source.

As a result of the particular arrangement of atoms in a single crystal and the concentration of a positive charge in the nucleus of each atom, the electrons move in a non-uniform electrostatic field. Depending on the character of this intrinsic field and diffraction effects on certain electrons, forbidden regions may occur at one or more energy levels, producing similar forbidden regions or gaps in the electronic energy bands. A diagram of these bands with a wide gap, equivalent to several electron-volts and so too large for electrons to be thermally accelerated across it, typifies an insulator diagram. On the other hand, a diagram having overlapping bands or no forbidden region, allowing electrons to move freely under influence of an electric field, characterises a metal conductor. In the case of a semi-conductor, the band diagram will have a forbidden gap so short and of the order of one electron-volt or less, that electrons can be accelerated across it from the filled to the empty conduction band by thermal excitation.

The semi-conductors employed in electrical circuitry are crystalline materials, which owe their peculiar conductivity properties to the presence, in extremely minute quantities, of selected impurities. The introduction of these impurities results in either an excess or a deficiency of electrons in the crystal structure. This process of 'doping' materials like germanium and silicon, the most widely used at the present time, gives rise to additional and distinct energy levels in the normally forbidden gap between the filled and the empty conduction band. This facilitates the passage of electrons across the forbidden gap, by reducing the energy difference between the bands.

Any transfer of electrons to the conduction band creates vacancies in the filled band which, although they have no mass, act like positive electric charges. Thus, we have the current in semi-conductor devices carried by positive carriers as well as negative electrons. It is perhaps difficult to appreciate how these vacancies give rise to conduction, as in effect they do, and to facilitate comprehension of this anomaly, the concept of conduction by 'holes' has been devised and this is basic to the present theory of semiconductors.

The operation of semi-conductor devices depends on the use of two kinds of semi-conductor, the *n*-type and the *p*-type, having net negative and positive charges respectively. In both types, electrons and holes co-exist with a preponderance of electrons in the former and holes in the latter. The impurities introduced in these two cases are designated 'donor' and 'acceptor' respectively. For germanium and silicon semiconductors arsenic and antimony are representative of the former, whilst aluminium, boron, gallium, and indium are acceptor impurities. The higher conductivity resulting from the presence of the impurity is attributed to the relatively small energy difference between the impurity energy level and the bands. For donor impurities the level lies nearer to the conduction band and for acceptor materials, nearer to the filled band, as depicted by the energy diagrams in Figure 1.16 for the two types of semi-conductor material under consideration.

Semi-conductor devices

There are two main groups of semi-conductor devices, namely diodes and triodes. More prevalent among the former are the rectifier diodes used mainly for rectification of an alternator output to direct current for battery charging in automobiles. Voltage reference devices known also as Zener diodes are used in generator output control circuits. In the triode group, the transistor is a device suitable either for amplification, somewhat similar to a triode thermionic valve or as a solid-state switch performing the same function as an electromagnetic relay or regulator.

(a) Rectifier diodes

Rectifier diodes, whether of germanium or silicon, are units in which an *n*-type and a *p*-type element together constitute a fused junction device.

Connecting a d.c. supply across the junction so that the positive supply is connected to the *p*-type element and the negative to the *n*-type element will result in current flow across the rectifier from the *n*- to the *p*-region.

With this 'forward' current connection, the positively charged holes are repelled by the positive supply polarity towards and across the junction to combine with electrons similarly repelled by the negative polarity in the *n*-region. This combination of electrons and holes permits the entry of fresh electrons from the negative side of the supply and new holes are injected into the *p*-region from the positive supply side. With the supply polarity reversed, that is for a 'reverse' current connection, the electrons and the holes move away from the junction, being attracted to the positive supply and negative supply poles respectively. In consequence, the donor and acceptor impurities no longer have balancing hole and electron charges available and a voltage builds up across the junction to counteract the applied voltage. The reverse current is not entirely zero, but is extremely small compared with the forward current.

It will be evident from the foregoing that when an a.c. voltage is applied across a semi-conductor diode, current will flow in a forward direction for one half-wave only of each cycle, since for the other half-wave the polarity across the diode will be reversed. By a suitable circuit arrangement of rectifier diodes, both halves of each cycle can be rectified.

Silicon rectifiers can operate safely up to 250° C, whereas the limiting junction temperature for germanium rectifiers is 75° C. At these operating temperatures, the performance characteristics of the two materials are very similar. The forward voltage drop of the silicon rectifier is rather higher than that for the germanium device, but it has the advantage of a notably higher permissible reverse voltage when operated at temperatures below 200° C. The conversion efficiency of both silicon and germanium rectifiers is very high, being of the order of 97%.

(b) Zener diodes

The Zener diode differs somewhat in structure from the rectifier diode as it is designed primarily to use the reverse current characteristics of an *np* junction, in the vicinity of the turnover in the reverse voltage-current curve.

Beyond the turnover in the curve small increments in the inverse voltage result in relatively large increments in current or as it is customarily expressed, as the inverse current is increased the slope resistance

(dV/dI) decreases. This characteristic renders the Zener diode particularly suitable for use in output voltage control circuits and also for use as a voltage reference device.

Whilst it is desirable to use as high an inverse current as possible to obtain the best regulation, the maximum permissible inverse current is determined by the thermal dissipation of the diode. The maximum permissible inverse current decreases with increase in operating temperature. The inverse current for silicon diodes decreases almost linearly from a maximum at 25° C to practically zero at the specified maximum temperature which can be as high as 225° C. Where the application demands, both the inverse and forward current characteristics may be used as reference sources. Zener diodes are available for voltage output control purposes, ranging from ¼–10 V and with inverse turnover or Zener voltages from 3–100 V in approximately 10% steps.

(c) Transistors

The transistor as now manufactured in considerable numbers is a back to back arrangement of two *np* junctions and can be either an *npn* or a *pnp* structure, the latter being the more prevalent. The middle member of the sandwich which is termed the 'base' is a thin single-crystal wafer about 2 mils thick of germanium or silicon. The *p* regions on either side of the base are called the 'emitter' and 'collector' according to their respective functions.

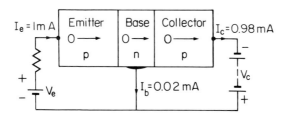

Figure 1.17   A pnp transistor diagram illustrating current flow with appropriate emitter and collector biasing

A *pnp* transistor is shown diagrammatically in Figure 1.17 and the operation of this type of transistor is briefly summarised. By appropriate doping of the three elements, free electrons are induced in the base and positive carriers or holes in the collector and emitter. On the initial formation of the transistor, holes from both collector and emitter migrate into the base to combine with the free electrons. Also electrons from the base migrate to the collector and emitter to combine with holes in these elements. The base losing electrons and the other two elements losing holes, a potential is built up at the junctions resulting in the base becoming positive with respect to both collector and emitter. Further migration of carriers across the junctions is ultimately precluded when the repulsive action across the junctions, due to the electric fields associated with the internal potentials, become balanced and a state of equilibrium is attained.

Operation of a transistor is brought about by the disturbance of this state of equilibrium by the application of external potentials across the junction as shown in Figure 1.17. Considering first the emitter-base junction only, the applied potential reduces the junction potential, thereby permitting the passage of carriers across the junction. In actual transistors, the impurities introduced are purposely designed to induce a much higher concentration of holes in the *p* regions than there are electrons in the base or *n* region. The carriers that then cross the emitter-base junction to produce current flow, are therefore mainly holes passing from the emitter to the base, with but very few electrons crossing the junction in the opposite direction. The holes exist in the base for a finite but extremely short time period before re-combining with free electrons and this is termed the 'life-time' of the base material. Electrons lost in the base as a result of this re-combination with holes are replenished from the battery negative. A positive bias of the emitter as shown gives rise to a current I_e from the battery positive, through the emitter and base to the negative of the battery.

Biasing only the collector-base junction negatively increases the junction potential and intensifies opposition to carrier movement from base to collector. In practice an almost negligible current does flow due to thermally freed electrons around the junction. With both junctions biased as illustrated in Figure 1.17, the base-collector junction field is in such a direction as to induce holes that have crossed from the emitter to the base, to cross also the base-collector junction. Providing the base is sufficiently narrow for the holes to be influenced by the base-collector junction field before they have time to re-combine with electrons in the base, holes from the emitter become the main carriers of current from the emitter across the base to the collector. As already explained, the base thickness is very small and is intentionally so, not only for this reason but also to obtain high frequency response.

The operation of the *npn* transistor is similar in principle to that of the *pnp* type except that electrons and not holes are the main carriers of the emitter-collector current. Moreover, as the carriers are of the opposite polarity, the operating voltages and currents are reversed as compared with the *pnp* transistor.

The foregoing will suffice to indicate that the emitter-collector current may be controlled by the base current I_b. Variation of the base current varies the voltage across the emitter-base junction and in this way controls the emitter-collector current. With suitable base drive, the emitter-collector impedance can be varied in an extreme from about 100 kilohms to 1 ohm, corresponding to virtually open and closed circuit respectively and this can be accomplished as far as the transistor is concerned in a fraction of a μs. Here then, we have the essentials for an extremely rapid switching device or an oscillator capable of operating, with modern transistors, over a frequency range from a few hundred hertz to a frequency of the order of 10 MHz. Such an oscillator would be very much superior to any electromagnetic vibrator. In Figure 1.17 typical currents for an amplification circuit are given and from these, it will be seen that the transistor current gain, which is collector/emitter current ratio, is 0.98 and also that a change of base current of 0.02 mA can produce a change of 0.98 mA in the collector current, thus giving a current amplification ratio of 49:1.

The main advantages of the transistor are robust construction; no moving parts or contacts needing maintenance; near-immunity from vibration and shock; negligible power consumption; small dimensions and light weight. All semi-conductor devices are, however, sensitive to temperature change and due allowance must be made for this in their application. Transistors commercially available range from low output units for radio circuits to devices giving very high collector currents for power output control circuits.

Another noteworthy semi-conductor development is the photo-transistor which increases its conductivity on exposure to light rays. Just as electrons can be thermally excited from the filled to the conductance band, so also can they be similarly excited optically, providing the wavelength is sufficiently short to be absorbed by the semi-conductor and not pass through it. In the case of the phototransistor, current induced by the incidence of light on the emitter side of the crystal can be amplified by normal transistor action to a value of several milliamperes, sufficient to operate a small relay. Phototransistors already find numerous applications, including photo-electric counting and speed indication and have a potential application for automatic parking and headlight control. Available phototransistors have a wavelength response ranging from about 0.8 to 1.6 microns or millionths of a metre and have a peak response in the near infra-red. The response continues to a lesser degree through the visible spectrum to the near ultra-violet.

CHAPTER 2

THE COMPLETE ELECTRICAL EQUIPMENT
INCLUDING INSTRUMENTATION

On the modern car we have, in effect, a miniature electrical system, possessing many of the essentials of a public electric supply system. Whereas the public supply system receives constant attention so that a continuous supply is always available, the automobile electrical system may be completely neglected as regards regular servicing or checking over. In spite of this it is expected to operate under world wide climatic conditions and also conditions of severe vibration and damage by corrosion or contamination from water, oil, dust, mud and corrosive fluids such as brake fluids, battery acid and engine coolants. The system may, on occasions such as starting under conditions of extreme cold, be called upon to provide power for short periods thirty to fifty times greater than the normal output.

The function of the electrical system is to convert the mechanical energy derived from the engine into electrical energy and to distribute this to various points where it may be used for starting, ignition, lighting and the various other uses which provide for the safety, comfort and convenience of driver and passengers.

The various sections of the electrical equipment may be classified under some seven main headings:

1. Generation, storage and distribution.
2. Ignition.
3. Lighting.
4. Starting.
5. Heating and air conditioning.
6. Fuel injection systems.
7. Accessories.

Item (7) includes such devices as electric horns, instrumentation, windscreen wipers, signalling devices, switches, relays, electric petrol pumps.

In this chapter it is proposed to deal briefly with the broad principles underlying the use of electricity on the automobile. Subsequent chapters will give more detailed consideration to the design, operation, performance and practical construction of the important components.

It will be evident from the foregoing that the duty of the automobile electrical equipment is very exacting and many intricate problems arise in its design and manufacture which demand for their solution specialised knowledge and techniques from all branches of modern science and technology. Figure 2.1 shows the diversity of electrical devices which may be employed on the modern passenger service vehicle. The functioning of these devices depends on an adequate supply of electrical power from the alternator via the battery. It is important to appreciate that the battery merely 'floats' across the alternator or generator output and primarily provides electrical power for functions such as ignition, starting and lighting prior to the engine reaching idling speed.

Figure 2.2 illustrates schematically the electrical equipment of a popular British car and the complete wiring circuit to all the electrical

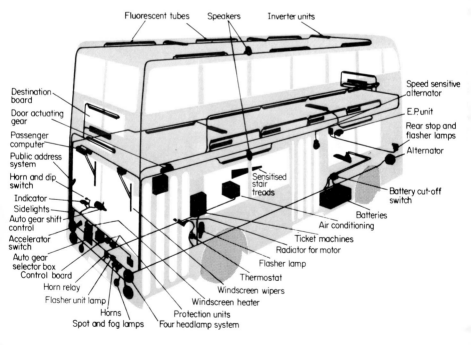

Figure 2.1   Diagram showing variety of electrical devices used on a passenger service vehicle

38

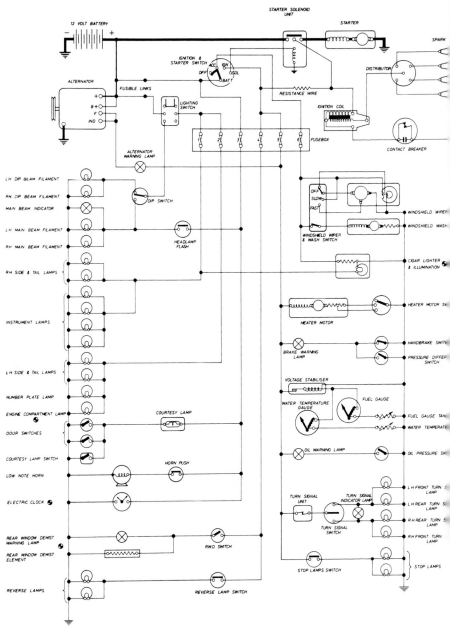

Figure 2.2 Typical car wiring diagram. Vauxhall Chevette (courtesy General Motors)

items including those required by law in the interests of road safety and those required for the efficient operation of the vehicle.

GENERATION, STORAGE AND DISTRIBUTION

The primary source of electrical energy on any vehicle is the alternator or d.c. generator. This is the means whereby mechanical energy derived from the engine is converted into electrical energy. The design and operation of the alternator or generator is described in Chapter 3, also the methods of controlling the output in accordance with the electrical load of the electrical equipment on the vehicle.

So that electrical energy shall be available when the engine is stationary, a secondary source is provided in the form of the *battery*. The battery may be regarded as the storage tank of the electrical system. When the voltage produced by the alternator or generator, as it is driven by the engine, exceeds the battery voltage a flow of current occurs in the alternator-battery circuit. This current flowing through the battery results in a chemical change in the active constituents of the battery, the process being termed *charging* the battery. Upon *discharging* the battery, as for example when switching on the lights, the chemical action taking place in the battery is reversed. In this way electrical energy is converted into chemical energy on *charge* and reappears as electrical energy on *discharge*.

As discussed in Chapter 1 it is necessary to provide an adequate conducting path for the current from the alternator to the battery and suitable conducting cables to distribute the electrical energy to the various components of the system. Details of the various forms of wiring and methods of connection and the practical problems involved are discussed in Chapter 12. Control and switching of various components are also dealt with in Chapter 11.

TYPES OF IGNITION SYSTEM

The ignition of the gaseous charge compressed within the cylinder of a petrol engine is effected by a tiny spark which occurs at the proper moment between the points of the sparking plug. It is the function of the ignition apparatus to produce a regular succession of sparks at the plug points. This may at first appear to be a relatively easy matter, but further consideration will show that the duty of the ignition apparatus is both difficult and exacting.

Present-day means of ignition may be divided into the following classes:

1. Battery-coil ignition.
2. Electronic ignition systems.
3. High-tension magneto.

Electronic ignition systems are now widely used and their adoption is rapidly increasing. The battery-coil system is still in use on a very large number of vehicles. The magneto now has a very limited use in racing and competition work and on some marine and motor cycle applications.

Magneto ignition

This ignition device is a self-contained unit driven by the engine. It draws energy from the engine and converts it first into low-tension electrical energy which is then transformed into high-tension electrical energy, appearing in the form of a succession of sparks at the plug points.

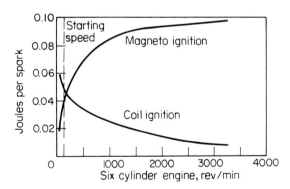

Figure 2.3   Coil ignition spark energy curve

A magneto is basically a small permanent magnet generator capable of generating electricity at very high voltages over a wide speed range. By rotating the magneto shaft, extremely rapid changes of the magnetic flux linking the armature are produced resulting in a series of high-tension voltage rises which cause a succession of sparks.

The primary source of the high-tension electrical energy is the permanent magnet and since the spark energy is produced by rotation it will be evident that this will increase with speed. Actually the spark energy increases rapidly from zero to medium speeds and it then remains fairly constant as shown by the magneto spark energy curve in Figure 2.3.

Battery and coil ignition

Whereas the magneto derives its energy from a permanent magnet, the battery-coil system derives its energy from the battery. The coil system is therefore not self-contained but is interlinked with the battery, which in turn is dependent on the generator. The conversion of mechanical energy into low-tension electrical energy is performed by the generator which supplies current to the battery.

An induction or ignition coil, consisting of windings mounted on a soft-iron core, operates in conjunction with a contact-breaker to transform the low-voltage energy available from the battery into high-voltage energy necessary for the production of the ignition sparks.

As the coil system depends upon the battery for its low-voltage energy, it is clear that, providing the battery is in a reasonable state of charge, this energy is available at any speed of the engine. Theoretically, therefore, it may be stated that even at a speed of 1 or 2 rev/min, the coil system is capable of producing a spark. Whilst this is true in theory, in practice it is not quite so, although most coil systems will produce a spark at the plug points at an engine speed as low as 15 rev/min. It is at low speeds that the coil ignition system gives its maximum spark energy. As will be explained in a subsequent chapter, the spark energy of this system decreases with speed in a manner shown by the coil ignition energy curve in Figure 2.3.

The nature of the two curves shown in Figure 2.3 clearly indicates the advantage of the battery-coil system over the magneto for starting and low speed running. This advantage coupled with the high cost of manufacturing the magneto was the reason for the general adoption of coil ignition systems.

Electronic ignition systems

In the conventional form of ignition distributor described in Chapter 7 the contact breaker operates as a high speed switch in the primary circuit of the ignition coil. The spark voltage and energy are limited by the mechanical and electrical capabilities of this form of switch. As a result of the higher speeds now obtained, particularly from six-, eight- and twelve-cylinder engines, much higher demands are placed on the conventional contact-breaker assembly which cannot be met by a mechanical switching device which has a maximum sparking rate of about 400 sparks per second, which is equivalent to 6000 rev/min on an eight-cylinder engine. This drawback has been overcome by the use of electronic components to replace the mechanical switching elements, (cam, rocker arm, spring mechanism). The semi-conductor (solid-state) switches have no moving parts, hence operate without inertia and have a longer service life.

The electronic system is no longer based on the traditional energy storage principles of the ignition coil, but takes energy for the spark directly from the battery in the form of a high amplitude current pulse of short duration. This higher ignition energy remains almost constant up to the highest engine speed and sparking rates up to 1000 sparks per second are possible.

Distribution of the accurately timed sparks produced by the electronic switching system is via a rotor arm, distributor cap and h.t. cables as in the conventional coil ignition system.

The function and various forms of electronic ignition systems at present in use are also described in Chapter 7.

THE SPARKING PLUG

The sparking plug is an essential component of the ignition system, as it is the means whereby the electrical energy generated by the ignition unit is utilised to ignite the explosive mixture in the cylinder.

It is fundamentally a simple spark gap in which the points or electrodes are insulated from each other and separated by a gap of the order of 0.8 mm. The spark which ignites the mixture takes place across this small gap. The plug consists of a central stem or electrode surrounded by insulating material which is encased and sealed within an outer metal shell which serves to fit and secure the plug in the cylinder head. Welded or otherwise secured to the outer shell is another electrode, which is brought into close proximity to the central stem to form the spark gap.

Ionisation

Immediately prior to the application of the high-tension voltage to the sparking-plug gap, the gas between the electrodes is almost a perfect insulator and as such offers a very considerable resistance to the passage of an electric current. Upon the application of the voltage, a constitutional change occurs in the gas, which greatly reduces the resistance, with the result that current flows across the gap in the form of a spark. This change or process is termed *ionisation* of the gas and is the result of the electrical stresses imposed on the gas. Ionisation is the splitting up of the gas molecules into positively and negatively charged elements or *ions*. Since these ions are the vehicles of an electric current, the conductivity of the gas between the electrodes is dependent upon the degree of ionisation. This in turn is dependent upon the applied voltage, the width of the gap, the shape of the electrodes, and the temperature of the gas. Just as mechanical stresses are more highly concentrated at small sections, so also are electric stresses at fine edges or points. Ionisation is therefore more intense with pointed electrodes than with rounded or ball electrodes and the voltage necessary to produce a spark across a given gap is less for pointed than for rounded electrodes.

Apart from electrical considerations there are other factors which affect the design of the plug electrodes and the plug in general. Not only is the sparking plug subjected to high electrical stresses but also to extremely varying mechanical and thermal stresses. This will be quite evident if we consider the operating conditions. During the propagation of the explosion, the plug is subjected to a temperature approximating to $2000°$ C and on the next induction stroke is rapidly cooled by the incoming charge at a temperature little higher than normal atmospheric temperature. Further, when the explosion occurs, the pressure in the cylinder rises with extreme rapidity and may be as great as 800 lb/sq in.

Internal design

In order that the sparking plug shall be capable of withstanding these widely varying stresses, it must necessarily be of robust construction. The spark-gap electrodes are made of nickel or some alloy having a high melting point. Whilst the size of the electrodes must be fairly substantial to resist burning, it is desirable that the electrode temperature shall be sufficiently high to burn off rapidly any sooty or oily deposit. These considerations set a limit to the size of the electrodes, which will vary for different types of engine. The shape of the electrodes should also be such as to allow any oil to drain away from the spark gap.

Problems of structure

With regard to the insulator, this has to withstand high pressures and temperatures without distortion, shrinkage or cracking. Further, adequate leakage surface must be provided so that when the surface is coated with an oily carbon deposit, the spark will not leak over the surface instead of jumping across the gap.

Effective sealing of the central electrode within the insulator and the insulator within the outer shell is essential at all operating temperatures, if the plug is to remain gastight. This is a very difficult problem and is effected in a variety of ways by different manufacturers. The more usual method is to fit gland washers to take care of the difference in the expansion of the metals and insulator at various temperatures.

Sparking plug requirements and constructional details are discussed in Chapter 9 together with the various factors affecting the performance and service life of sparking plugs.

AUTOMOBILE LIGHTING REQUIREMENTS

To provide satisfactory road illumination it is essential that the headlamps should project a strong and uniformly distributed beam of light which will clearly reveal the road surface for a distance ahead. This will enable adequate braking time to be available at whatever speed the vehicle is travelling should obstacles be encountered on the road. Both sides of the road immediately in front of the vehicle must also be clearly illuminated.

This has been achieved by modern headlamps which incorporate bulbs giving the maximum amount of light for a given watts consumption and highly efficient reflectors, and lenses which control the beam pattern with great accuracy. The overall shape and form of modern headlamps are designed to give this efficient illumination, at the same time blend with the overall styling of the vehicle. The efficiency of rear and signal lamps must also be maintained in the interests of road safety, particularly in bad weather conditions and to ensure that a slow moving vehicle can be easily seen at night by oncoming or faster moving traffic.

Light measurement, designs of bulbs and reflectors and other forms of light signalling systems are discussed in detail in Chapter 6 together with the standards and statutory regulations which must be met in their design, manufacture and installation.

ESSENTIALS OF ENGINE STARTING

The function of the starting motor is to rotate the engine at a sufficiently high speed to produce satisfactory carburation or fuel injection and sparking at the plug electrodes under all climatic conditions. To do this the internal friction of the engine both from the engine bearings and the resistance of the first compression stroke must be overcome. These frictional forces are greatest when the engine is very cold and added to this there may be considerable moisture on the sparking plugs, cables, and ignition unit which renders the production of sparks more difficult.

The power output of the starting motor for a given engine is therefore decided by the extreme conditions just mentioned. The size of the starting motor on the other hand is limited by the space available for its mounting. The starting motor is usually mounted on the engine casting so that a pinion on the end of the motor shaft can be readily engaged with the gear on the flywheel. This engagement is effected automatically when the starter switch is operated, usually by turning the ignition key switch.

Probably the most common form of engaging gears is the Bendix pinion. The teeth of the pinion and also the ring gear on the flywheel are also chamfered on their end faces to facilitate meshing. Provision must also be made for the pinion to demesh automatically as soon as the engine has started, otherwise the motor could be destroyed by being driven at too high a speed, bearing in mind that the gear ratio between the starter pinion and flywheel gear ring is usually about 1:10 but may be as high as 1:20.

Details of starting motor designs, starting systems and starter switching are discussed in Chapter 4.

Aids to starting

On heavy commercial and passenger vehicles, notably those powered by diesel engines, starting under extreme climatic conditions is very

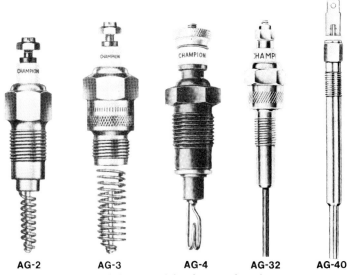

AG-2   AG-3   AG-4   AG-32   AG-40

Figure 2.4   Typical diesel engine glow plugs

difficult and auxiliary means, such as a booster battery or rectified booster current plugged into the starter circuit, are employed. Such means are, normally, only available at the vehicle garage.

Pre-heating the engine water cooling system by means of a mains energised electric heater unit is another method of facilitating starting and this may be supplemented by over-night low rate charging of the battery from a mains rectifier unit. When independence of the mains or garage equipment is necessary, an induction heater permanently fitted to the engine air intake and operated from the vehicle battery is an effective device.

Some engines are also provided with heater plugs called 'Glow Plugs' which are energised from the vehicle's battery immediately prior to starting. The construction of this type of plug is similar to a sparking plug except high voltages are not utilised. Such plugs are exemplified in the range manufactured by Champion Sparking Plug Co. as shown in Figure 2.4.

HEATING AND AIR CONDITIONING

Although climatic conditions in this country are not so severe in the extremes as for example in the USA or in tropical countries, it is now general practice for some form of interior heating and windscreen de-misting and de-frosting to be provided.

The simplest form of interior heating is provided by a small radiator heater supplied with hot coolant from the engine cooling system, with a small electric fan to draw air through the heater element and distribute the heated air via ducting to the car interior or on to the windscreen for the purpose of de-misting or de-frosting. The electric fan is driven by a small permanent magnet motor (details of which are described elsewhere). The speed of the motor can be varied by the use of switching or a rheostat in order to control the temperature of the inside of the car and the degree of de-frosting necessary. On some vehicles, in addition to the heating and ventilating systems an air extraction system is also incorporated.

In complete air conditioning systems the air inside the car is cooled and the humidity also reduced, dust and fumes are also absorbed. The functioning and constructional details of typical systems in use on various types of vehicles are discussed in Chapter 13.

FUEL INJECTION SYSTEMS

Improvements in engine performance in terms of power, torque, flexibility and fuel economy have been obtained by the use of petrol injection instead of the more orthodox carburettor or multiple car-burettors. It is also claimed that a fuel of lower octane rating may be used for a given compression ratio.

Mechanical forms of injection pumps and injectors have been in use for many years. This type of equipment injects directly into the engine cylinders against the compression pressure and, to meet the exacting requirements of both metering the fuel and timing the injection, has to be manufactured to very high standards of precision. Consequently this type of equipment, whilst possessing many advantages over the carburettor, is very costly.

A type of equipment more recently developed is controlled electronically. Whilst this form of control may appear complex, the results are achieved much more economically. In this form of injection equipment, the fuel is injected into the manifold at a much lower pressure and under less arduous conditions as regards temperature and corrosion. The actual pump elements need not be manufactured to such a degree of precision as with the high pressure system and the engine manifold design can be greatly simplified.

Where high performance is required, such as racing and some sports cars, the cost of a fuel injection system is considered justified. A detailed description of various forms of fuel injection equipment is given in Chapter 14.

ELECTRICAL ACCESSORIES

A wide range of electrical equipment comes under this heading. Some items are essential to the running of the engine such as electric fuel pumps, electrically driven cooling fans (where fitted), coolant temperature indicators, oil pressure switches, etc.

Other devices are often fitted in the interests of road safety, complete control and operation of a vehicle and passenger comfort. These comprise instrumentation, audible signalling systems (including horns), windscreen washers and wipers, electrically-operated door and seat adjusting gear and car radio. These form part of the electrical circuit and must be catered for in the complete wiring system which should include the requisite switching and protecting from short circuits and overloads.

Instrumentation is a subject of great importance in the interests of overall control of a vehicle and also to minimise driver fatigue. Considerable instrument development work has been done to meet the essential control and legal requirements also to ensure that the instruments are arranged on the facia panel in a way which blends with the design of the interior trim. Light signals, illuminated identification of controls, tell-tales, and indicators are in widespread use. These devices must conform to national and international regulations.

Items referred to above will now be considered in detail.

Electric fuel pumps

The type of electric petrol pump in very widespread use consists of a solenoid provided with a centre rod attached to an armature which is

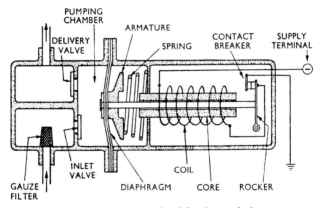

Figure 2.5   Basic principle of the electric fuel pump

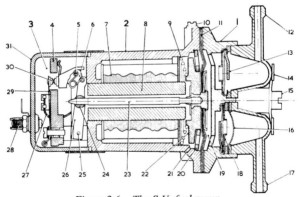

Figure 2.6   The S.U. fuel pump
See text for key to numbers

pulled towards the coil when the coil is energised. The centre rod actuates a contact-breaker mechanism located at the end of the coil remote from the armature. The armature is attached to a flexible diaphragm which moves in an enclosed cylindrical chamber fitted with simple valves so that on one stroke or movement of the armature and centre rod, fuel is drawn from the tank into the diaphragm chamber and on the next stroke it is expelled along a pipe to the engine.

Figure 2.5 shows a schematic sectional view of such a pump, the basic principle of which is that when the coil is energised it becomes magnetic and the armature and diaphragm are drawn towards the coil. The centre rod then opens the contacts and the armature is returned to its original position by a return spring and the contacts are again closed. This oscillatory movement ensures a continuous fuel supply. A highly successful example of such a pump from a range of similar pumps for specific purposes, produced by the SU Carburettor Company is shown in Figure 2.6.

The type AUF 200 pump, in common with previous types of SU high-pressure electric fuel pump, is designed to be mounted in the vicinity of the fuel tank and at a level not appreciably above that of the top of the tank. This situation ensures freedom from vapour generation troubles, even under the most severe conditions of high ambient temperature and high altitude operation. Mounted in this position and provided with fuel lines of approximately ¼ in bore, the pump is capable of supplying approximately 75 pints of fuel per hour at a delivery point about 3 ft above the level of the tank. It is thus capable of supplying fuel at this rate when ascending the most severe gradient liable to be encountered even by a vehicle of exceptionally long wheelbase.

The pump is normally provided with a Lucar connecting tag to the terminal screw 28, which can, however, be replaced by other types of connector tag if required. It is essential that a sound connection should be made to earth on the vehicle from the earthing screw 10. The most suitable method of mounting, particularly with regard to inaudibility of operation, is by a circular mounting clip surrounding the coil housing 2, from which it is separated by a soft rubber packing strip. The pump should be mounted in a horizontal position with the outlet nozzle 12 uppermost.

When a pump is being connected, it must be primed by disconnecting the fuel pipe at the carburettor until a flow is obtained.

Construction

The pump comprises three main assemblies:

1. The body casting.
2. The diaphragm, armature and magnet assembly.
3. The contact-breaker assembly.

The body

The body 1 is a casting into which the clamp plate 14, retained by two screws 15, holds the inlet 17 and outlet 12, moulded nozzles and both valve assemblies, all of which are arranged to be accessible from the outside of the pump. The inlet valve 18 consists of a thin plastic disc permanently assembled into a pressed steel cage. The outlet valve 13 is an identical assembly, but reversed in direction. A dome shaped filter 16 is provided on the entry side of the inlet valve 18. The valve allows passage to the pumping chamber 19, a shallow depression formed on the face of the body casting and bounded by the diaphragm 11.

The diaphragm, armature and magnet assembly

The diaphragm 11 is clamped at its outer edge between the coil housing 2 and the body, and attached at its centre to the iron armature 22. The armature spindle 23 passes freely through the magnet core 8 and is screwed into a trunnion 24 carried by the inner rocker 26. Eleven spherically-edged rollers 20 are fitted between the coil housing and the armature; these centralise the armature in the housing and allow freedom of movement in a longitudinal direction. An atmospheric vent 21 may be fitted to the coil housing.

The contact-breaker assembly

This consists of a bakelite-pedestal moulding 4 which carries two rockers, outer 6 and inner 26, both hinged to the moulding at one end by the rocker spindle 25 and interconnected at their top ends by two small toggle springs 5 arranged to give a 'throw-over' action. The inner rocker, as mentioned carries a trunnion into which the armature spindle is screwed. The outer rocker 6 is fitted with one or two tungsten points 30 which contact other tungsten points carried by the spring blade 29. One end of the coil 7 is connected electrically to the spring blade and the other end is connected to the terminal stud 28. A short length of flexible wire 27 connects the outer rocker to one of the screws securing the pedestal moulding to the coil housing, thus providing an earth return. This must then be thoroughly earthed to the body or chassis of the vehicle via the earthing screw 10. The contact breaker is contained in an end-cover moulding 31 secured to the pedestal by a nut and lock washer on the pedestal stud.

Action of the pump

When the pump is at rest, the outer rocker 6 lies in the position shown and the tungsten points 30 make contact. When switched on, current passes from the terminal stud 28 through the coil, back to the spring blade 29, through the points and so to earth, thus energising the coil and attracting the armature 22. The armature, together with the diaphragm assembly, moves towards the coil, against pressure from the armature spring 9, drawing fuel through the inlet valve into the pumping chamber 19. When the armature has travelled well towards the end of its stroke the 'throw-over' mechanism operates and the outer rocker moves rapidly backwards, thus separating the contact points and breaking the circuit. The armature and diaphragm will now move away from the coil under the influence of the armature spring, thereby expelling the fuel through the outlet valve at a rate determined by the requirements of the engine. As the armature approaches the end of its stroke, away from the coil, the 'throw-over' mechanism again operates, the tungsten points re-make contact and the cycle of operations is repeated.

A similar design of SU pump, the type AUF 400, is provided with two solenoids mounted back-to-back and the two pumping chambers with their associated pumping units operate simultaneously from a common inlet connection.

The AUF 500 is designed for use on vehicles on which a fuel-reserve system is provided to give a reserve of 1 to 2 gallons is a 'double-entry' pump constructionally somewhat similar to the AUF 400 'dual pump'. It differs, however, in that the two pumping chambers with their associated pumping units are not normally intended to operate simultaneously, the one being provided with an inlet connection supplied by a feed pipe which terminates short of the bottom of the fuel tank while the other feed pipe draws fuel from the bottom of the tank. Thus the pump is provided with two inlet passages marked, respectively, 'Main' and 'Reserve'. The two pumping units are alternatively energised by a two-way switch with corresponding 'Main' and 'Reserve' markings.

Motor driven electric fuel pumps

A motor driven fuel pump is used for special applications in which the motor is used to drive an impeller, or the pump is of the roller vane type as used on electronic fuel injection systems. A permanent magnet d.c. motor supplies the motive power for the pump and in the case of the impeller type the fuel envelops the whole motor and in the roller vane type (described in detail in Chapter 14) the armature operates totally immersed in fuel. This is possible because petrol is an insulating fluid and not ignitable when enclosed in a space without air.

Electrically-driven engine cooling fans

An overall increase in engine efficiency and power output can be achieved by the use of an electrically-driven fan instead of the orthodox belt-driven fan and increasing numbers of vehicles are fitted with this type of cooling as original equipment. The fan motor is provided with a thermal control which is operated by a sensor or bimetal switch sensitive to engine coolant temperature and therefore provides maximum cooling when this is required. Advantages obtained over the belt driven fan are:

1. A more rapid 'warming up' of the engine.
2. Use of choke is reduced.
3. Reduces the fuel consumption of a cold engine.
4. Cuts out fan noise.
5. Eliminates unnecessary power loss, unavoidable with the conventional belt driven fan.
6. Improves interior heating.

Figures 2.7 and 2.8 show performance characteristics for the Lucas 9GM multi-purpose motor used for radiator cooling fans and also for other applications such as interior heating, cooling and air replacement systems, water pumps, fuel pumps, mechanisms associated with automatic gear change, electric window lift and seat adjustment.

The model 9GM is of totally enclosed construction and incorporates a two-pole ceramic permanent magnet field system. It is intended for use on either 6 V, 12 V or 24 V d.c. supplies and is supplied complete with supply cables to any required length. These normally terminate in a non-reversible terminal block, but other forms of connector can be provided. The armature is carried in porous bearing bushes which are lubricated for life. The brushes are positioned on the neutral magnetic axis to allow the output shaft to be rotated in either direction according to the polarity of the electrical connections, the motor running with equal efficiency in either direction.

Warning devices

An audible warning of approach is a legal requirement on automobiles, and therefore the electric horn is an essential fitment on all types of

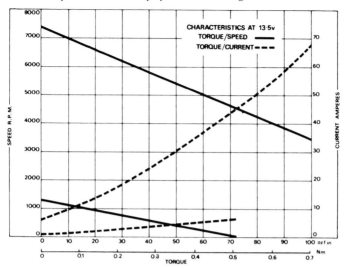

Figure 2.7   Performance characteristics for the Lucas 9GM motor

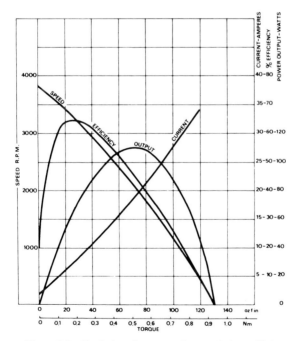

Figure 2.8   Typical performance characteristics at 13.5V

vehicle. The penetrating notes of the present-day electric horn command immediate attention without in any way annoying or irritating the public in general.

The designs of electric horns in general use for some years past are of the high-frequency impact and windtone horn types.

High-frequency horn

A good example of the high-frequency horn, which was first introduced as far back as 1929, is the Lucas horn, which is clearly illustrated by Figure 2.9. On pressing the horn button, current flows through the magnet coil and induces a magnetic field which atracts the soft-iron armature to the magnet core. This movement of the armature (and the diaphragm to which it is attached) causes the contacts to separate and thereby the current through the magnet coil is interrupted. The armature then reverts to its original position and, closing the contacts in the process, initiates the same sequence of events, which is repeated with extreme rapidity so long as the switch button is depressed. In consequence the diaphragm emits a note of low frequency and sets up in the tone disc supported on the central boss of the diaphragm vibrations of a much higher frequency to provide the harmonics or overtones.

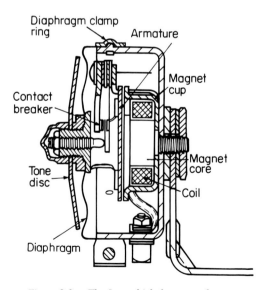

Figure 2.9   The Lucas high-frequency horn

Windtone horn

This type of horn is extensively used and the general practice is to fit them in pairs of different pitch in order to provide high and low notes, blended to give a pleasing warning.

Figure 2.10   Windtone horn with cover removed

In principle, the windtone horn is similar to an orchestral wind instrument the note being produced by setting up resonance in an air column. The air column is in the form of a passage in the die-cast base of the horn and for the 'low note' horn the passage is somewhat longer than is the case with the 'high note' instrument.

The air is set in motion in the throat of the air column by the vibration of a diaphragm actuated electrically by means of an electro-magnetic solenoid, make-and-break contacts and a push rod. The construction and general appearance of this design of horn is well illustrated by the Lucas windtone horn shown in Figure 2.10. In order to prevent extraneous vibration impairing the purity of the tone, a resiliently-mounted bracket is provided which serves to secure the horn to some rigid member of the chassis frame.

Air horns

Air-operated horns producing a strident and powerful note are popular on sports cars and some commercial vehicles in Europe. The frequency of the note produced is a function of the length of the trumpet and kits are available comprising a motor-driven compressor and two or three trumpets each producing difference notes.

There are various legal requirements to be met in different countries regarding the use of such warning devices.

The regulation concerning audible warning devices for use in this country is contained in *The Construction and Use Regulations 1973.*

This is Regulation 27 which states that every motor vehicle shall be fitted with an instrument capable of giving audible and sufficient warning of its approach or position. The sound emitted by any instrument of the kind described fitted to a motor vehicle, being a motor vehicle first used on or after 1st August 1973, shall be continuous and uniform and not strident.

Because of their relatively high current consumption, matched pairs of large diameter horns should be connected through a relay to prevent damage to the horn push switch contacts.

INSTRUMENTATION

The speedometer may be regarded as the first automotive instrument. This was produced as early as 1899 and took the form of a centrifugal governor device and had a full scale speed of 60 m.p.h. Various mechanical devices followed known as chronometric instruments and later the magnetic speedometer, all operated by a flexible drive generally driven from the gearbox.

About 1936 a remotely operated electric speedometer was produced for commercial and public service vehicles. This instrument had no mileage counter and consisted of an a.c. generator and a voltmeter scaled in m.p.h. By the late 1940s d.c. was used to operate a magnetic speedometer via a synchronous motor incorporating a mileage counter operated from the vehicle's battery.

In 1967 an electronic speedometer with mileage counter was developed in which an inductive signal proportional to speed is used to trigger a circuit connected to the vehicle's d.c. supply.

Instrument grouping

Individually-cased instruments are now rapidly disappearing on the mass produced car and instruments are grouped in small 'panels' combining

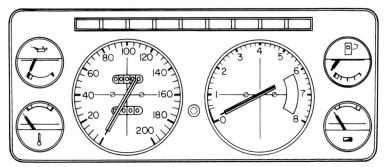

Figure 2.11   Typical instrument grouping (Courtesy Veglia Borletti Ltd.)

Figure 2.12   Instrument panel (Courtesy Veglia Borletti Ltd.)

speedometer, engine rev. counter, fuel gauge, coolant temperature
indicator, clock and several warning lights or 'tell-tale' indicator lights.
Such self-contained instrument panels are produced as a complex plastic
moulding with a very attractive facia which also has a high standard of
appearance under night driving conditions. This is achieved by the
uniform level of illumination possible in one integral unit. The dimming
of the lighting on the one unit can also be more uniformly carried out
than with separately mounted instruments. The all-electric content of
this type of instrument grouping is usually connected to a flexible
printed circuit which in turn is linked to the complete vehicle harness
by means of a multi-way plug. A typical example of an instrument group
is shown in Figure 2.12.

Impulse tachometers (rpm indicators)

A compact design of impulse tachometer has been developed by Messrs
Smiths Industries Ltd, which is in general use where a suitable alternator
drive is not available for the equipment described in the previous
paragraph.

58

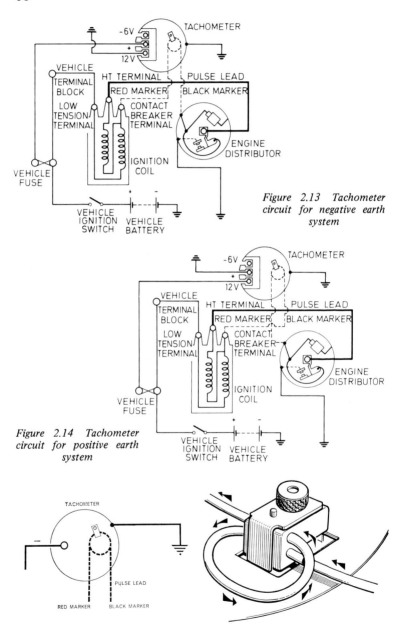

Figure 2.13 Tachometer circuit for negative earth system

Figure 2.14 Tachometer circuit for positive earth system

Figure 2.15 Pulse lead core (left) and iron core assembly (right) of impulse tachometer

This equipment consists of an indicator head and a pulse lead. The pulse lead when connected in series between the contact breaker terminal on the engine ignition coil and the CB terminal on the distributor, will transmit voltage pulses to the indicator head. The equipment employs a d.c. voltmeter movement of quite small dimensions which is controlled by a two transistor circuit contained within the indicating instrument.

Figures 2.13 and 2.14 show this external circuit arrangement for negative and positive earth systems respectively and Figure 2.15 (a) and (b) shows the arrangement of the pulse lead and iron core assembly on the back of the indicating instrument. The induced impulses in the indicator circuit are directly proportional to the engine speed and the engine ignition is in no way impaired by this operation of the speed indicator or even by the failure of the indicator. Another feature of this design is that for all practical purposes the indicator scale calibration is linear.

Electrical speed indicators

For road speed indication a mechanical speedometer driven by a flexible drive from the gearbox is still in general use on many types of vehicles. On commercial and public service vehicles an electrical system has been in use for some thirty years.

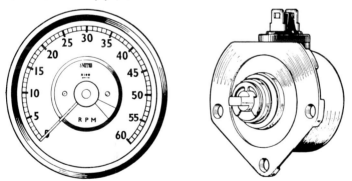

Figure 2.16   (left) indicator head, (right) generator unit

This system consists of a small permanent magnet a.c. generator (the induced voltage of which is proportional to speed) and a voltmeter scaled to read mph/kph. The generator may be mounted in any position convenient for the drive and connected to the indicating instrument (voltmeter) by a twin conductor cable. This arrangement has a distinct advantage on large vehicles where a long flexible drive would be necessary with a mechanical speedometer and also maintains its accuracy over prolonged periods of service.

The same equipment can also be used to indicate engine speed in which case the generator is driven from either the engine camshaft or crankshaft, whichever is more convenient. The voltmeter is then scaled to read rev/min. A typical indicator is shown in Figure 2.16. The voltage output of such a generator is 1 V per 100 r.p.m. and the current of the order of 25 mA.

Electronic speedometer – odometer

This system comprises a pulse generator and indicating instrument connected electrically as shown in the block diagram Figure 2.17. An example of this design is the Veglia Borletti electronic speedometer-odometer. In this design the pulse generator is mounted on the gearbox and its drive is identical to an orthodox speedometer flexible drive. This generator produces constant amplitude electrical pulses whose frequency is proportional to the gear drive. These pulses provide the information on speed and mileage and the vehicle battery provides the power to display this information on the indicating instrument.

The construction of the generator is shown in Figure 2.18. It comprises a die-cast body containing a ferromagnetic vane wheel mounted on a shaft which rotates on self-lubricating bearings and an insulator cap supporting the electronic circuit and the output connections.

The electronic circuit figure 2.19 is a high-frequency oscillator in which positive feedback is provided by two coils mounted on the same ferrite core. The ferromagnetic segment of the vane wheel reduces the energy transfer from one coil to the other thereby stopping oscillation. Oscillation resumes when the segment is drawn out of the magnetic field of the coils. Since the core of the coils is assembled opposite to the vane wheel segments, when the vane wheel turns the circuit oscillates and rests alternatively.

The circuit current in these two conditions varies considerably and this variation is the signal of the pulse generator. This operation is independent of the speed of rotation of the vane wheel; therefore there is no lack of output at low speeds, as with magnetic generators, where signal amplitude depends on speed. The high frequency oscillation is suppressed by a bypass capacitor to provide radio interference suppression.

The circuit is designed to operate over a wide temperature and voltage range. The bias of the oscillator transistor obtained by the base emitter of a matched transistor ensures satisfactory operating conditions from -40° C to $+150^{\circ}$ C and from 2 V to 35 V. The generator is completely sealed and will operate submerged in water and is vibration and shock proof up to 80 g.

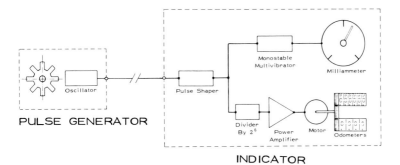

Figure 2.17   Block diagram of generator and indicator system

Figure 2.18   Pulse generator, wheel and body

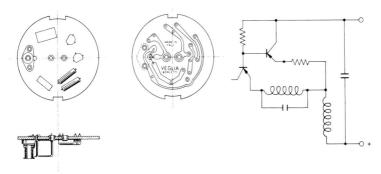

Figure 2.19   Electronic circuit of the pulse generator

The indicating instrument shows the following:

Vehicle speed
Total distance travelled up to 999,999 km/max.

Trip distance up to 9,999.9 max which may be re-set to zero. This comprises a moving coil milliameter movement mounted on a frame shown in Figure 2.20 which also carries the step motor, the two odometers C and the electronic components mounted on a control board E (see Figure 2.21) fitted to the side of the lower part of the frame.

Figure 2.20   Moving coil milliammeter

The moving coil milliameter has a high torque and wide angle and the reinforced moulded armature F (Figures 2.21 and 2.22) has moulded-in coil connections. The coil G is secured to the armature by epoxy resin and an orthodox magnetic circuit with recent modifications giving improved scale linearity compared with earlier electronic tachometer designs.

The electronic circuit Figure 2.17 consists of an integrated bipolar circuit which functions as follows:

An input pulse shaper (a conventional Schmidt trigger) which shapes the generator signal and suppresses any noise (stray signals) picked up by the cable connecting the generator to the indicating instrument. These shaped signals from the input trigger are then fed into a mono-stable vibrator and then into the milliameter instrument which indicates the mean value of the pulses. This mean value is directly proportional to the frequency of the current pulses, since these are of uniform shape and intensity and the frequency of the current pulses is proportional to the speed of the shaft driving the generator.

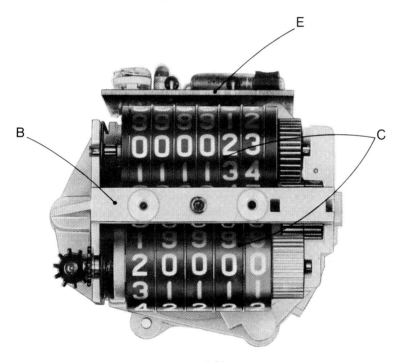

Figure 2.21

The frequency of the pulse generator is arranged to minimise pointer vibration at low speeds and, for distance counting, the output from the pulse shaper to fed to the odometers via a frequency divider circuit power amplifier and step motor powered from the vehicle battery. The dividing circuit consists of a number of cascaded flip-flops each of which divides the input frequency by two.

To make use of conventional integrated circuits, power is stabilised at 7 V on the whole circuit except the power amplifier, which can therefore provide the permanent magnet step motor with all the available voltage. A filter protects the stabiliser from any power surges in the vehicle circuit.

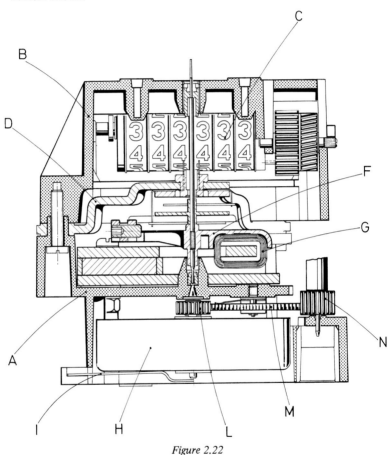

Figure 2.22

A principle of operation of an electronic speedometer developed by the Vehicle Instrumentation Division of Smiths Industries Ltd, for use on private cars is shown in Figure 2.23. From this diagram it will be seen the electronic speedometer receives input pulses whose frequency is proportional to road speed. These pulses are branched into two circuits, one for speed indication and the second for distance recording.

The speed indication pulses are shaped to controlled width and amplitude and then fed directly to a moving coil movement. Distance recording pulses are divided electronically by a factor of 64 and then amplified and fed into a stepped motor which drives the total and trip odometer via a gear reduction.

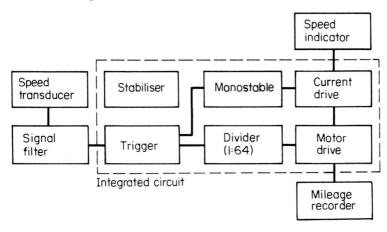

Figure 2.23   Block diagram of the speedometer operation

The speed indicator can be calibrated to any mean speed with the usual tolerances. The speedometer is a moving coil movement mounted as a sub-assembly with a thick film electronic circuit including a custom designed integrated circuit. The total and trip odometers are constructed as a separate sub-assembly complete with stepper motor. The two sub-assemblies are then united to form the electronic speedometer.

The speed transducer is a quill-driven device which fits into the gearbox take-off as a direct replacement for the conventional flexible drive.

Fuel level indicators

Designs in current use are the moving-iron resistance type and the more recently introduced bimetal resistance type. The latter was developed mainly to overcome a tendency for swinging in the moving-iron design, due to the surging of the fuel in the tank when the vehicle was moving. Both designs comprise two units, a transmitter with float arm mounted in the fuel tank, as shown in Figure 2.24, and an indicator on the facia board. The transmitters in both cases are variable rheostats but differ in construction and action.

In the moving iron-resistance type the float arm moves a sliding contact over a linear wound resistance, so that as the fuel level falls, the resistance in the circuit is reduced. The indicator, as shown in the circuit diagrams of Figure 2.25, consists of two coils, a control and a deflecting coil, which influence a pivoted soft-iron armature to which is attached the indicator pointer.

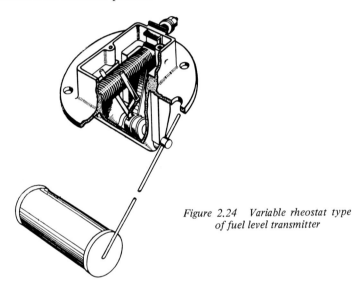

Figure 2.24   Variable rheostat type of fuel level transmitter

When the tank is empty, the transmitter resistance is a minimum and the circuit current flows almost wholly through the control coil and, in consequence, the armature is attracted to the control coil pole. With rising fuel level and the resultant movement of the float arm, the transmitter resistance is increased, diverting current from the control to the deflection coil. The armature then comes under influence from the deflecting coil and the pointer moves over the indicator scale until it ultimately registers a full tank.

The indicator in the later design is a bimetal device, the high thermal inertia of which provides a heavily damped pointer indication, not subject to variation due to normal fuel disturbance in the tank caused by vehicle movement. The bimetal-resistance combination is, however, significantly sensitive to voltage variation and for this reason a voltage stabiliser is introduced into the supply circuit.

Prior to the introduction of transistorised stabilisers, voltage stabilisers consisted of a bimetal strip sensitive to the heating effect of a winding around the strip. One end of the strip is fixed and the other end carrying a contact is free to move. The strip contact is biased to press against a fixed contact and so close the supply circuit to the heating winding. Current flowing in this winding heats the strip causing it to deflect and open the contacts, which close again on interruption of the current when the strip cools down. In this way the strip is made to oscillate and does so at a frequency depending upon the applied voltage. A drop in voltage results in a fall in the rate of oscillation and consequently an

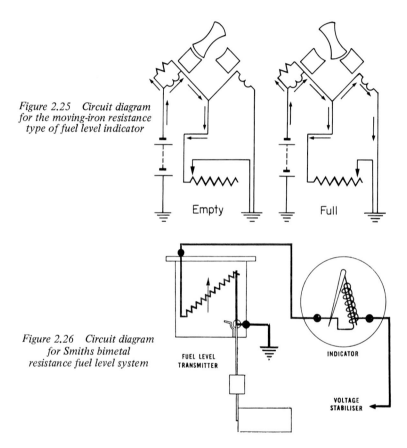

Figure 2.25 Circuit diagram for the moving-iron resistance type of fuel level indicator

Empty   Full

Figure 2.26 Circuit diagram for Smiths bimetal resistance fuel level system

FUEL LEVEL
TRANSMITTER

INDICATOR

VOLTAGE
STABILISER

increase in the period of closed contact between the strip contact and a fixed contact. This operates to maintain the energy supply to the indicator and it has been found necessary to ensure an energy supply to the system equivalent to the steady d.c. voltage. To accommodate normal supply voltage variations, the indicator calibration voltage is less than that of the supply, being 10 V for a 12 V battery system.

voltage. To accommodate normal supply voltage variations, the indicator calibration voltage is less than that of the supply, being 10 V for a 12 V battery system. This form of stabiliser can, where necessary, be used to control the energy input to a group of bimetal indicators.

To provide the required scale form for the indicator, the transmitter resistance is wound on a graded former. The float arrangements also differ from the earlier design, inasmuch as the float is turned through 90° and the float arm is loaded with a weight to vary the float submersion in the fuel throughout its travel. The transmitter resistance also increases with lowering fuel level and float arm fall and is a maximum when the tank is empty.

The bi-metal resistance system is shown diagrammatically in Figure 2.26 from which it can be seen that the indicator consists of a bimetal strip wound with a heater coil and linked to a pointer. Current flowing

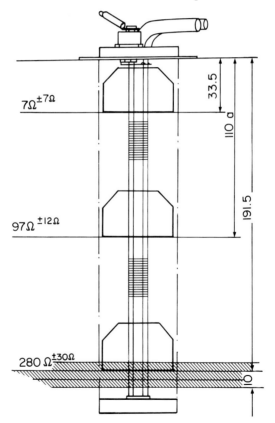

Figure 2.27   Veglia fuel level indicator

through the coil causes deflection of the strip which in turn deflects the indicator pointer, as determined by the transmitter float arm position and the magnitude of the current in the heater coil. The particular form of the bimetal strip renders the indicator particularly sensitive to variations in the ambient temperature. Since the principle of a bimetal-resistance system is applicable to temperature and pressure indication, identical bimetal indicator elements can be used for these applications, providing the characteristics of the transmitter are matched to suit.

A form of fuel level indicator developed by Veglia which is not affected by surging of the fuel in the tank is shown in Figure 2.27. From this it will be seen that the transmitter consists of a cylindrical body mounted vertically in the fuel tank, which contains an axially mounted wire resistance element on which an annular form of float moves with the level of fuel. Variation in fuel level produces a corresponding variation in resistance and voltage applied to the indicating instrument.

Coolant temperature indicators

The electrical alternatives to the Bourdon gauge are the thermal and semi-conductor types.

A circuit diagram for the thermal type is given in Figure 2.28 from which it will be seen that both the indicator and the transmitter are of the bimetal design. The latter is similar in form to the voltage stabiliser device already described. The heater windings of both units are in series and when the transmitter contacts are closed, the current flows through both windings. When the ignition switch is closed, the transmitter oscillates at a frequency depending on the temperature surrounding the transmitter. Increase in this temperature reduces the tension between the transmitter contacts and reduces the closed period of the contacts. The energy transmitted to the indicator is thereby reduced and since the energy transmitted is proportional to the oscillation frequency of the transmitter, the indicator will register the coolant temperature.

The maximum energy is supplied to the indicator when the coolant temperature is lowest and therefore, this type of indicator zeros at the hot end of the scale. Periodic interruption of the circuit current due to the transmitter oscillation is not observable on the indicator owing to the thermal lag of the bimetal.

From the circuit diagram in Figure 2.29 it will be seen that the semi-conductor type of transmitter may be used with either a moving-iron or a bimetal indicator. The availability of semi-conductor thermistors, suitable for the water boiling range and automobile voltages, has made possible this design of transmitter. A thermistor is a thermally-sensitive

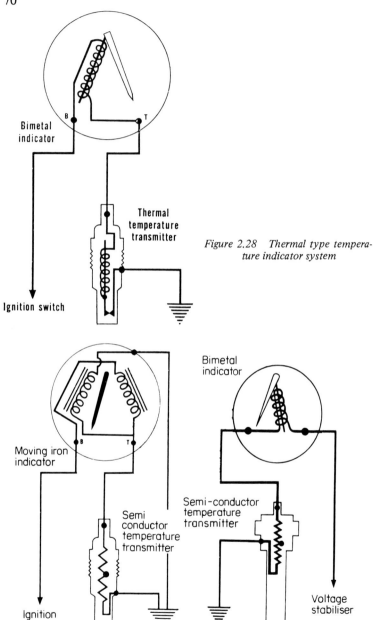

Figure 2.28   Thermal type temperature indicator system

Figure 2.29   Circuit diagrams for Smiths semi-conductor systems of temperature indicator

resistor having a high negative thermal coefficient of resistance and for the purpose under review, provides a variation of the order of 220–20 ohms over a temperature range of 50–115° C. The semi-conductor temperature transmitter comprises a brass bulb housing the semi-conductor in the form of a pellet, provided with a brass heat sink and surrounded with grease having good thermal conduction to dissipate, at a controlled rate, the self-produced heat due to current flow through the thermistor. Whether used with a moving-iron or a bimetal indicator, the thermistor acts as a variable resistor to influence the indicator reading, in accordance with the coolant temperature, in similar manner to the fuel level transmitter. When the bimetal indicator is employed, a voltage stabiliser will be required in the supply circuit.

Pressure indicating systems

Examples of the electrical pressure gauges are the thermal and the bimetal-resistance types. The similarity of the circuit diagram in Figure 2.30 for the thermal type pressure indicator, to that in Figure 2.28 will be readily apparent. The essential difference is that for pressure indication, the tension between the transmitter contacts is controlled by the pressure on a diaphragm instead of by the temperature surrounding the bimetal element.

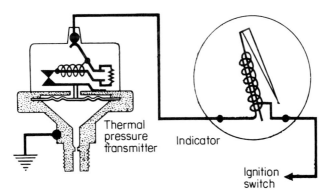

Figure 2.30 Circuit diagram for Smiths thermal type oil pressure indicator

In the case of the bimetal-resistance design, the indicator is similar to that shown in Figure 2.30 and is used in conjunction with a variable resistance transmitter actuated by pressure on the transmitter diaphragm. A voltage stabiliser is also necessary with this type of indicating system.

Battery condition indicator

This is a voltmeter with a specially calibrated dial which indicates the condition of the battery. The movement is very similar to that in the bimetal type of instruments.

Air cored instruments

For service in environments which are too severe for bimetal instruments, such as for heavy tractor applications where high vibration levels are encountered, an air cored instrument has been developed by Messrs. Smiths which may be used for measuring:

Fuel level
Temperature
Pressure
Battery condition

It also replaces bimetal instruments where rapid response is required, such as oil pressure measurement. (A new pressure transmitter has been developed for this application – the Resistance Pressure Transducer).

The air cored indicator can be considered as a compass in a magnetic field, that can be made to change its direction with the compass always remaining in alignment with the surrounding magnetic field. This field is produced by electric current through several coils and changing the relative value of currents in the coils, alters the direction of magnetic field. The method of operation is as follows.

A bar magnet, pivoted about its centre, is mounted at the centre of three coils (Figure 2.31(a)). To cause the magnet and hence, the pointer to move, the relative currents in the three coils is changed. There are two types of instrument:

(a) Fuel, temperature and pressure.
(b) Battery condition

As it is easier to understand the operation of the battery condition indicator, this will be considered first (Figure 2.31b). The direction of magnetic field that would be produced by electric current being passed through each coil, individually, is indicated by the arrow adjacent to that coil. The direction of magnetic field that would be produced by each coil is fixed, but field intensity can be varied by altering the current.

When the three coils are energised simultaneously, these interact to produce a magnetic field with a direction that can be found as shown in Figure 2.31c. The magnetic field that would be produced by each coil can be considered as proportional to the number of turns in that coil, times the current passing through it (ampere turns).

The length of the arrows shown in the diagram is equal to the ampere turns and from this, it can be seen that, by altering any one, the direction of the resultant field will change, e.g. by increasing the ampere turns in coil C, the direction of resultant field can be made to rotate anticlockwise (Figure 2.31d).

In the battery condition indicator (which is a voltmeter), at all voltages above the minimum scale reading, the current in coils A and B is constant and the current in coil C increases as voltage increases. Hence, referring to the previous figure, the direction of the resultant magnetic field will rotate anticlockwise, as voltage is increased and the magnet and pointer will follow. The full circuit diagram for the battery condition indicator is shown below (Figure 2.31e).

The resistances are used to reduce the power dissipation in the coils, to prevent overheating and provide adjustment for calibration. The zener diode ensures that coils A and B receive constant current irrespective of battery volts (assuming battery volts are equal to, or exceed minimum scale reading of indicator). To understand how the zener diode works in this application, it can be considered as described below.

Starting at a low battery voltage (below bottom chaplet) and gradually increasing the volts, the zener diode will, at first, pass no current and the current through all three coils will increase proportionally; the direction of the resulting magnetic field will not alter. At about the bottom chaplet voltage, the zener diode will start to conduct and, as the battery voltage is increased, the zener diode will pass more and more current, keeping the voltage across coils A and B constant, hence keeping the current in these coils constant. The current through coil C will increase, as described previously, as voltage increases, hence causing the direction of magnetic field to change and the magnet and pointer to move.

The circuit diagram for fuel, temperature and pressure indicators is shown in Figure 2.31f. The transmitters are all variable resistors and as these vary, the current through coil C will change. For instance, for a fuel gauge, the 'empty' resistance value is approximately 250 ohms and this will change to about 20 ohms at 'full'. The current in coil C will be at a maximum when the transmitter resistance is about 250 ohms and will reduce to a minimum value of current when the transmitter resistance falls to about 20 ohms (Figure 2.31g). Whilst the current through C is altering, the currents in A and B will also vary, but to a lesser extent. The result is that the direction of magnetic field alters with transmitter resistance and the magnet and pointer move rotationally.

In all types of indicators, when the supply is switched off, the magnetic field from the coils disappears and the magnet and pointer would remain at the value indicated before switch-off. To prevent this, a permanent magnet is placed near the coils to produce a small magnetic field that returns the moving magnet and pointer to the Zero end of the scale upon switch-off. This permanent magnet which is called the 'pull-off' magnet is fitted to all types of air cored indicators.

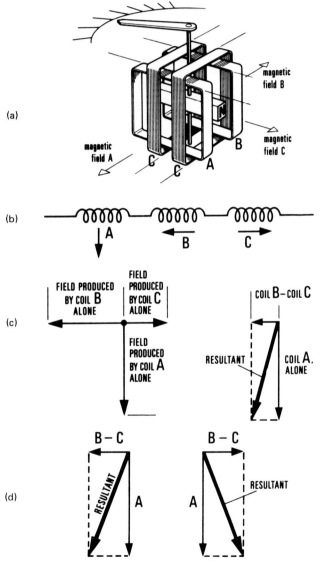

Figure 2.31   Showing the operation of Smiths air-cored indicator

The above description on how the indicators function has been kept as simple as possible and many minor factors have been ignored. For instance, the effect of the pull of magnet on the magnetic field produced by the three coils has been ignored, but this must be taken into account when calculating indicator characteristics, voltage errors, etc.

The action of the zener diode has also been over simplified to describe its function more clearly.

The advantages of air cored indicators (other than battery condition), are that they are self-compensated for voltage changes and no voltage stabiliser is required. Response is instantaneous. The indicator is rugged

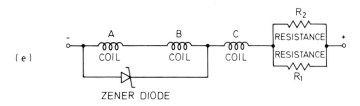

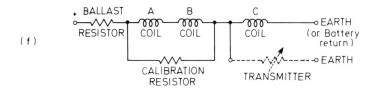

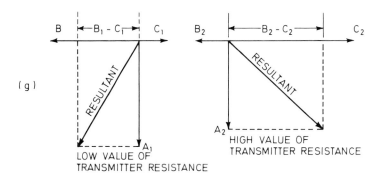

Figure 2.31 (continued)

and able to withstand high levels of vibration, cause no electrical interference with radios etc., and the indicators can be edge-reading.

Dual level monitoring

A dual level monitor has been developed by Smiths Industries Ltd for use on both private cars and on commercial vehicles. The unit has two channels, as shown in Figure 2.32, each capable of monitoring level in any system of the vehicle which appears as a variable resistance or

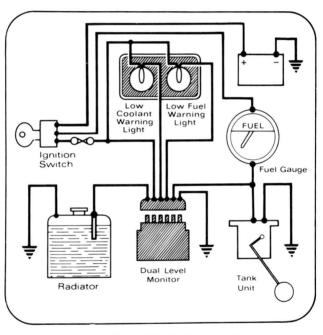

Figure 2.32   Smiths dual level monitor

variable voltage. A visual warning is given to the driver by a flashing light indicating a change beyond the stipulated level within the service, but if both channels are triggered then the separate warning lights will operate in antiphase. The response time can be varied to suit particular service requirements from less than one second for low coolant to above thirty seconds for low fuel reminder.

The dual level monitor operates by a pulse sampling technique which allows a high sampling current to give greater accuracy whilst maintaining very low average current. The service being monitored is unaffected by the unit and in the case of an electrically conductive liquid, corrosion, due to electrolysis is inhibited. The dual level monitor is suitable for 12 V or 24 V d.c. electrical systems either negative earth return or insulated return.

Typical services which could be monitored on motor vehicles using the dual level monitor are:

1. *Coolant low level.* Using a probe or probes in the coolant.
2. *Coolant high temperature warning.* Reducing resistance (RR) operated in conjunction with electric temperature indicator system, the signal obtained from the existing transmitter.
3. *Liquid low level.* Any liquid having an electrical resistance between the two probes or probe and earth of less than 500 K (possibly brake fluid or clutch fluid but engine and gearbox oils are unsuitable).
4. *Oil pressure low.* Operated in conjunction with electric pressure gauge systems, the signal obtained from the existing transmitter.
5. *Fuel low level.* Operated in conjunction with the electric fuel gauge system, the signal obtained from the existing tank unit.
6. *Vehicle supply voltage.* Indicating either high voltage (faulty voltage regulator) or low voltage (RV) (undercharged battery).
7. *Windscreen washer low level.* Operated with probes immersed in liquid.

Combination of any two of the above services may be incorporated in one dual level monitor providing that both signal channels are not triggered by reducing voltage (RV) or reducing resistance (RR) action of the transmitter.

Warning light and indicator colours and markings

It is general practice to incorporate group indicators within the instrument cluster. The colour and form of marking with appropriate symbols is covered in an EEC Directive (78/316/EEC).

Solid state instrumentation

A recent development of great interest for the future is the illumination of the instrument panel by electroluminescence. This phenomenon occurs when some forms of phosphor are subjected to the action of an electric field.

A sandwich or laminar type of construction is used in making up an electroluminescent panel. Figure 2.33 shows an example of this type of instrumentation developed by Smiths Industries Ltd, to whom the author is indebted for the technical data and relevant diagrams.

The construction is shown in Figure 2.34. When a direct current is passed from the rear electrode (−) through the phosphor to the transparent conductive pattern (+) the molecular layer in immediate contact with the conductive pattern on the glass emits light. A 'television screen' type of display is produced which can be offered in various colours by filtering and can include functions such as speedometer, tachometer,

Figure 2.33   An electroluminescent panel

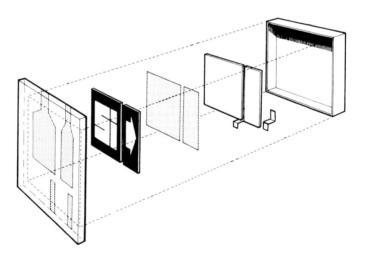

Figure 2.34   Construction of electroluminescent panel

clock, odometer, tell-tale indicators as well as fuel, coolant temperature and battery condition indicators. The display area of the device theoretically has no limitations and because all the functions displayed can be accommodated on one panel, this automatically leads to a more compact display than that achievable with conventional instruments. There are no moving parts in the system and the use of integrated circuits in which many active components are on one silicon chip makes for both efficiency and reliability.

Electric clocks

Various forms of electric clock are in use on automobiles. In an early design the balance wheel was in effect a motor armature driven from the vehicle battery. In a later design the mainspring of an orthodox mechanical movement was rewound electrically at 1−2 minute intervals. When the main spring approached run-down, a pair of contacts closed a solenoid circuit. The movement of the solenoid armature rewinding the spring and opening the contacts before the end of its travel.

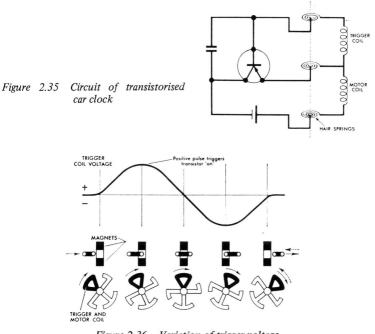

Figure 2.35   Circuit of transistorised car clock

Figure 2.36   Variation of trigger voltage

Contactless solid state switching led to the development of the transistorised car clock. Two version are in use; in one, a permanent magnet is stationary and triggering and motor coils are mounted on the balance wheel, and in the other version, the reverse arrangement is used.

In both cases, when the balance wheel is given a starting swing a voltage is induced in the triggering coil. This causes the transistor to switch thus energising the motor coil which in turn provides a force pulse on the balance wheel. The gear train from the armature to the clock hands may be regarded as a conventional escapement driven in reverse.

A typical example is the Smiths transistorised car clock which is essentially a powerful oscillating coil motor in which iscillation is maintained at the required accurate frequency by a transistorised circuit (Figure 2.35) and hair springs. Once the clock is started a current pulse is generated in the trigger coil which switches a transistor; this causes the motor coil to be energised helping the balance through the magnetic field against the hair springs. When the trigger coil emerges from the field the transistor switches off and the hair spring starts to return the balance through the field and the above procedure is repeated. Figure 2.36 shows diagrammatically this operation.

An efficient magnet system and the use of high energy magnets incorporated in the motor ensure that it is capable of maintaining operation against the increase of frictional resistance caused by very low ambient temperatures. The hair spring material is also closely batch controlled to ensure good temperature compensation.

The transistorised car clock is manufactured in two basic forms, one type is powered from the vehicle battery and another type is energised by a small mercury cell housed spearately in a special holder.

Quartz clocks for automobiles

Miniaturisation in electronics and improvement in the techniques of quartz manufacture have enabled a quartz car clock to be produced at a price acceptable to the vehicle manufacturer and providing an accuracy hitherto found only in master and laboratory clocks.

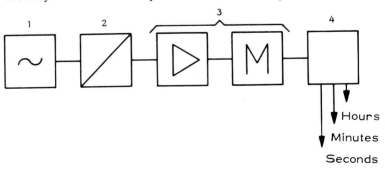

Figure 2.37   Block diagram sequence for a quartz car clock

Figure 2.38   Assembly and reduction gear arrangements of quartz clock (Courtesy Veglia Borletti Ltd.)

The operating principle of quartz clocks is very simple. The high stability of quartz oscillations, due to the piezoelectric properties of the crystal, is utilised to pilot a synchronous motor at a constant frequency. The quartz crystal is kept oscillating by an electronic circuit; another circuit steps the quartz frequency down to pilot the motor and drive the clock hands through a reduction gear. The speed of the clock hands, hence clock precision, is rigidly controlled by the quartz crystal extremely stable oscillation frequency, the frequency step down circuit, the synchro motor and the reduction gear.

An example of a quartz clock for automobile use is the quartz car clock produced by Veglia Borletti. This consists of the following main parts:

1. Quartz crystal oscillator.
2. Frequency step down circuit.
3. Motor and motor pilot circuit.
4. Mechanical assembly which includes reduction gearing and time hands (seconds, minutes, hours).

These parts are arranged in the sequence shown as a block diagram in Figure 2.37. The assembly of the electronic components and reduction gear arrangement is shown in Figure 2.38.

It will be noted that the mechanical assembly does not contain such delicate components as balance wheels, hair springs, or escapement assemblies. This simplification in the 'clockwork mechanism' greatly assists in maintaining accuracy in vehicle service. The complete assembly is unaffected by vehicle vibration under the most severe conditions and its accuracy is unaffected by vibrations up to 300 Hz and 10 g, also over a temperature range of -40° C to $+85^{\circ}$ C.

Electronic hour meter

The electronic hour meter for use on tractors and earth-moving machinery is basically one of the clock units already described which is not affected by vibrations of low frequency and high amplitude, geared to a mechanical hours counter. The counter mechanism is usually in the form of an assembly of plastic counter wheels geared to each other by means of external pinions. The unit is self-starting when electrically energised.

Windscreen wipers

Electric windscreen wipers have the advantage that they are able to operate at a regular speed, irrespective of the speed of the car. These are operated by a small electric motor working through gearing and some form of eccentric pin with rack and pinion action to provide steady oscillatory movement of the wiper arm.

The deeper screens of modern vehicles, together with the trend towards increased convexity, has necessitated increased actuating torque and power from the driving motor. For heavily curved screen surfaces even articulated wiper blades have become necessary. The manually parked wiper has now been superseded by a design with a self-switching device, which automatically parks the wipers in the rest position on switching off the control switch.

There are two types of wiper in general use, the flexible rack and the link types. The former type has the advantage of ease of installation but is less efficient than the link type and requires greater power output from the driving motor, the current consumption on a 12 V system being 3 A. Although very efficient, the link type has two disadvantages, the first being that the linking mechanism requires appreciable operating space and therefore is not so readily accommodated as the flexible rack type. The second drawback is that the angle of wipe is restricted by the need to avoid approaching too closely the dead centre of the wiper crank member, at either end of the oscillation travel. Noise is a factor which presents quite serious difficulties and is largely overcome in the flexible rack type by using single worm-gear reduction. This consists of a steel worm meshing with a moulded nylon gear wheel, to which is attached the crank actuating the wiper cable racks.

The forms of screen wiper motors so far referred to have been wound field machines. It is now more general practice to use motors having permanent magnet fields, such as the Lucas Model 14W which comprises a self-switching power unit which drives two wiper arm wheel-boxes by means of a flexible cable rack running through a rigid tube. The two-pole motor has a permanent magnet field consisting of two ceramic magnets housed in a cylindrical yoke. A worm gear formed on the extended armature shaft drives a moulded gearwheel within the die-cast gearbox. Motion is imparted to the cable rack by a connecting rod and crosshead actuated by a crankpin carried on the gearwheel.

Associated with the terminal assembly is a self-switching limit switch unit. Two-stage contacts inside the switch are operated by a plunger, which in turn is actuated by a cam on the underside of the moulded gearwheel inside the gearbox. When the manually operated control switch is moved to 'OFF' (or park) the motor continues to operate under the automatic control of the limit switch. When the wiper blades reach the parked position, the first-stage contacts open and the motor is switched off. A momentary period follows during which no contact is made by the switch, then the second-stage contacts close causing regenerative braking of the armature which maintains consistent parking of the blades.

The motor is produced in single and two-speed form. To provide the latter requirement, the brush-box plate is fitted with a third brush to which the armature positive feed is switched when the second (higher) speed is required. An exploded view of this motor is shown in Figure 2.39.

A later design of windscreen wiper by Messrs Lucas is the 16W model for cars and light commercial vehicles, with a recommended maximum length of arm and blade of 16 in (406 mm). The unit comprises an electric motor 3 in (76 mm) yoke diameter with integral gearbox. The motor field system is provided by two high energy permanent magnets. Reciprocating motion is transmitted from the gearbox through a cable rack running in a rigid tube, formed to suit individual layouts, to two scuttle-mounted wheelboxes and hence to the wiper arms.

Single and two-speed motors are available, the higher speed being achieved by the addition of a third brush, a self-parking switch mechanism is usually incorporated in the gearbox: when the motor is switched off the self-parking action results in the wiper blades being parked unobtrusively below the normally wiped area of the screen. Alternatively, a self-switching version is available in which the blades are parked at the limit of the normal wiping arc. The speed of the operation for two-speed

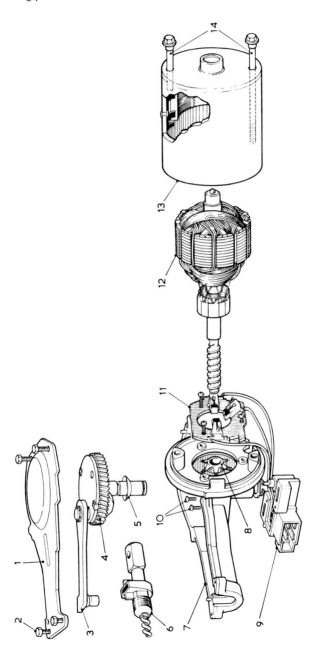

Figure 2.39 Exploded view of Lucas windshield wiper motor model 14W (two-speed)

1. Gearbox cover
2. Screw (gearbox cover fixing)
3. Connecting rod
4. Shift and gear
5. Dished washer
6. Cable rack with crosshead and outer casing ferrule
7. Gearbox
8. Set-aligning bearing brush
9. Limit switch assembly
10. Screws (limit switch fixing)
11. Brushgear, comprising: insul. plate and brushboxes, brushes, springs, fixing screws
12. Armature
13. Yoke assembly, comprising: two permanent-magnet poles and retaining clips and armature bearing brush
14. Bolts (yoke fixing)

models is 50 and 70 wiping cycles per minute and for single speed models 50 wiping cycles per minute.

A typical wiper for commercial vehicles is the AC-Delco unit. This is a powerful wiper and the motor operates in conjunction with a two-stage reduction gear, providing an overall reduction of 50:1 and a final drive speed of 60 rev/min. To reduce fluctuations in the motor torque requirements, a balancing system is provided comprising a tension spring attached to the linkage rods. This spring absorbs power on the downward stroke of the blades and gives up power on the upward stroke.

It is a statutory requirement for windscreen wipers to operate correctly.

Headlamp wipers and washers

Headlamp wipers and washers generally similar in design to the units described for windscreen use, are now being fitted to many vehicles particularly to meet the demand in Scandinavian countries for operation in heavy snow and mud.

Windscreen washers

It is now also a statutory requirement for vehicles to be fitted with a means of applying a jet of water or cleaning fluid onto the screen in front of each wiper blade to enable the blades to clean the glass effectively. This was originally achieved by a very simple mechanical pump operated by a push button on the dashboard. This function is now carried out electrically. A typical example of this is the Lucas 'Screenjet' model 5SJ.

This unit consists of a 12 V permanent magnet motor on the top of the screw cap of a plastic water container. The motor drives a small centrifugal pump which delivers water, through plastic piping and a nozzle, to give two jets on the windscreen. The operation is controlled by a push-button switch and electrical connections are via 'Lucar' connector blades.

The pump and water container are located in the engine compartment but a position must be chosen in which it will not be subjected to undue heat from exhaust pipes or manifolds and where it will be cooled by airflow. The tubing between the container and nozzle should be as short as possible.

A wide range of such washers is on the market and the latest practice is to have a motor/pump unit separately mounted from the water container. An example of this type is the Lucas Model 10SJ windscreen washer. This is fully insulated and therefore suitable for either positive or negative earth electrical systems.

The windscreen washer comprises a small electric motor coupled directly to a self-priming pump which transfers stored water, via plastic tubing, to the jet nozzle(s) and thence to the windscreen. A permanent magnet motor, controlled by a push-type switch or via a steering column mounted switch drives a pumping element incorporating a multi-bladed rubber impeller. The motor/pump assembly is available either with an integral bracket for mounting to the vehicle body (the water being stored in a separately mounted container) or fitted to the cover of the water container. A filter is incorporated in the submerged end of the inlet tube. The inlet and outlet ports are indicated.

If the water supply is exhausted during washer operation, the motor will run for a further period not exceeding 20 seconds without damage to the pump.

The operating current is 2.8 A (at 13.5 V); 1.4 A (at 27.0 V) and the operating voltage range is 11 to 15 V (12 V units); 22 to 30 V (24 V units). The pump delivers water at a minimum rate of 10 gallons (45 litres) per hour at a pressure of 10 lbf/in^2 (68.95 kN/m^2).

The operating temperature range over which the pump will operate satisfactorily is up to 80° C down to $-23°$ C (with approved anti-freeze additions in the container).

C.A.V. electrically-operated door gear

On public service vehicles the use of power-actuated doors has obvious advantages. The C.A.V. electrically-operated door gear, shown in Figure 2.40, comprises a reversible electric motor with a worm reduction drive, operating folding doors through the medium of suitable levers and rods. The double lever above the gear shaft operates the door shafts at each end by means of rods through cranked levers. An overload slipping clutch is fitted between the reduction gear and the double lever.

The overload clutch is provided for two purposes; first, to permit the opening of the doors by hand in the event of electrical supply failure and second, in the event of any obstruction in the doorway or runners, the clutch will slip and so prevent damage to the mechanism. Similarly, it will also prevent injury to any passenger caught by the door.

Figure 2.40   CAV electrically-operated door gear

The doors are operated by push buttons under the control of the driver. For use in emergency, additional 'OPEN' control buttons are provided inside and outside the vehicle. The power consumption is very small, amounting to 0.01 Ah approximately for each cycle of opening and closing. A time period of about three seconds is required for opening or closing the doors.

CHAPTER 3

THE ALTERNATOR AND D.C. GENERATOR

A vehicle electrical system is subjected to ever increasing loads as more sophisticated electrical equipment is developed. This is particularly the case with equipment having relatively high current consumption such as air conditioning systems, heated rear windows, power operated windows, power-operated seat-adjusting mechanisms, high wattage driving and fog lamps, etc.

It is essential for the battery to be fully charged by an engine-driven alternator or generator which automatically maintains the battery fully charged so that it can provide reliable starting and supply all electrical services under severe cold conditions. The generator should also be capable of quickly re-charging a 'flat' battery after a difficult start and also prevent over-charging and possible damage to the battery in hot climatic conditions.

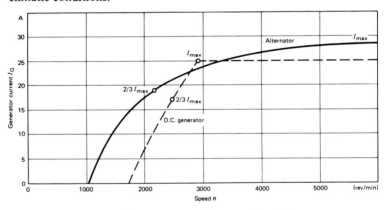

Figure 3.1   Current as a function of rev/min

For many years the d.c. generator or dynamo with its electro-magnetic regulator and cut-out relay has fulfilled these requirements. As engine speeds and battery loads increased the inability of the dynamo to provide sufficient charge at low and idling speeds and commutation problems at

88

high speeds have resulted in its replacement by the alternator with its built-in rectifier and electronic regulator and its ability to run at higher speeds and cut-in at lower engine speeds. Other advantages offered by the alternator are its light weight in relation to output, and that it requires little or no maintenance; also less wear occurs giving longer life. As a result of the extended speed range possible with an alternator, the power requirements of a vehicle can be met even when the engine is idling and the battery size may also be reduced since rapid charging is guaranteed.

Figure 3.1 shows comparable output characteristics for an alternator and a d.c. generator of about the same maximum output. It will be seen that the alternator starts to supply current at lower speeds which means battery charging begins at lower engine speeds.

THE ALTERNATOR

Principle of electromagnetic induction

Alternator design is based on the principle of electromagnetic induction and the fact when an electrical conductor cuts through the lines of force of a magnetic field, an electrical potential (electromotive force – e.m.f.)

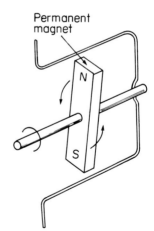

Figure 3.2   Alternator principle. The magnet rotates within a stationary loop of wire

is induced in the conductor. This principle is shown in its simplest form in Figure 3.2 which shows how a permanent magnet rotated in a loop of wire will cause an e.m.f. to be induced across the ends of that loop of wire.

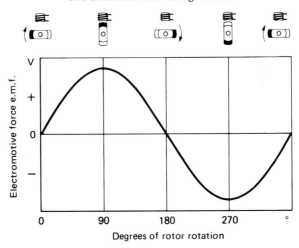

Figure 3.3   Curve of induced alternating current during one turn of the rotor

In an automobile alternator the permanent magnet is replaced by an electromagnet; the rotor (which rotates) and the loop of wire is replaced by a large number of loops or turns which remain stationary and form the stator winding in which the e.m.f. is induced and the output current is generated. This induced voltage will be alternating in both magnitude and direction, and if the rotor is turned at a uniform rate will take the form of a sine wave, as shown in Figure 3.3.

ALTERNATOR DESIGNS

The claw pole alternator

In most practical applications the stator coils are placed in slots around the laminated iron stator core and the rotor has a single winding co-axial with the shaft and embraced by two claw-shaped iron pole pieces to form a multi-pole rotating magnetic rotor with North poles derived from one end of the winding and South poles derived from the other end. To obtain the maximum output from a given weight it is common practice to use a claw pole rotor with 12 poles and a three phase stator winding as indicated in Figure 3.4 which shows a sectional view of the Bosch K1 claw pole alternator.

Figure 3.5 clearly indicates the claw pole principle and the way in which each half of the claw poles having six fingers or pole pieces

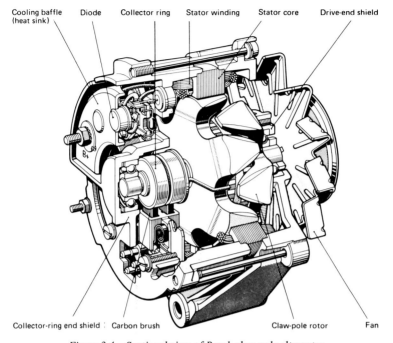

Cooling baffle (heat sink) | Diode | Collector ring | Stator winding | Stator core | Drive-end shield

Collector-ring end shield | Carbon brush | Claw-pole rotor | Fan

Figure 3.4   Sectional view of Bosch claw-pole alternator

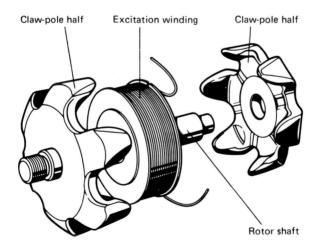

Claw-pole half | Excitation winding | Claw-pole half

Rotor shaft

Figure 3.5   Parts of a claw-pole motor

embraces the excitation winding which is wound co-axial with the rotor shaft. The complete rotor becomes a rotating electro-magnet when energised via slip or collector rings mounted on the shaft and carbon brushes mounted on the stator and shield. The stator winding consists of three separate windings or phases in which an alternating current is induced. This form of winding utilises the available winding space in the most efficient manner. The three-phase current generated in the stator windings is then rectified by six diodes (three positive and three negative) mounted in heat sinks fitted in the collector ring endshield.

The excitation current for the generation of the magnetic field is tapped from the stator winding and is rectified by three separate exciter diodes and the three negative power diodes. Excitation of the rotor is initiated from standstill by the effect of residual magnetism in the iron core with the assistance of current drawn from the battery via the ignition switch and the ignition warning lamp. It is essential that the current taken by the warning lamp is sufficient when passing through the excitation winding, to produce a magnetic field strong enough to start self-excitation of the alternator. Details of the circuitry required for excitation and output control are given later in this chapter under the heading 'Electronic Regulators'.

Another type of the claw pole alternator is designed for high output and arduous duty such as that required on buses. The basic rotor/stator design is similar to the claw pole arrangement described earlier except the iron circuit is of heavier section. The collector ring compartment is enclosed for dust protection, ensuring the longest possible life for the brushes. Constant lubrication of the bearings is provided by the grease channels and grease cups. The suppression capacitor and the air intake vent are fitted to the cover of the collector ring end shield.

Claw pole alternator with exciter (without collector rings)

A further alternative form of construction of the claw pole alternator has been developed to give still longer periods without maintenance. This machine has neither carbon brushes nor collector rings and maintenance intervals are determined only by the ball bearings.

Figure 3.6 shows the construction of such a machine. It will be seen from this sectional drawing that the collector ring section is replaced by a special exciter which is in itself a miniature alternator having a rotating three-phase winding and a stationary excitation winding. The current induced in the three-phase winding of the excitor is rectified by diodes and used to energise the excitation winding of the main rotor

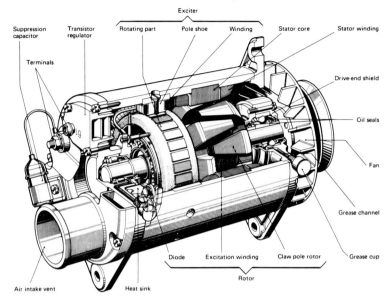

Figure 3.6   The Bosch T4 claw-pole alternator without collector rings

of the alternator. This machine has a transistor regulator built into a cavity in the end shield on the air intake side.

Alternator construction

The claw pole design has been adopted by many British, Continental and American companies and is manufactured in large quantities for worldwide use. This design is exemplified in the Lucas range of alternators.

In standard form, these alternators are intended for use on cars and light commercial vehicles. Alternative versions are available for agricultural and industrial applications where larger section driving belts are usually employed or where there is increased risk of contamination. The alternators are suitable for negative earth systems only.

Nominal rated outputs (hot) at 6000 rev/min are:

15ACR	28A
16ACR	34A
17ACR	36A
18ACR	45A
23ACR	55A
25ACR	65A

Nominal machine voltage is 14.2 V (at 20% rated output).

The first four models listed above employ a common frame diameter, the variations in rated output being achieved by differences in rotor and stator windings. Also the principal mechanical features are identical, except that stator length on the 17ACR and 18ACR models is increased by approximately 6 mm and the 18ACR model has a larger diameter fan and drive end bracket to provide adequate cooling. Apart from the exceptions mentioned, the component parts employed in the construction of the alternators are in general interchangeable, so making for ease of servicing.

Models 23ACR and 25ACR employ a stator diameter approximately 19 mm greater than that for the 15–18ACR range. Except for the use of a larger fan and drive end bracket on the 25ACR the two models are of the same basic physical size and construction, the alternative rated outputs being achieved by different rotor and stator windings.

In all models, a voltage regulator is incorporated on the slipring end casting, to provide a composite generating and control unit. The system is arranged for direct connection of a charge indicator light.

Alternator construction

The alternator is of rotating field, ventilated design. It comprises principally:

1. A laminated stator on which is wound a 3-phase output winding; delta-connected or star-connected.
2. A 12-pole rotor carrying the field winding. The rotor shaft runs in ball race bearings which are lubricated for life.
3. Natural finish aluminium die cast end brackets, incorporating the mounting lugs.
4. A rectifier pack for converting the a.c. output on the machine to d.c.
5. An output control regulator.
6. A surge protection diode.

Figure 3.7 shows external and exploded views of the 15–18ACR range and Figure 3.8 similarly indicates the construction of the 23 or 25ACR alternators.

The rotor is belt driven from the engine through a pulley keyed to the rotor shaft. A pressed steel fan adjacent to the pulley draws cooling air through the machine. This fan forms an integral part of the alternator specification and has been designed to provide adequate air flow with a minimum of noise. The alternator can be driven in either direction provided the appropriate fan is used. Maximum continuous rotor speed is 15000 rev/min with a short duration peak speed of 16500 rev/min.

In standard form, the alternator is fitted with normal duty ball race bearings, that at the slip-ring end being of shielded type. On the alternative versions, a heavy duty sealed bearing is employed at the drive end and the slipring end bearing is also of sealed type.

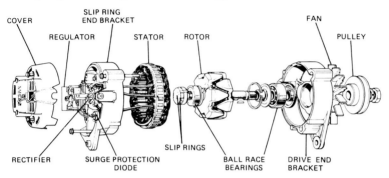

Figure 3.7   Typical Lucas 15 to 18 ACR range alternator

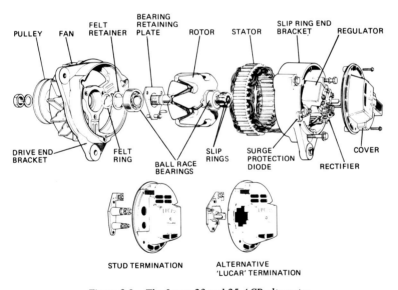

Figure 3.8   The Lucas 23 and 25 ACR alternator

The standard and recommended form of installation is that in which the alternator is swing-mounted from holes in the drive-end and slip-ring end brackets; with this configuration, vibrational stresses are minimised. An alternative form of mounting by means of a single lug on the drive-end bracket is available if required. With both methods of mounting a further lug provides belt adjustment. Alternative left and right hand versions are available for this belt adjusting lug, to cater for various engine mounting positions.

Rectification of alternator output is achieved by six silicon diodes housed in a rectifier pack and connected as a 3-phase full wave bridge. The rectifier pack is attached to the outer face of the slip-ring end bracket and contains also three 'field' diodes which, together with three of the output diodes, form a second bridge through which rectified current from the stator windings passes to give self-excitation of the field at charging speed.

The face-type slip-rings are carried on a small moulding attached to the rotor shaft, outboard of the slip-ring end bearing. The inner ring is centred on the rotor shaft axis, while the outer ring has a mean diameter of 19 mm approx. The slip rings are connected to the rotor field winding by wires carried in grooves in the rotor shaft.

A moulding screwed to the outside of the slip-ring end bracket carries the brushgear and encloses the slip-ring assembly. On standard alternators, this provides protection against the entry of dust and moisture. Alternators intended for use in highly contaminated environments incorporate further protective measures which ensure that the slip-ring/brushgear is fully sealed.

The output control regulator is of electronic pattern, having in-built radio interference suppression. It is set during manufacture and requires no further attention.

The purpose of the surge protection diode is to protect the regulator output transistor against over-voltage surges which may arise in the event of external wiring failure or battery disconnection.

The voltage signal required for the control system may be provided either from within the alternator ('machine sensing') or from the battery via a separate voltage sensing lead ('battery voltage sensing'), as described later in this chapter under 'Electronic regulators'.

Electrical connections to external circuits are brought out to 'Lucar' connector blades, these being grouped to accept a non-reversible moulded connector socket on the cable harness. Spring-clip retention of the connector is available if required. Models 23ACR and 25ACR are available also with stud terminals for use with eyeleted cables.

Charge indicator light

Provision of the three additional 'field' diodes makes possible the use of a simple charge indicator light.

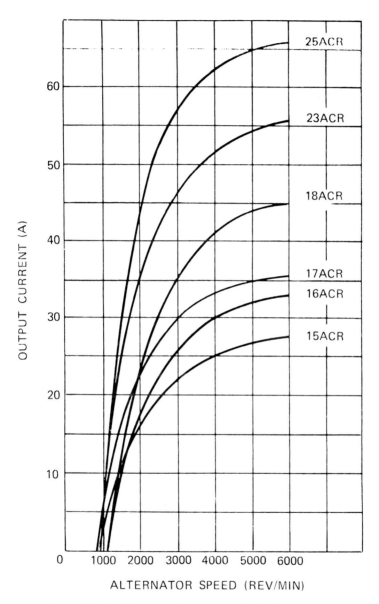

Figure 3.9 Typical output characteristics (hot) of an alternator

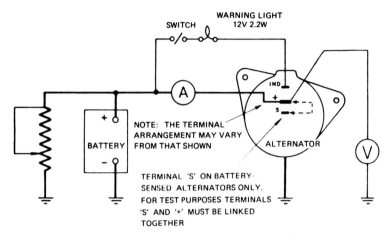

Figure 3.10   Alternator output test circuit

When the ignition switch (or its equivalent on diesel-engined vehicles) is turned on, the 12 V 2.2 W indicator light is connected to the battery, the circuit being made by way of the alternator field winding and the regulator and the indicator light glows brightly. The small current flowing in the field sets up a flux which supplements the residual flux in the rotor and aids the initial build-up of stator winding voltage as the rotor begins to revolve.

As the rotor speed increases and the voltage generated in the stator winding rises, the field current supplied by the stator winding through the 'field' diodes increases correspondingly until finally the alternator becomes self-excited. During the rise in the stator winding voltage (reflected at alternator terminal IND) the brilliance of the indicator light is reduced. At about the speed at which the alternator commences to charge, the voltage at the IND terminal approximates to that at the battery side of the indicator light and the latter is extinguished. In the event of drive belt breakage, the voltage will not build up within the alternator and the charge indicator light will remain on to indicate failure.

With the battery in a fully charged state, abnormal functioning of the charge indicator light may occur if the standing electrical loading on the system is zero or very low. On petrol-engined vehicles the ignition system provides sufficient electrical loading to prevent this occurrence, but on diesel-engined vehicles it may be necessary to fit a suitable resistor to provide a continuously connected load of at least 0.5 A whenever the engine is running.

Installation (Mechanical)

Location

Factors to be considered when choosing the site for alternator mounting are as follows.

The alternator must be sited where it will not be exposed to undue contamination. Airborne particles, sand, dust, etc, are normally drawn through the machine and therefore present no special problem, although where the alternator is likely to be subjected to a higher degree of contamination from this source (as may be the case with agricultural vehicles) fully sealed bearings should be specified. A special derated model 17ACR alternator is available where any severe contamination hazard (e.g. as from chaff) exists.

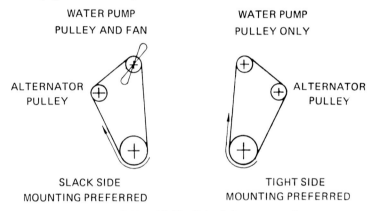

Figures 3.11 and 3.12   Drive belt arrangement

The alternator should be mounted on the engine in a cool position where the flow of air through the unit will be unobstructed and where there will be no risk of damage due to radiant heat from the exhaust system. On in-line engines the side of the engine opposite to the exhaust outlet is therefore to be preferred. If exhaust side mounting is unavoidable, as on many vee engines, heat shields should be considered.

The mounting should be preferably on the slack side of the belt drive unless the engine cooling fan is either remote (electrically operated) or crankshaft mounted, when the tight side is preferred (Figures 3.11 and 3.12).

Mounting

Figure 3.13 illustrates methods of mounting the alternator on to alternative engine mounting brackets and shows the sequence of washers, etc. If a two-lug fixing bracket is employed (Figure 3.13a and b) its design must ensure that no pre-stressing due to misalignment of the lugs or fixing bolt holes. The lugs should be an integral part of the bracket rather than one on the bracket and the other on the engine. The lugs are fitted with a sliding sleeve to prevent the lugs being damaged due to misalignment.

The risk of pre-stress can be eliminated by the use of a full-width boss (Figure 3.13c) which may either form part of the engine casting or

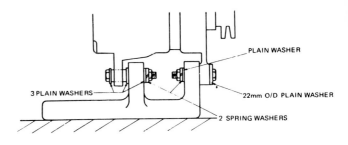

(a) STEEL BRACKET

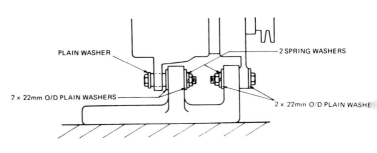

(b)   ALUMINIUM BRACKET

*Figure 3.13
Alternator mounting
details*

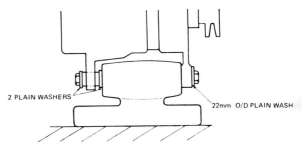

(c)   FULL-WIDTH MOUNTING BOSS

be rigidly fixed to it. The alternator is then secured to the boss by one full length bolt (where space permits) or by set screws inserted from each end into the tapped boss. When a separate fixing bracket is employed, a three-point mounting on the engine block is desirable, with staggered fixing bolt holes. Flexible fixing brackets must be avoided, since these may suffer short life and may also serve as dynamic magnifiers causing damage by vibration at the alternator. A load spreading washer, 1.6 mm thick and 22 mm outside diameter, must be fitted between mating steel and aluminium surfaces where the steel surface is of less area than the 22 mm washer.

Care must be taken in the design of the tensioning strap to avoid pre-stress during installation; the tensioning strap is not a mounting bracket. The strap should preferably be straight, if necessary using a spacing collar to achieve this. For fixing straps less than 127 mm in length, a thickness of approximately 1.75 mm is recommended. For longer fixing straps, or if the strap has to be cranked more than 6 mm a thickness of 3 mm is necessary.

Driving media

Pulley ratios should be chosen to utilise fully the maximum permissible alternator rotor speed without risk of being continuously driven beyond this.

Lucas alternators are designed for use with premium grade V-belt drives of either AV10 or AV13 section according to duty. Engine manufacturers should consult the leading makers of drive belts and follow the calculations in BS AU150, in order that a grade and type of belt to give satisfactory service may be chosen. Features such as 'notching' to preserve belt life should be considered. In some instances it may be beneficial to consider cut' or 'raw edge' V-belts. Tensioning of drive belts is considered later under the heading 'Stress'.

Bearing life

Bearing life is a function of pulley overhang and radial pulley load. The calculation of drive-end bearing life for any given application can be supplied by Lucas. In general, the engineering of a radial pulley load which will provide acceptable bearing life means that satisfactory belt life will automatically follow.

Stress

The stress on the alternator mounting bracket is basically in two parts, a 'static' stress introduced by the belt tension and fixing bolts, and an 'alternating' stress arising from relative motion due to vibration. The combined effect of these stresses must not exceed the fatigue limit stress of the bracket material.

The static stress is, in general, much higher than the alternating stress and since the static stress will vary with belt tension, the latter factor becomes of vital importance. Hitherto, with the wrapped V-belts widely employed, belt tension has to some extent been self-protective; irrespective of the value of initial installed tension, there is a rapid reduction during the first 1000 miles of operation, stabilising at values between 30 and 40 lbf. However, with poly-V or raw edge belts, which do not have this self-relieving feature, it is important that belt tension must not exceed the quoted value and this is equally applicable to maintenance in service.

Noise

During product development, alternator noise measurement and 'sound jury' techniques were employed to ensure an acceptable level of alternator noise. As in the case of vibration, this factor may be affected by the characteristics of the installation.

Maintenance in service

The equipment has been designed for the minimum amount of maintenance in service, the only items subject to wear being the brushes and bearings. Brushes should be examined after about 75 000 miles and renewed if necessary. The bearings are lubricated for life and should not require any attention if mounting and driving recommendations are followed closely.

Service precautions

Polarity

The alternator is for use on negative earth electrical systems only.

Refitting battery

Reversed battery connections will damage the alternator rectifiers. When refitting a battery, therefore, it is essential to ensure that the negative terminal is connected to the earthing cable or strap.

The cable(s) are live at all times while the battery is connected. When disconnecting the alternator, care must be taken to keep cable terminals well clear of earthed parts.

Single-pole alternator with collector rings

An alternator of this design has been developed by Bosch, for special vehicles with very high current demands (in excess of 85 A). The

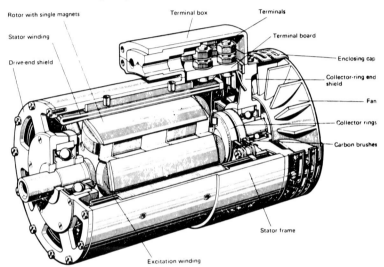

Figure 3.14   Sectional view of Bosch U2 model single-pole alternator (6 magnets, without built-in rectifier)

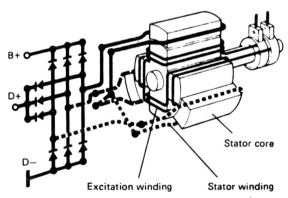

Figure 3.15   Basic construction of a single-pole alternator

machine is of very robust construction as will be seen from the sectional view shown in Figure 3.14. The rotor/stator arrangement and method of winding is shown in Figure 3.15. The rotor supplied with this basic design has four or six radially disposed single poles (instead of claw pole pieces) on which are wound the excitation windings. Each pole has its own excitation winding which replace the one common co-axial excitation winding for all the poles as in the claw pole design. The Bosch U2 machine has a built-in fan and the diodes are externally mounted in a separate water cooled rectifier housing.

Terminal arrangements in a box mounted on the outside of the alternator are such that by changing the terminal strips in the terminal box the machine may be easily converted from 'star' to 'delta' configuration according to the current output required. If, for example, the alternator has a maximum current output of 320 A with a star connection, the maximum current for the delta connection is:

$$I = I_p \sqrt{3} = 320 \times 1.73 = 550 \text{ A}.$$

ELECTRONIC REGULATORS

To maintain the battery in the ideal state of charge and to ensure the correct vehicle system voltage at all engine speeds and under all load conditions, some form of automatic voltage regulation of the alternator must be provided. Transistor regulators are now generally used for this purpose. As these units have no moving parts or contacts and the principal semi-conductor components in their construction are transistors and Zener diodes, such regulators provide long and trouble free service without attention.

Transistors

Figure 3.16 shows how a transistor (left view) can assume the functions of a relay (right view). If the switch in the control circuit of the relay is closed, the relay contacts also close, completing the main circuit (power circuit). By this method, a comparatively small control current can be used to control much larger working currents. This can, however, also be accomplished with transistors, as shown in the left hand side of the diagram.

If the switch in the control circuit is closed, a comparatively small control current flows from the positive terminal of the battery through the emitter (E) and base (B) of the transistor and returns to the negative terminal of the barrery via the resistor and switch. The control current through the path emitter to base, causes the path emitter to collector

(E–C) to become conductive, which completes the circuit for the much larger main current. This is the distinguishing characteristic of the transistor.

As compared to a relay, a transistor is far superior in terms of size and weight. Although its size depends upon the strength of the exciting

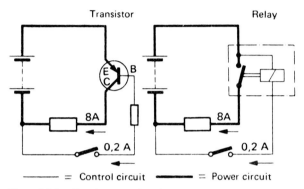

Figure 3.16   Circuit comparison between a transistor and a relay

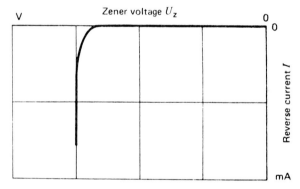

Figure 3.17   Characteristic curve of Zener diode

current it must conduct, the transistor, like the diode, is only approximately the size of a thimble. The main transistor in alternator regulators is used to cut the exciter field on and off in rapid succession. The main transistor does not function as an amplifier, as for example in a transistor radio, but serves solely as a switch or relay.

Z diodes

Another very important semiconductor element in transistor regulators is the Zener diode (named after its discoverer), see Figure 3.17. It is used only in its blocking state region due to its characteristic of suddenly increasing its reverse current once a specific voltage (Zener voltage) is reached. Consequently, the Zener diode is very suitable as a reference source. It is used for this purpose in the transistor regulator: once its reference voltage (Zener voltage) has been reached the Zener diode triggers another transistor.

Construction of transistor regulators

The transistor regulator has no contacts. Voltage is regulated electronically by the diodes, transistors, resistors and capacitors, which are all mounted on a printed circuit board. This means that all the moving, subject-to-wear, regulator components have been eliminated.

Voltage regulation

The simplified transistor regulator shown in Figure 3.18 operates in the following manner.

Control current flows from the exciter diodes through terminals D+ to the emitter E of the main transistor T1. From the emitter the current flows via the base B of T1 through resistor R3 to ground. This current causes the emitter to collector path (E − C) of T1 to become conductive and current flows through E − C to the excitation winding via terminals DF. The alternator is thus fully excited and output voltage increases. This voltage is also applied to the voltage divider R1/R2 which supplies the reference voltage for the Zener diode. When the specified voltage is reached the voltage across Z equals the Zener voltage and the Zener diode conducts. The Zener diode triggers the control transistor T2 which conducts and connects the base B of the main transistor T1 to terminal D+. Base current no longer flows, causing transistor T1 to interrupt the flow of exciting current. The alternator is now no longer excited. The output voltage falls below the specified value and the Zener diode switches back to the non-conducting state and interrupts the base current of transistor T2. This again connects the base of the main transistor T1 to the resistor R3 and terminal D−: T1 switches the exciting current back on. This process repeats itself in rapid succession, resulting in a very precisely regulated voltage.

The Lucas regulator (model 14TR) has been designed for use with alternators having nominal field currents up to 3 A maximum, as an inbuilt fitment to the Lucas 15, 16, 17, 18 and 20ACR alternator

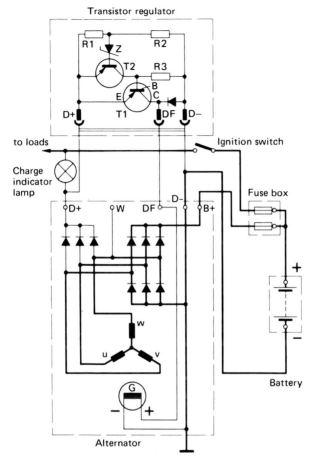

Figure 3.18   Installation with claw-pole alternator T1 and transistor regulator ED

T1. Main transistor   R1/R2. Voltage divider
T2. Control transistor   R3.   Resistor
Z.   Zener diode

range already described and is basically interchangeable with earlier 8TR and 11TR regulators. Features of the design are:

1. Improved reliability by a reduction in the number of separate semi-conductor devices.
2. The ability to function as a multi-purpose unit for application in a variety of charging system arrangements.

In standard form, the regulator is suitable for use with machine sensing of system voltage. For use in alternator applications where external battery voltage sensing is employed, a version of the 14TR regulator is available incorporating special circuit features to eliminate reliability hazards which would otherwise result from open circuits occurring in the main output or voltage sensing cables (see later reference to battery voltage sensing). For each of these applications, the temperature characteristic is nominally level.

In addition, a special version of the 'external voltage sensing' regulator is available for use when the battery is subjected continuously to very low or very high ambient temperatures, necessitating the use of a battery temperature sensor (see later under 'Temperature sensing version').

In all alternator installations employing the 14TR regulator, additional protection against surge voltage arising from wiring failure in the alternator output circuit or battery disconnection, and from ignition transients, is provided by the use of a protection diode.

Machine-sensing regulator

The regulator (see Figure 3.19) comprises discrete components (resistors, capacitors and semiconductors) flow soldered to a printed wiring board. The capacitors are of polyester type and the resistors, with the exception of the setting resistor, are of carbon film pattern. The setting resistor is a thick film resistor screenprinted onto a miniature alumina substrate. Abrasion of this resistor permits automatic setting of the regulating voltage (so avoiding selection of a resistor of suitable value and its subsequent assembly, as is necessary in the case of the earlier 11TR design).

The semiconductor devices consist of a silicon diffused-junction field recirculating diode and two plastic encapsulated devices. The first of these contains the voltage regulator diode and the input transistor combined in a single package: the second – the output stage – comprises an integrated Darlington Pair. The use of these combined devices simplifies assembly and reduces the total number of circuit connections.

The Darlington Pair device has a heat dissipating tag which is riveted to the metal heat sink housing the complete regulator assembly. The printed wiring assembly is encapsulated in epoxy resin as further protection against vibration and adverse environmental conditions.

Two flexible wire connections are provided, the yellow lead being connected to alternator IND and the black lead to alternator earth. The F connection to alternator field takes the form of a metal strip connector

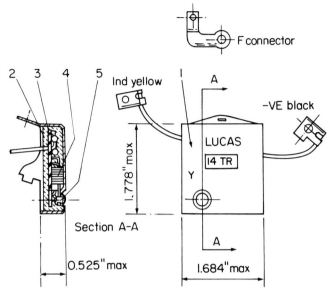

Figure 3.19   Lucas 14 TR regulator assembly (machine sensed version)
1. *Case heat sink*
2. *Encapsulant*
3. *Printed circuit board*
4. *Darlington pair transistor*
5. *Darlington pair retaining rivet*

located under the regulator mounting screw, the metal case of the regulator being common to the collector circuit of the Darlington Pair.

The semiconducting devices

Planar processes are used throughout the manufacture of the input and output devices employed in the 14TR regulator, this technology lending itself to a high level of quality control.

In the Planar process, oxide films are used to delineate the areas of silicon into which diffusion of impurities is to take place. This has two advantages; first, all junctions are protected by this oxide film and second because of this protection it is possible to measure the electrical parameters of the devices being manufactured prior to assembly, which enables a much closer control to be maintained on production. Contacts for the devices are made by evaporation of aluminium on to which are attached gold wires in the case of the input device and aluminium wires in the case of the output device. In the output device, two bonding wires (each wire being rated for full duty) are used for both the emitter and the base connections to give a higher level of reliability.

During the course of manufacture a continual check is made of the electrical parameters, the results being recorded on punched tape to provide a daily computer analysis of performance of the line.

After final encapsulation by transfer moulding in Polyset 300 epoxy resin, all devices are thermally cycled four times between -30° C and $+175^\circ$ C before being electrically tested at both room temperature and 175° C.

Before batches of devices are released for building into regulators, a random sample is taken and subjected to high stress testing. These tests include thermal cycling between -40° C and $+175^\circ$C, thermal shock by liquid immersion for 200 cycles between 15° C and 190° C and voltage stress testing at 60 volts and 175° C for 6 days followed by a 6-day humidity test. Failure rates in excess of the permitted limit in any of these tests result in quarantining of the batch and if an explanation of the deviation which would permit selection is not found, the entire batch is discarded.

The field recirculating diode is taken from a standard production line which is operated in conformity with CV and BS 9000 quality assurance procedures.

Alternative version for battery-voltage sensing

This is similar to the standard version described above but having an additional resistor and diode, as described later in this chapter.

Four flexible wire connections are provided. The yellow and black wires are connected as for machine-sensing. The red lead is connected to the alternator output terminal and the white lead to the sensing terminal.

Operation of the machine-sensing regulator

The regulator circuit (Figure 3.20) is similar to that used in 8TR and 11TR regulators. Its operation is briefly as follows.

When the ignition switch is switched on, the supply from the battery through the warning light provides base current via R4, switching on transistor T2 which in turn switches on transistor T3. With T3 switched on, field current can flow and the alternator output will build up until full excitation is obtained. Dependent upon the state of charge of the battery, and other electrical loads, the system voltage will rise until the regulating voltage is reached. At this point the voltage regulator diode D1 will conduct and commence to turn on transistor T1, reducing the base current of T2 which will turn off, causing T3 to turn off.

With T3 off, the instantaneous field current is diverted via the field recirculating diode D2 and commences to fall.

Over the regulating band the regulator will oscillate at a frequency dependent upon the internal time constant of the circuit and the voltage is controlled by modulation of the mark-space ratio brought about by the variations in the base current T1.

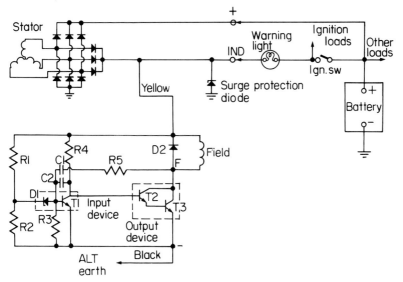

Figure 3.20   Circuit diagram for Lucas 15–18 ACR alternators with 14 TR regulator (field diode-sensing) and surge protection diode

It is desirable that T3 is never held in a high power dissipation state and is switched quickly to either the fully-on or the fully-off condition. This is achieved by the positive feed-back circuit composed of R5 and C1, which ensures that a voltage change at the collector of T3 is made cumulative by applying a signal of the correct phase to the base of T1.

The field recirculating diode D2 is connected across the field winding to prevent high voltage being applied to T3 as a result of sudden collapse of field flux due to rapid switching of this transistor.

In this circuit arrangement, in the event of output lead failure the alternator voltage is controlled to the regulator setting voltage.

Operation of the alternative version

The regulator circuit (Figure 3.21) is similar to that already described with the addition of components R6 and D3. Their purpose is to overcome reliability hazards which would otherwise occur on a battery-voltage sensed system due to:

1. Voltage runaway and consequent damage to the regulator in the event of an open circuit in the output lead between alternator and battery.
2. Uncontrolled output and consequent battery overcharging in the event of an open circuit in the voltage sensing circuit.

The first of these has been eliminated by a form of dual sensing so that, while under fault-free conditions the system voltage is sensed

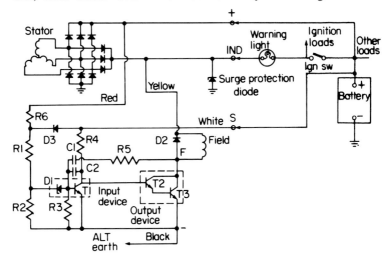

Figure 3.21   As for Figure 3.20 but with external voltage sensing

externally as designed, a secondary machine-sensing circuit is provided which in the event of output lead failure prevents the alternator voltage from rising above a pre-determined value. The second hazard has been overcome by arranging the circuit so that, in the event of an open circuit in the sensing circuit, the alternator is shut down.

The operation is as follows.

Under normal operating conditions, D3 is forward biased and the regulator is controlled by the voltage at the external sensing point. With a break in the main output lead, the voltage at the output terminal will rise and the current in R6 will increase. Under this condition, D3 will be reverse biased and the voltage at the R6-R1 junction will reach the level at which T1 switches on. The alternator voltage will then be clamped at a value equal to the setting voltage plus the additional voltage drop across R6. With a break in the voltage sensing lead, D3 is again reverse biased and in the absence of base drive to T2 the machine shuts down.

Temperature sensing version

In applications where the battery is continuously subjected to extreme temperatures, it may be desirable to modify the regulated voltage dependent upon battery temperature. For this purpose, a version of the 14TR regulator has been developed for use with a remote battery temperature sensor.

Two examples of circumstances where it may be desirable to adopt this system are:

1. Where a vehicle is required to operate continuously in low ambient temperatures and the battery is located remote from any source of heat.

2. Where the battery is subjected to exceptionally high ambient temperatures due to its close proximity to a source of heat (e.g. engine, torque converter, or exhaust system) or where abnormally high underbonnet temperatures are experienced.

Figure 3.22 shows the circuit arrangement. To simplify wiring, it is advantageous to adopt battery voltage sensing with this type of installation, the battery temperature sensor being fitted in the immediate vicinity of the battery and electrically connected in the voltage sensing lead to the in-built regulator.

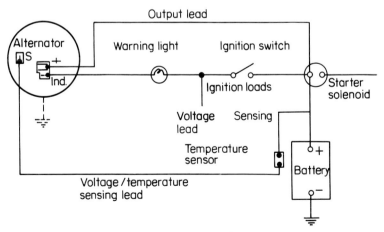

Figure 3.22   Circuit of system with temperature sensor

The 14TR regulator for use in these instances is therefore of the type described for battery-voltage sensing, but the combined temperature coefficient of regulator and sensor is −28 mV/deg.C. It is not necessary to 'pair' the regulator and sensor.

Ideally, the sensor should monitor the battery internal temperature but since in practice this cannot readily be arranged, the sensor should be in contact with the battery case or sired in the immediate vicinity of the battery.

Table 3.2 PERFORMANCE SPECIFICATION

Field current	Suitable for use with alternator having a nominal field current not exceeding 3 A with a system voltage of 14.0 V after stabilisation at 25° C ambient temperature. Maximum current (cold starting) 4.9 A at −40° C.
Regulating voltage	14.2 V nominal 20% available alternator output at 5000 rev/min. Tolerance ± 0.2 V max. ± 0.1 V typical
Voltage load characteristic	Change in system voltage between 20% and 80% alternator output measured at the voltage sensing point. Nominal 0.26 V Typical value 0.1 V
Temperature characteristic	Nominally level. The slope of the setting voltage temperature graph lies between 1.7 mV/deg C. to −3.6 mV/deg. C. over the temperature range −20° to +120° C. *Note* The combination of special version regulator and battery temperature sensor gives a temperature coefficient of −28 mV/deg. C.
Saturation voltage	Saturation voltage at a field current of 3 A and regulator excited at 13.5 V. 0.9 to 1.5 V (Typical value 1.1 V).
Operating ambient temperature range	−40° C to +110° C.
Voltage speed characteristic	Change in regulating voltage with alternator speed 0.01 V max. per 1000 rev/min over range 2000−12 500 rev/min.
Radio interference (Line born)	When fitted to an alternator with the recommended suppression capacitor across its output terminals, the radio interference generated by the regulator and the rectifier diodes combined, measured at any point in the external charging system, at any load/speed condition, is 70 dB max. relative to 1 μV within the range 0.15−5 mHz.

THE DYNAMO OR GENERATOR

An electromotive force may be *statically* or *dynamically* induced by electromagnetic means. Transformers and ignition coils are examples of electrical apparatus in which e.m.f.s are statically induced, whilst the dynamo or generator is an electrical machine for dynamically inducing electromotive force. A dynamo is a machine in which a winding moves relative to a magnetic field, and consists of a rotary member carrying the winding which is termed the *armature*, and a stationary magnetic field system.

Simple generator

We can best illustrate the principle of the standard generator by considering a single coil or turn of wire rotating between the poles of a magnet, as shown diagrammatically in Figures 3.23 and 3.24. Here we have a two-pole generator in its simplest form. By rotating this coil in the magnetic field produced between the N and S poles of the magnet, an e.m.f. is

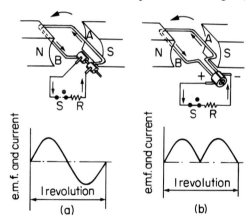

Figures 3.23 and 3.24   Principle of a simple generator

induced in it. The direction of the induced e.m.f. will be as indicated by the small arrows, when the coil is rotated anti-clockwise, and in the reverse direction if rotated in a clockwise direction. The total e.m.f. induced in the coil will be the sum of the e.m.f.s induced in the two sides (conductors A and B) of the coil which are parallel to the axis of rotation.

The direction of a dynamically induced e.m.f. may be readily ascertained by the use of Fleming's right-hand rule. Holding the right hand as shown in Figure 3.25, so that the thumb is pointing in the direction of motion of the conductor and the first finger in the direction of the magnetic field, that is towards the south pole of the magnet, then the second finger, if held at right angles to the first finger, will indicate the direction of the induced e.m.f.

Figure 3.26 shows that the induced voltage in the coil will be a maximum when the conductors are moving at right angles to the magnetic field, and zero when moving at right angles to the magnetic field, and zero when moving parallel to the field.

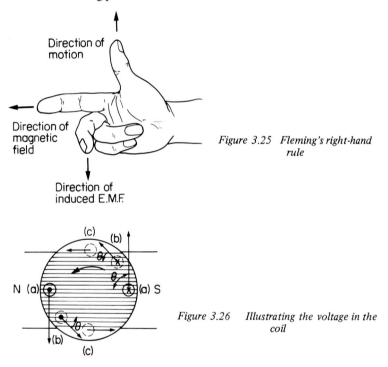

Direction of motion

Direction of magnetic field

Direction of induced E.M.F.

Figure 3.25   Fleming's right-hand rule

Figure 3.26   Illustrating the voltage in the coil

The induced voltage is proportional to the rate at which the conductors are traversing or *cutting* the magnetic lines of force. When the conductors are in position (*a*) their direction of motion is at right angles to the magnetic field, consequently the rate at which they cut the magnetic lines is a maximum.

Sine-wave induction

In position (*c*), since the conductors are moving parallel to the field, they are not cutting the lines of force, and the induced voltage is zero.

The induced voltage in each conductor at any instant is proportional to the sine of the angle θ between the direction of motion and the magnetic field. Therefore the voltage induced in the coil will follow a sine wave as shown in Figure 3.23(a).

A little consideration will show that there are two reversals in the direction of the voltage during each revolution. An *alternating* voltage is therefore induced as the curve in Figure 3.23(a) shows. If then we connect the rotating coil to an external circuit through brushes and slip rings, the current generated will likewise be alternating in character.

Function of the commutator

For the purpose of charging a battery it is essential that the current shall be uni-directional. For automobile application we must provide some means of reversing the connections of the external circuit to the generator, so as to make the current uni-directional in that circuit. This is the function of the commutator, and in the simplest case under review the commutator would consist of a metal ring divided into two segments as shown in Figure 3.23(b), the ends of the single turn coil being connected to these segments.

It will be seen from the illustrations that each conductor as it is moving in a downwards direction will be connected to the positive (+) brush. Similarly, the upward moving conductor is always connected to the negative (−) brush. The current in the external circuit will therefore always flow in the same direction. The brushes are set in relation to the magnet poles to bridge the gaps between the segments when the induced voltage in the coil is zero − that is, when the coil is in a vertical position and the induced voltage is about to change its direction. The effect of the commutator, so far as the external circuit is concerned, is to transfer the negative half-wave to the positive side.

Eliminating voltage fluctuation

Now whilst we have an induced voltage that is uni-directional, it is, nevertheless, widely fluctuating in value. Suppose instead of one coil we had two coils A and B connected in series to a four-part commutator and disposed at right angles to one another. At any instant the voltage at the brushes would be equal to the sum of the instantaneous voltages in the four conductors of these two coils. As we see from Figure 3.27, at no instant would the generator voltage be zero. When the voltage in coil A is zero that of coil B is a maximum, and by adding the ordinates of the two voltage curves A and B we obtain curve C, which gives us the voltage variation at the brushes. The addition of the second coil has reduced very considerably the fluctuation in voltage at the brushes. If we increase the number of coils to say four or six, the fluctuation in voltage would be still further reduced. With twenty or more coils in series, the variation in voltage at the brushes is so small as to be practically negligible.

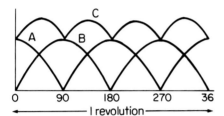

Figure 3.27   Voltage fluctuation

In the case of automobile dynamos, the number of coils in series is always sufficient to ensure that the voltage fluctuation is negligible.

Brushes and brush gear

Electrographitic brushes of rectangular shape are generally employed for automobile generators. This type of brush has a working current density of 50 to 70 A/in² and also has a relatively high contact resistance of 0.02 to 0.04 ohm/in², which is desirable from the point of view of commutation. Brushes used on starting motors are usually of the copper-carbon type, owing to the necessity for a higher working current density. In referring to generator or motor brush dimensions it is useful to remember that the *width* of the brush is that dimension *parallel* with the axis of the commutator, whilst the *thickness* is the dimension at *right angles* to the commutator axis.

The brushes are provided with flexible leads or pigtails, the ends of which are secured to terminal posts on the brush gear endplate. Box-type brush holders are generally employed, the brush being an easy but by no means sloppy fit in the box. The brushes are maintained in contact with the commutator by springs of various shapes, the pressure exerted by the spring depending upon the size of the brush. Generally, automobile dynamo-brush pressures are greater than for ordinary lighting and power units, in order to maintain good contact between the brush and commutator under conditions of vibration customary for road vehicles. The brush pressures will also vary with the design and size of dynamo and may be as high as 4 lb/in^2. in the case of the standard Lucas 3.9 in and 4.5 in dynamos, the brush pressures are 1½ and 2½ lb/in^2 respectively.

Types of wound field generators

The performance of an electric generator will depend mainly upon the manner in which the magnetic field is excited. In self-excited wound field generators there are three ways of connecting the field windings as shown in Figure 3.28. According to the manner of connecting the field windings, the generator is defined as a series, shunt or compound wound machine.

The field of the series wound machine is excited by the armature or load current, the field winding consisting of a relatively few turns of heavy-gauge wire. It will be evident that under open-circuit or no-load conditions the voltage will be zero, since no current is flowing in the field winding. The generator voltage will increase as the load current increases, until the field current is sufficient to saturate the iron circuit, when any further increase in load will cause a diminution of the generator voltage. This type of generator finds but a very limited application.

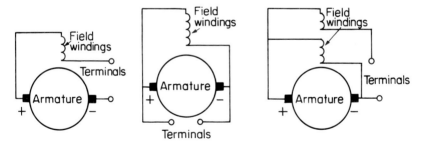

Figure 3.28   Types of wound field generators

In the case of the shunt wound generator having field windings of many turns of fine-gauge wire connected directly across the brushes, the generator voltage is a maximum on open-circuit and diminishes slightly as the load is increased for the following reasons

1. IR drop in the armature
2. Demagnetising effect of the armature current
3. Reduction in the field current as a result of the lower generator voltage.

This type is suitable for steady load conditions, where the variation in voltage is not of great importance, as for example battery charging. It is a type of generator which is very suitable for supplying lighting and power for domestic and industrial purposes.

Generators for motor vehicles are required to maintain a fairly constant output over a very wide speed range. Since the series type will not generate a voltage under open-circuit conditions, it is clearly unsuitable for battery charging. Before the generator can be connected to the battery, the generated voltage must be greater than that of the battery, otherwise current would flow from the battery through the generator windings. This reverse current through the field windings could only excite a field of opposite polarity to that actually required, with the result that the generator voltage would also build up in the wrong direction. Therefore a series wound generator could never charge the battery and in fact, even with a compoind wound generator, there would always be a risk of the battery reversing the polarity of the generator.

With a shunt wound generator, on the other hand, the battery voltage would tend to maintain the field current and polarity in the right direction, should the generator voltage be less than the battery voltage. Automobile generators are therefore of the shunt wound type. The voltage characteristics of the shunt wound generator are not, however, satisfactory for battery charging under the widely varying speed conditions which obtain on motor vehicles, where it is desirable that the full charging current shall be generated at as low a speed as possible. If, for example, the low speed conditions were met, we should find that at normal and high speeds the charging current would be very much in excess of what was required. The generated voltage is proportional to the product of the speed and the field strength, and if the voltage is satisfactory at the lower speeds, it must necessarily be too great at the higher speeds.

As the speed increases, some means must therefore be provided for reducing the field strength proportionally in order to keep the product (speed × field strength) constant, if the generator voltage is to be maintained constant. Modern practice is to control the field strength, although in the very early designs of generator slipping clutch drives were employed to maintain the speed constant after it had attained a predetermined value.

Control of the field strength may be effected externally with some automatic means of *inserting resistance in series with the field windings*, or it may be effected by utilising the *armature reaction*. Both these methods are employed, the former being known as the *constant voltage* and the latter as the *constant current* method.

Output regulation

As mentioned earlier, to meet the battery changing requirements, some form of output control or regulation must be provided. This is effected by two methods known as constant current and constant voltage systems. The constant current system which necessitated a third brush on the generator need be mentioned only as of historical interest, as it is now completely replaced by the constant voltage system.

From Figure 3.29 which shows the essentials of the constant-voltage system, it will be observed that a shunt wound generator is employed with a resistance in series with the field winding; it will be noted that there is no third brush. This resistance is normally short-circuited by an alternative circuit which is closed by a pair of contacts actuated by a spring. An electromagnet, wound with many turns of fine wire to keep the operating current low, is connected directly across the generator brushes. At a predetermined voltage the magnetic pull of the electro-magnet attracts the hinged or vibrating contact arm and in separating the contacts, inserts the resistance in the field circuit.

When the generator is first run up to speed, the voltage generated is insufficient to cause the electromagnet to open the contacts, consequently the main field builds up rapidly since the resistance is short-circuited. With increasing speed the generator voltage increases until it is high enough for the electromagnet to separate the contacts. Immediately this happens, the main field is materially reduced owing to the reduction of the field current as a result of the resistance being inserted. The generator voltage then falls and the electromagnet can no longer keep the contacts open. Short-circuiting the field resistance

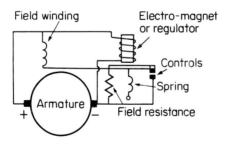

Figure 3.29   The constant-voltage system

increases the main field, causing a rise in the generator voltage, which again results in a separation of the contacts. This sequence of operations is repeated and continues at a very rapid rate. The generator voltage is thus automatically maintained within closely defined limits through an automatic variation in the ratio between the open- and closed-circuit contact periods. It will be noted that the regulation of the generator voltage is quite independent of the battery.

The automatic cut-out – its function and operation

It will be apparent that if an automobile dynamo were permanently connected to the battery, there would be a continual discharge from the battery through the generator windings, when the generator was stationary or the generated voltage was lower than that of the battery. Therefore some automatic means must be provided to disconnect the generator from the battery when the generated voltage is too low for charging. This is the function of the automatic cut-out, though, of course, it also serves to connect or 'cut-in' the generator when the voltage of the latter is a little higher than that of the battery.

The automatic cut-out is actually an automatic switch and it operates in quite a simple manner. Referring to Figure 3.29, the device consists of a soft-iron core on which are wound a series winding and a shunt winding. As its name implies, the series winding is connected in series with the generator and battery, and carries the full charging current. In order to keep the resistance low, this winding consists of a few turns of heavy-gauge wire. The shunt winding is connected directly across the generator terminals and is wound with a large number of turns of very fine-gauge wire. As the shunt current is a constant drain on the generator, the winding has the maximum number of turns possible, commensurate with the required magnetising effect, in order to keep the shunt current as low as possible.

Flexibly mounted, by means of a hinge or leaf spring, is a soft-iron armature which carries at one extremity a silver contact. Immediately under this contact is another contact mounted on a fixed and insulated bracket. It will be observed that these contacts are in the main generator and battery circuit. The soft-iron armature is actuated by a spring so that normally (that is when the soft-iron core is not magnetised by current in the shunt winding) the contacts are separated. The action of the automatic cut-out is as follows:

When the generator is running the induced voltage is impressed across the shunt winding. Current therefore flows in this winding and produces a flux proportional to the generator voltage. As a result a magnetic force will be exerted on the soft-iron armature, pulling it towards the core against the tension of the spring. With increasing current, the pull exerted on the armature will increase until, with a given value of current in the shunt winding, the armature will be pulled near enough to the

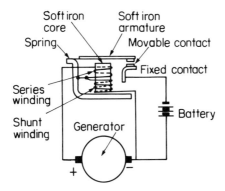

Figure 3.30 The automatic cut-out

core to cause the contacts to close. The shunt winding current necessary to close the contacts will bear a definite relation to the generator voltage.

Automatic cut-out adjustment

The automatic cut-out is adjusted so that when the generator voltage is slightly in excess of the battery voltage, the shunt winding current is sufficient to cause the contacts to close. This voltage is usually 6.5, 13 and 27 V for 6-, 12-, and 24-V systems respectively. When the contacts close, the battery is then connected to the generator and the charging current will flow through the series winding and the contacts. The current flowing through the series winding augments the action of the shunt winding and increases the pull on the armature and the pressure between the contacts.

When the generator speed falls and the voltage is too low for charging the battery, current will then flow from the battery through the series winding to the generator. This current, being in a reverse direction to that in the shunt winding, creates a field which opposes the magnetic flux induced by the shunt current, thereby operating to weaken the pull on the armature. Further reduction in the generator speed and voltage results in an increase in the discharge current from the battery. The magnetic pull on the armature is still further reduced until the control spring operates to open the contacts. This disconnects the battery from the generator and prevents any further discharge of current from the battery.

Car and commercial vehicle generators

Examples of d.c. generators together with their associated forms of control are described in the following pages. Whilst most vehicles are now fitted with alternators, very large numbers of vehicles are still operating with d.c. generators.

Their generator, which is of two-pole shunt wound design is capable of an output of 22 A at 12 V. The armature is carried in two bearings, a ball race at the driving end and a porous bronze bearing bush at the commutator end. The brushes are fitted in box holders secured to the commutator end casting. The magnet field assembly comprises a steel tubular yoke to which the pole shoes are secured by screws. Through-bolts secure both endplates to the steel yoke member. Connections to the armature and field are made by means of terminals mounted on the commutator end cover plate. Holes in the endplates allow a stream of air to be drawn through the generator by a fan mounted integrally with the driving pulley, thus ventilating the machine and so allowing an increased output to be achieved.

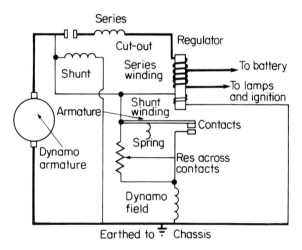

Figure 3.31   Circuit diagram for Lucas c.v.c. dynamo and equipment

Both the compressed voltage and the current-voltage systems of output control are used in conjunction with Lucas dynamos. The current voltage system is similar to the CAV system next described. For the compensated voltage control the regular unit is combined with the automatic cut-out, although electrically they are independent units.

Reference to Figure 3.40 will show that the regulator is provided with a voltage winding connected directly across the dynamo and two current windings, one of which carries the current from the dynamo to the battery, while the other winding carries the battery load current. These windings assist each other in energising the magnet system and effecting movement of the armature.

At a predetermined dynamo voltage, the magnetic field due to the voltage winding is sufficiently strong to attract the armature, thereby opening the contacts and inserting resistance in the dynamo field circuit. The resultant reduction in the field current lowers the dynamo voltage, which is turn weakens the regulator magnetic flux and in consequence the contacts close again to permit a building up of the dynamo field current and voltage. The repetition of this sequence of events causes the regulator armature to vibrate at a frequency sufficient to maintain the dynamo voltage within certain fixed limits.

As the dynamo speed increases above that at which the regulator comes into operation — about 20 mph — the periods of interruption of the contacts increase, with the result that the mean value of the dynamo output undergoes practically no increase once the operating speed has been attained.

In order to prevent overloading the dynamo by delivering an unduly heavy current to a battery in a discharged state, the series windings on the regulator operate to assist the shunt winding. Thus with the dynamo charging at a high rate and a full load on the battery, the regulator comes into operation at a sufficiently reduced dynamo voltage to limit the dynamo output to a safe value.

A bi-metal spring suspension for the regulator armature affords means of temperature compensation, whereby the regulator operating voltage is increased in cold weather and reduced in hot weather in order that the voltage characteristic of the dynamo shall conform more closely to that of the battery under all climatic conditions.

Generators for commercial vehicles manufactured by CAV operate either on the compensated voltage control system or on the current-voltage system. In both systems a plain shunt wound generator works in conjunction with a vibrating regulator. The operation of the compensated voltage control system is substantially the same as the Lucas system just described.

The current-voltage system differs from the compensated voltage control in that independent controls are used for the current and voltage, involving the use of two regulators. The winding of one regulator is responsive to the dynamo output current, whilst that of the other regulator is responsive to dynamo voltage. A connection diagram of the system, including the automatic cut-out, is shown in Figure 3.32, whilst in Figure 3.33 the output characteristics of the two systems of control

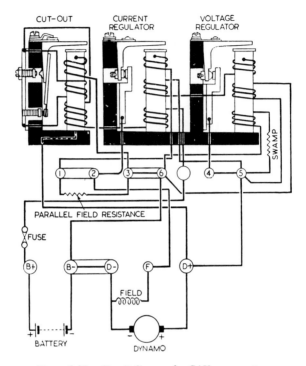

Figure 3.32   Circuit diagram for CAV c.v.c. system

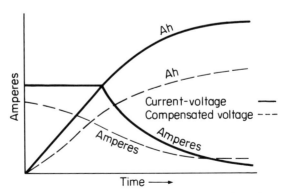

*Figure 3.33   Comparison of current-voltage and
compensated voltage regulators*

for similar operating conditions are compared. From this comparison it will be noted that with the current-voltage system the battery is charged at a constant current, until a predetermined voltage is reached, when the voltage regulator takes over and gradually reduces the charging current until trickle charge conditions obtain.

In the case of the compensated voltage control system, the charging current falls steadily from the commencement of charging. It is evident from Figure 3.33 that the ampere-hour input to the battery in a given time period is much greater in the case of the current-voltage system. This is a very desirable feature on road vehicles carrying equipment for radio communication.

A further advantage of the current-voltage system is that it provides much closer control of the dynamo output. Through the agency of the current regulator, the dynamo maximum output is restricted within the accuracy of the regulator setting. External influences, such as a short-circuited battery cell, short-circuited wiring or excessive lamp load cannot overload the dynamo. Consequently, a fuse in the dynamo circuit is unnecessary. Briefly, the operation of the system is as follows:

When the dynamo is running just above the cut-in speed, the current through the current regulator series winding (see Figure 3.32) is insufficient to attract the regulator armature and the contacts remain closed, providing full field current for the dynamo. With increased speed, full dynamo output is generated and the series winding current is then sufficient to attract the armature and open the contacts. This diverts the field current through the bucking resistance coil on the current regulator, thereby weakening the field current and reducing the dynamo current through the series winding. The contacts then re-close and this cycle of events is repeated at the rate of approximately 100 per second.

When the battery voltage through charging attains a predetermined value, the voltage regulator is sufficiently energised to open the voltage regulator contacts. The field current is then diverted through the 'points resistance winding' on the current regulator and the bucking coil on the voltage regulator. The dynamo field is weakened in consequence and the voltage regulator contacts re-close, the cycle being repeated rapidly at approximately 100 per second. In this way, the voltage is maintained at a steady value and the charging current gradually diminishes to a trickle charge rate.

To take advantage of the higher ampere-hour output often obtainable by accepting a higher cut-in speed and the elimination of interpoles rendered possible by the consequent reduction in dynamo diameter, a range of high speed generators is manufactured by CAV.

A single field system without interpoles is employed and the brush gear is mounted directly into the inside of the yoke. The yoke is of the extended type, with disc-pattern endshields carrying grease-packed bearings — roller at the drive end and ball at the commutator end. A ventilation fan is incorporated.

Dynamos for commercial vehicles manufactured by Simms Motor Units, Ltd., normally range in size from 5- to 7-in diameter, with the addition of an 8-in dynamo for extra heavy loads. Supplied for 12- or 24-V systems, Simms dynamos are of the four-pole single or double shunt field design for operation generally with an external voltage regulator of the compensated type. To meet conditions of short running times compared with the number of starts an automatic double voltage system of control has been devised.

A characteristic feature of the Simms compensated voltage control system is the diverter strip switch which controls the current flowing through the regulator series winding. By altering the thickness of the diverter strip, the dynamo output may be modified as desired.

The primary purpose of the double-voltage control system is to reduce the overall time of the charging cycle and to obtain this the initial high charging current is maintained for a longer period of time than normal.

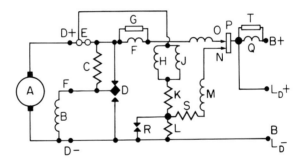

Figure 3.34   Circuit for Simms auto double-voltage control

The control circuit is illustrated in Figure 3.34 and in this the dynamo consisting of armature A and shunt field B is controlled by a standard type of regulator with field resistance C and contacts D. The regulator and cut-put shunt winding H and J respectively are in series with a swamping registance K and L whilst the regulator series winding F and diverter strip G are in the main dynamo output line. The automatic contactor consists of a shunt coil M in series with a swamping resistance S and operating points R, connected across the combined regulator and cut-out resistance portion L. The shunt coil M is connected into circuit by the operation of the cut-out points. The contactor series coil Q and the diverter strip T are in the battery positive line, while all resistance loads in the external circuit are fed directly from the cut-out and do not pass through the contactor series circuit. The output from the control board is thereby split and the method of operation is as follows:

The standard regulator is set for 12-V systems at an open-circuit voltage of approximately 17 V and the series coil and diverter strip arranged so as to drop the voltage to about 13 V with maximum dynamo output. The rise in battery voltage, therefore, during the charging cycle, reduces the charging current only to about two-thirds of its initial value.

The contactor series circuit Q and T is arranged in such a way that with the initial charge a battery voltage of about 15.5 V is required to operate the contactor points. As the charging cycle proceeds, however, the current is slightly reduced and the battery volts rise and these two factors combine to cause the contactor points R to close, when the battery is four-fifths fully charged and the specific gravity for a lead acid battery has risen to about 1.23. When this occurs and that part of the regulator swamping resistance L is short-circuited, the open-circuit voltage setting is reduced to 14.5 V and the remainder of the charging cycle is completed at a trickle charge rate.

Proportioning of the series and shunt values in the regulator and contactor have been so arranged that application of load to terminal LD+ in no way affects the operation of the contactor, which will always close when the battery is four-fifths fully charged. The contactor shunt coil circuit is fed from an insulated section N of the cut-out so that the points R will always be re-set whenever the cut-out opens.

AC/DC MOTOR CYCLE GENERATOR SYSTEMS

Alternator systems are now in general use on motor cycles. The system most widely used comprises a rotating magnet alternator, a bridge connected full wave rectifier, an ignition coil; a contact breaker with automatic timing device (and also a distributor when used for multi-cylinder motor cycles) a combined ignition and lighting switch and a 6- or 12-V battery. Variations in the system are provided to take care of special applications, emergency starting and different forms of ignition.

Such equipment is exemplified by the Lucas RM series of alternators and battery charging components, as shown in Figure 3.35.

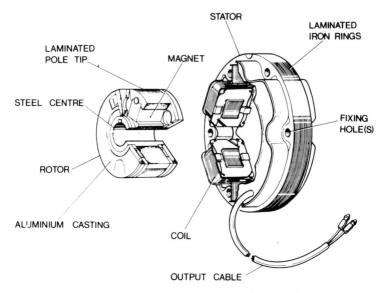

Figure 3.35   Exploded view of Lucas RM Series alternator for building into a motor-cycle engine

The alternator

This comprises two units, the stator and the rotor.

The Stator

The stator consists of six coils mounted on circular six-pole laminations. The assembly is encapsulated to protect the windings from damage by metallic swarf or vibration. There are two types of stators used:

    1. The three-lead stator, suitable for both 6- and 12-V systems.
    2. The two-lead type, which is similar in construction to the three-lead version, except that it is internally connected for full output and is only suitable for 12 V operation.

The rotor

The rotor has an hexagonal steel centre with a high energy magnet mounted on each face. Each magnet is keyed to a laminated pole tip and the complete assembly is cast in aluminium and machined to give a smooth external finish. The rotor is driven by an extension of the engine crankshaft and revolves inside the stator. The alternator requires no maintenance beyond checking that the connectors are clean and tight.

The rectifier

The rectifier consists of four semiconductor diodes connected in a bridge formation, each diode being mounted on its own heatsink. As the diodes get hot during operation, the rectifier must be mounted in a good airflow and kept clean and free from road dirt or corrosion.

If the rectifier has to be removed for any purpose, two spanners should be used, one for the head of the through bolt and the other for the securing nut. The nut clamping the plates together should not be disturbed, otherwise damage to the diode connections may occur.

The rectifier requires no regular maintenance beyond checking that the connections are clean and tight; especially the earth connection through the mounting bolt.

Zener diode

The Zener diode regulates the charge rate to the battery by diverting the excess charging current away from the battery, through the diode.

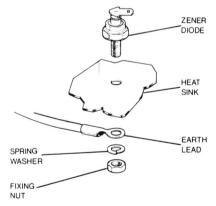

Figure 3.36   Zener diode and heatsink

During this process a large quantity of heat is produced. This would destroy the diode if it were not conducted away. The diode is consequently mounted on a heatsink supplied by the manufacturer of the machine.

To ensure efficient operation the diode base must make good metal-to-metal contact with the heatsink as dirt or corrosion under the diode, or a loose diode, would cause overheating and failure, see Figure 3.37.

Controlling the alternator output

Some form of output control is necessary, otherwise the generator output would remain at a maximum irrespective of load requirements and the battery would become overcharged. The alternator stator carries six coils, one on each pole. These are three pairs of coils, each connected in series. The first pair have a green/black cable connected to them, the other two pairs are linked together and connected to a

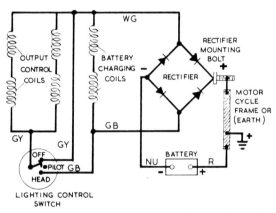

Figure 3.37   Charging circuit with lights switched off

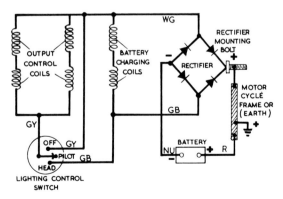

Figure 3.38   Charging circuit in the pilot light position

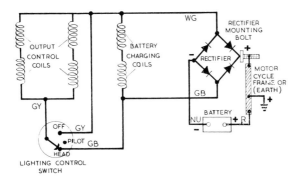

Figure 3.39   Charging circuit in the headlight position

green/yellow cable. All the coils have the remaining terminals connected to a common white/green cable.

The single pair of coils connected permanently across the bridge rectifier provide some charging current for the battery, whenever the engine is running.

Connections to the remaining coils vary according to the positions of the lighting and ignition switch, with the ignition key in the IGN position, for coil ignition the basic output control circuits are shown in Figures 3.37, 3.38 and 3.39.

With the lighting switch in the OFF position, the output control coils are short-circuited, as shown in Figure 3.37 and the alternator output is regulated to its minimum value by the interaction of the coil flux, set up by the heavy current circulating in the short-circuited coils, with the flux of the magnet rotor. Trickle-charging is provided by the permanently connected charging coils.

In the PILOT position, Figure 3.39, the control coils are disconnected and the regulating fluxes are consequently reduced. The alternator output therefore increases and compensates for the additional parking light load. On certain applications the short circuit lead is omitted in the OFF position giving the higher output for both OFF and PILOT positions.

In the HEAD position, Figure 3.39, the alternator output is further increased by connecting the control coils in parallel with the charging coil. Maximum output is now obtained.

SPECIAL APPLICATIONS – MOTOR CYCLE ALTERNATORS

Police machines and those used by the Automobile Association fitted with two-way radio are not fitted with a short circuit lead in the OFF position. They incorporate a separate 'boost' control switch, which can be used at any time, irrespective of the position of the main lighting switch. When in the 'boost' or closed position, maximum output is obtained from the alternator, see Figure 3.40. When the switch is open,

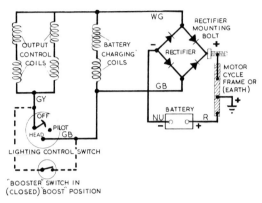

*Figure 3.40   Charging   circuit   for   motor   cycles
equipped with a radio*

the output from the alternator depends upon the position of the lighting switch.

EMERGENCY STARTING WITH MOTOR-CYCLE ALTERNATORS

Motor cycles fitted with coil ignition and the alternator-rectifier battery charging system are normally provided with a means of starting the engine when the battery is badly discharged. For this purpose, a three-position ignition switch is used, marked IGN, OFF and EMG. On switching to EMG and kick-starting the engine, the battery receives a high charging current and after running for a while, the ignition switch should be turned back to the normal running position IGN. (In the case of single-cylinder machines and twins fitted with two ignition coils, the appropriate time to change back to normal ignition is indicated by a tendency for the engine to misfire. This is because the rising battery voltage is in opposition to the alternator voltage, and consequently the amount of energy available for transfer to the ignition coil is reduced.)

The emergency starting feature also enables short journeys to be made (if absolutely essential) without battery or lighting. This is done by connecting the cable normally attached to the battery negative terminal to an earth point on the machine and kick-starting the engine with the ignition switch in the EMG position. Thus a rider can return home even if his battery has failed completely.

It must be emphasised, however, that continuous running in these conditions would result in badly burned contacts in the distributor or contact-breaker unit and cannot therefore, be recommended. Also the lighting system must not be switched on.

Single cylinder machines

When current flows through the windings in the direction shown in Figure 3.41 and the contacts are closed, the main return circuit to the alternator is through one arm of the rectifier bridge. As the contacts separate, the built-up electro-magnetic energy of the alternator windings quickly discharges through an alternative circuit provided by the battery and the ignition coil primary winding. This rapid transfer of energy from the alternator to coil causes h.t. current to be induced in the ignition coil secondary windings and a spark occurs at the plug.

Twin-cylinder machines (single ignition coil and distributor)

Figure 3.41 shows that the ignition coil primary winding and the contact-breaker are connected in series and not in parallel as for single machines. This enables a simpler harness and switching system to be used on

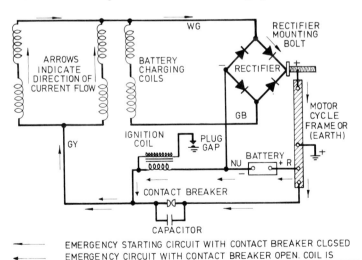

EMERGENCY STARTING CIRCUIT WITH CONTACT BREAKER CLOSED
EMERGENCY CIRCUIT WITH CONTACT BREAKER OPEN. COIL IS ENERGISED BY TRANSFER OF ENERGY FROM ALTERNATOR. BATTERY IS ALSO BEING CHARGED BY CURRENT GENERATED IN BATTERY COILS.

Figure 3.41   Emergency start circuit for single-cylinder motor cycles

twin-cylinder machines, but it is unsuitable for use with single-cylinder machines due to 'idle' sparking before the contacts separate. Twin engines, fitted with a distributor are unaffected by this premature sparking.

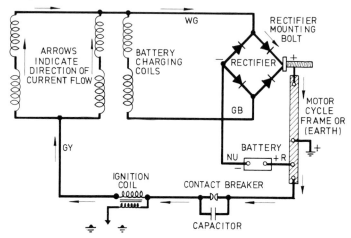

Figure 3.42   Emergency start circuit for twin cylinder motor cycles with a distributor

With single-cylinder machines connected as shown in Figure 3.42 'idle' sparking occurs after the contacts have separated and so does not affect these engines.

The machine should not run continuously with the switch in the emergency start position, because the rising voltage of the battery opposes that of the alternator and gradually the energy available for transfer to the ignition coil is reduced.

The engine will misfire, reminding the rider that he has not returned the ignition key to the IGN position.

Twin-cylinder machines (twin ignition coils and twin contact-breakers

When the ignition switch is in the normal running position IGN, each coil, with its associated pair of contact-breaker contacts, serves one of the cylinders, each functioning as an ordinary battery coil ignition circuit. On switching to EMG, however, one of the ignition coils functions on the energy transfer principle.

Figure 3.43 shows the circuit used for emergency starting. The No. 1 contact-breaker is arranged to open when the alternating current in the windings reaches a maximum in the direction shown by the large arrows.

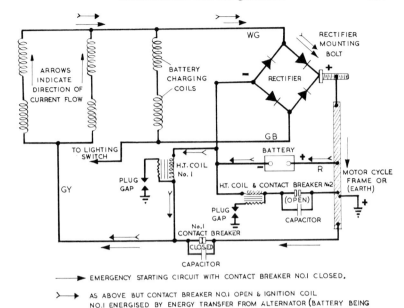

Figure 3.43   Emergency start circuit for twin cylinder motor cycles

The system functions as follows.

With the contacts closed, the main return circuit to the alternator is then via one arm (diode) of the rectifier bridge and the closed contacts. In effect the four output control windings have been short circuited allowing a heavy current to build up and circulate through them.

As soon as the contacts separate, this built-up energy discharges through an alternative circuit provided by the battery and primary winding of the No. 1 or EMG ignition coil. The rapid transfer of current from alternator to the ignition coil primary winding results in h.t. current being induced in the secondary winding and an efficient spark at the plug.

The efficiency of the energy transfer ignition is quite high because the alternative path through the battery, when the contacts are opened, is virtually a short-circuit. The 'flat' battery has little or no potential difference across it and consequently very little energy is lost at this point.

However, as the current surges pass through the battery and there are two charging coils also in circuit, a potential difference is formed across the battery terminals. After several current pulses (assuming the engine has fired and is running on one cylinder), the amount of energy available for the ignition coil is reduced, causing the engine to misfire. This will remind the rider to return the ignition key to the IGN position. The contact points will be badly burnt if the switch is kept in the EMG position for long periods.

Another feature of the system is that the coil No. 2 eventually comes into operation during emergency starting, so that after a few seconds running on one cylinder, number two cylinder cuts-in and the engine functions as a normal twin-cylinder unit. It will not operate on both cylinders after a few more seconds because the rising battery voltage causes misfiring on the one cylinder.

Although the No. 2 coil SW terminal is linked to the same feed cable as the SW terminal of No. 1 coil, it does not pass any of the energy transferred from the alternator, during the 'energy transfer' pulse. At this particular instant the No. 2 contact-breaker points are open, open-circuiting the No. 2 primary circuit. As the alternator current passes through the battery, the voltage rises. Further, while the No. 1 coil is fed by energy pulses from the alternator, the No. 2 coil, when its associated contacts close, will receive current direct from the battery which is gradually becoming charged. This results in the engine firing on both cylinders. It will not run at full power intil switched to the IGN position, because the energy now available for the No. 1 coil is being reduced and misfiring will still occur.

Actually, when both coils are functioning, their primary windings are being fed in opposite directions. The No. 1 coil is receiving pulses from the alternator via the battery, the insulated side of the circuit, through the primary from SW to CB and back to the alternator. The No. 2 coil is fed by a steady current direct from the battery, via earth, through No. 2 contacts to CB through primary to SW and back to battery —ve.

Ignition performance under emergency starting conditions should be equivalent to that of a magneto at kick-start speeds. Where a boost control switch is fitted, the switch must be in the off position before attempting an EMG start. As the switch short circuits the emergency starting system.

THE IMPORTANCE OF CORRECT TIMING FOR EMERGENCY IGNITION

Correct ignition timing, both electrically and mechanically, is a very critical factor with the a.c. sets, particularly in relation to emergency starting. As already stated, in the emergency start position the alternator supplies current direct to the ignition circuit. The timing is so arranged that the contacts are opened when the peak of the voltage wave coincides with each firing point of the engine, illustrated graphically in Figure 3.44a.

If the contacts do not open at the precise instant required, emergency starting performance will be affected. Electrically, the timing position is fixed by the manufacturer, i.e. the alternator rotor is keyed on to the crankshaft in a position consistent with peak voltage and cannot be altered. It is on the mechanical side that variations in timing can arise.

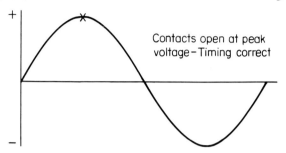

Figure 3.44(a)   The effect of correct timing

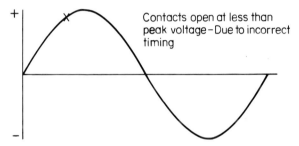

Figure 3.44(b)   The effect of incorrect timing

The engine ignition timing must be accurately set to the figures specified for the particular machine. The contact breaker gap must also be set and maintained to the specified figure as any variation in the gap setting will affect the timing position in relation to spark energy.

If the timing at the distributor or contact-breaker is advanced or retarded excessively either through a timing error, incorrect contact gaps, or weak automatic advance springs, the contacts will not open at the peak of the voltage curve, see Figure 3.44b, and consequently the spark will be weak. Remember ignition timing is critical.

A.C. ignition

The alternator designed for a.c. ignition has the ignition generating coils connected in series with each other and with the primary winding of a special ignition coil, model 3ET.

This special ignition coil employs a closed iron circuit and has a primary winding whose impedance is closely matched to that of the ignition generating coils of the alternator. As a result of this electrical matching, the ignition performance combines the good top speed characteristics of the magneto with the good low speed performance of the conventional ignition coil.

The system functions as follows.

The contacts of a contact breaker unit or distributor are connected in parallel with the ignition coil primary windings, since one end of

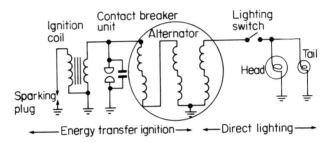

Figure 3.45   AC ignition circuit for single-cylinder motor cycles

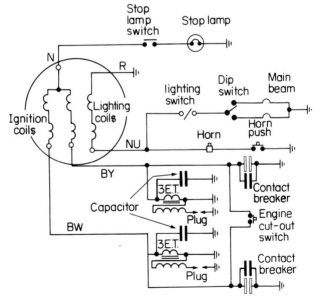

Figure 3.46   AC ignition circuit for twin contact motor cycles

the stator winding, one end of the ignition coil primary winding and one side of the contact-breaker is earthed, as shown in Figures 3.45 and 3.46.

When the contact-breaker contacts close, the primary winding of the ignition coil is short circuited and, at the same time, the stator ignition windings form a closed circuit. As the magnetic rotor turns, voltages are induced in the stator coils resulting in alternating currents while the contacts are closed. When the contacts open, a pulse of electromagnetic energy (developed in the stator while the contacts are closed) is discharged through the ignition coil primary winding. The effect of this energy pulse in the primary winding is to induce a high tension voltage in the ignition coil secondary winding which is then applied either directly or by way of a distributor to the appropriate sparking plug.

Timing considerations

Since the magnetic rotor of the alternator is keyed (or otherwise located) on the crankshaft, the magnetic pulse in the alternator stator, which produces the energy pulse to feed the ignition coil primary winding, must be timed to occur at the firing point of the engine.

The magnetic pulse occupies several degrees of crankshaft (and therefore the rotor) rotation. A fairly wide angular tolerance would thus be available for a fixed ignition engine.

However, it is desirable that most four-stroke engines should incorporate an ignition timing control (usually centrifugally operated) giving a range of advanced and retarded sparking. The magnetic relationship of the alternator rotor to its stator must therefore be governed by the fact that the engine firing point will vary by several degrees between the fully retarded starting condition and the fully advanced running condition.

This is exactly the same problem which obtains with a manually controlled magneto and gives rise to the same characteristics, i.e. the available sparking voltage for a given kick-start speed reduces progressively with the retard angle. A magneto, however, is a self-contained unit and will produce a spark even though seriously mistimed to the engine, because a magneto contact-breaker is always in the correct relationship to the magnetic geometry of the unit. With an alternator, however, the position of the magnetic rotor with respect to the stator and to the engine piston at the instant of firing, is pre-determined by its position on the engine crankshaft.

The range of retarded magnetic timing that can be used with a particular engine depends in part on that engine's starting performance, since the required plug voltage is influenced by many factors of engine design. The speed at which it can be kicked over in attempting to reach this voltage will depend on piston and bearing friction, kick-starter ratio, etc.

Figure 3.47 shows how the available plug voltage varies with different magnetic timing positions and for different speeds of rotation. The

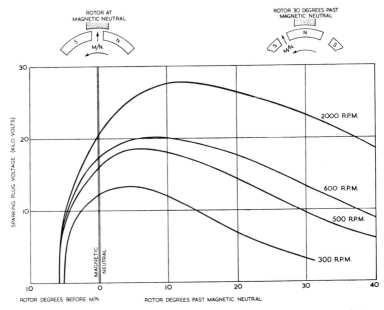

Figure 3.47   The effect of magnetic timing on plug voltage at various rev/min

reference point is known as the magnetic neutral position, where the interpolar gaps of the rotor are situated on the centre-lines of the stator limbs.

It will be seen that although the optimum magnetic position is just past the magnetic neutral at 300 rev/min, it changes to several degrees past at 2000 rev/min, due to distortion of the magnetic flux.

It will also be seen that the sparking performance deteriorates rapidly a few degrees before the magnetic neutral position. Hence commercial tolerances on keyways, etc., dictate the inadvisability of approaching too near to this critical point in the advanced or running position of engine timing.

As previously stated, the extent to which the retard timing can be used depends on plug voltage requirements at starting and on kick-starter speed. For example, if the required plug voltage is 6 kV, the retard timing would be restricted to about 20° (engine) if the kick-starting speed was to be limited to 300 rev/min – in practice, a fairly low speed. On the other hand, at the fairly normal kick-starting speed of 500 rev/min, the timing range could be widened considerably with plug voltages up to about 8 kV.

Accurate ignition timing is an important requirement in the operation of an energy transfer system. The optimum conditions are determined by the engine designers during the development and should always be maintained to ensure the highest performance, both from the engine and from the ignition system designed to work with it.

It will also be appreciated that amateur tuning will not improve the performance of highly developed engines. Indeed, in certain circumstances, this may be harmful. Indifferent sparking outside the prescribed range will almost certainly indicate tampering and may well serve as a warning to the would-be tuner.

12-volt charging system and capacitor ignition

The Lucas RM series alternator is voltage conscious, but, to avoid overcharging the battery, was used with 6-V equipment. However, with the introduction of the Zener diode it became possible to control the output for all conditions using 12 volt equipment. Many advantages were gained. Extra output from the alternator, more accurately controlled charge rates and better headlight efficiency.

The Zener diode is constructed from similar materials to the diode used in the rectifier and it will pass current in one direction only. However, when connected in the reverse direction, it will act as an insulator while the voltage is below approximately 14 V. Above this it will start to conduct current, slowly at first, until at approximately 15 V it becomes fully conductive.

If the Zener diode is connected in parallel with the battery in the charging circuit, as shown in Figure 3.48a, it will act as a regulator. When the battery is discharged, its terminal voltage is low and the system voltage is low and the Zener diode acts as an insulator. All the output from the alternator is then directed through the battery. As the battery is charged, its terminal voltage rises and the system voltage rises accordingly, until approximately 14 volts is reached. At this point the battery requires less charge, so the Zener diode starts to break down and conducts part of the charging current away from the battery. When the battery voltage reaches approximately 15 V, the Zener diode breaks down completely and conducts all the charging current away. When lighting equipment is switched on, the system voltage falls until it becomes an insulator again, at approximately 14 volts and the output from the alternator is available to balance the lighting load.

As the Zener diode conducts excess current a large quantity of heat is produced by the diode and in order to keep the temperature within the operating limits, a heatsink is required. The maximum current the diode can carry is limited by the efficiency of the heatsink. In order to remain within this limit the switching arrangements were as shown in Figures 3.48a and b. Later heatsinks have a surface area of at least

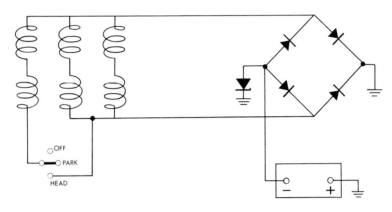

Figure 3.48(a)   12-V circuit for coil ignition

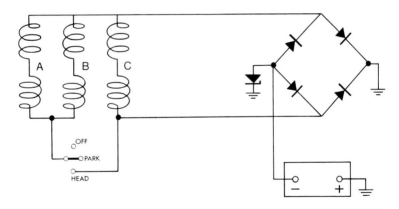

Figure 3.48(b)   12-V circuit for magneto ignition

72 square inches and are mounted in a good air stream. The full output from the alternator can now be permanently connected across the rectifier. The circuit for this system is shown in Figure 3.49 and as the charging system requires no switching arrangement, car type ignition and lighting switches are used.

When the RM series alternator is used for 12-V operation the maximum output is raised from approximately 60 W on 6 V to approximately 104 W. The increased output on kick-over is enough to charge the battery sufficiently to start the machine on third or fourth kick. Consequently, no emergency start is necessary allowing the simple coil ignition circuit to be used.

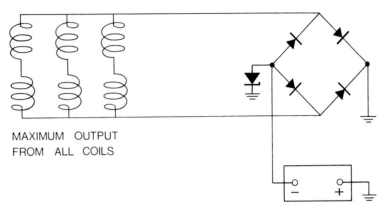

MAXIMUM OUTPUT
FROM ALL COILS

Figure 3.49   12-V circuit connected for continuous maximum output

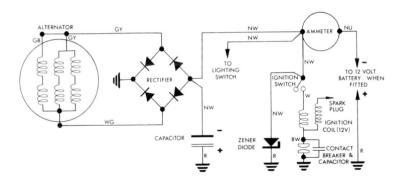

Figure 3.50   Basic capacitor ignition circuit

It may be preferable to remove the battery for sporting events. The Zener diode will then maintain the system voltage at a safe level, but there will be insufficient output from the alternator to start the machine on kick start. Consequently the 2MC capacitor is connected in parallel with the battery, as in Figure 3.50. When the battery is removed, the capacitor stores the impulses from the alternator when the points are

146

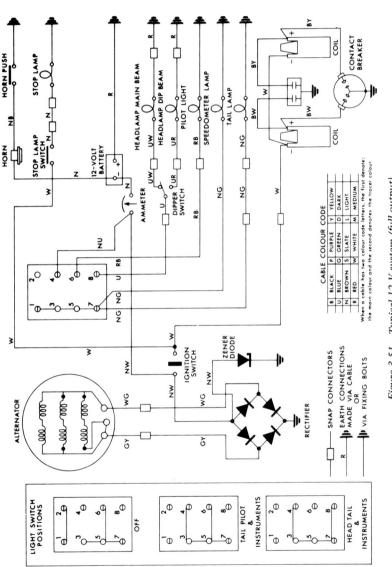

Figure 3.51 Typical 12-V system (full output)

open and discharges when the points close to enable other impulses from the alternator to start the machine. The machine will, therefore, operate without the battery and the full lighting load may be supplied while the engine is running. However, when the engine is not running there will be no lights available for parking etc.

Modern circuitry has now simplified the wiring system considerably and a typical 12-V charging and ignition circuit is shown in Figure 3.51.

CHAPTER 4

THE STARTING MOTOR

In the case of the generator, it has been shown that by rotating the armature we are able to transform mechanical energy into electrical energy. By applying a suitable electric pressure to the terminals of the generator, it is possible to reverse this process of energy transformation. Any d.c. generator can be operated as a motor by applying current to the armature and field windings. The dynamotor is an example of a single machine operating both as generator and motor. The essential components of the electric motor are therefore identical with those of the generator.

An example of a combined motor/generator in use on marine and stationary engines is the Siba Dynastart, details of which are given at the end of this chapter.

Necessity for heavy conductors

In the case of automobiles, electrical energy is derived from the battery for the purpose of producing the mechanical energy necessary for rotating the engine crankshaft when starting. The connections of the starting motor circuit are illustrated in Figure 4.1. The mechanical

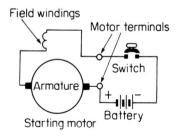

Figure 4.1   The starting motor circuit

148

power necessary to rotate the larger size car engine at starting speed is in the region of 1½ h.p., and the current required to produce the driving torque, allowing for a 50% overall efficiency, is of the order of 300 A for a 12-V system and on a large diesel engine as much as 10 h.p. may be required. It follows, then, that not only must the cables and switch connections be of substantial size to carry such heavy currents, but also the motor windings must be wound with conductors of much larger section than is the case with the automobile generator. This is one factor which renders it difficult to combine the two machines and obtain the best all-round efficiency.

Principle of operation

A current-carrying conductor, if transversely disposed in a magnetic field will experience a force urging it in a direction at right angles to the magnetic lines of force. The reason for this will be apparent if we consider the effect upon the main field by the magnetic field induced by the current in the conductor.

If we change the direction of the current or the magnetic field, the force exerted on the conductor will be in the opposite direction. If, however, both the current and field directions are changed, then the direction of the force remains unchanged.

The current flowing in the armature of a generator, when operating as a motor, is reversed as compared with the direction of a generated armature current, assuming the connections to the battery remain unchanged. Consequently the effect of armature reaction in a motor is to distort the main field in a direction *opposite* to the rotation. Any shifting of the brushes, therefore, to minimise sparking at the commutator, must be in the opposite direction to the rotation. This is just the reverse to a generator where it is necessary to move the brushes in the direction of rotation to reduce sparking under the brushes, making it difficult to operate one machine both as a generator and a motor.

Speed and torque

The force exerted upon a current-carrying conductor is proportional to the magnetic field intensity in the vicinity of the conductor. Since the self-induced magnetic lines are proportional to the current flowing, the force exerted on the conductor will also be proportional to the conductor current. Also, the longer the conductor, so proportionally greater will be the force urging it across the magnetic field.

Expressing this mathematically, we may say that:

$$f = \frac{Hli}{10} \text{ dynes} \qquad (4.1)$$

where f = force exerted on the conductor in dynes
H = magnetic field intensity in CGS units
l = conductor length in cm
i = current flowing in the conductor in amperes.

If then we have an armature with a total of Z active conductors (Z conductors actually under the magnet poles), the total force urging the armature to rotate will be:

$$F = Z \frac{Hli}{10} \text{ dynes} \qquad (4.2)$$

Now as these conductors will be rotating in a magnetic field, e.m.f.s will be induced in them, the direction of these being in opposition to the applied e.m.f. or voltage. The self-induced e.m.f. of a motor is termed the back e.m.f. and, as it opposes the applied e.m.f., it will tend to diminish the current flowing in the conductors. The armature current is given by the equation:

$$I = \frac{E - e}{R} \qquad (4.3)$$

where I = armature current in amperes
E = applied e.m.f. in volts
e = total back e.m.f. in volts
R = resistance of armature in ohms.

From this equation it will be evident that at the instant of applying a voltage to the terminals of a motor the armature current will be relatively great, since the back e.m.f. (e) is zero, the current being limited only by the resistance of the winding. As soon as the speed increases, the current will necessarily diminish to satisfy the equation. The back e.m.f. voltage is given by the formula:

$$e = \frac{\phi ZN}{60} \cdot \frac{P}{X} \cdot 10^{-8} \text{ V}$$

We can now rewrite equation (4.3) as follows:

$$I = \frac{E - \dfrac{\phi ZN}{60} \cdot \dfrac{P}{X} \cdot 10^{-8}}{R} \text{ A}$$

and from this evolve the speed formula:

$$N = \frac{E - IR}{\phi Z} \cdot \frac{X}{P} \cdot 60 \times 10^8 \text{ rev/min} \qquad (4.4)$$

where N = speed in rev/min
   E = applied volts
   I = armature current in amperes
   R = armature resistance in ohms
   ϕ = flux per pole
   Z = total number of armature conductors
   P = number of poles
   X = number of parallel paths through armature winding.

For lap winding $X = P$ and for wave winding $X = 2$.

The voltage drop IR in the armature winding under no-load or light-load conditions is very small, and if we assume that this can be ignored in equation (4.4), then we get:

$$\text{Speed} = N = k \frac{E}{\phi}$$

where k represents the constants of the equation.

It will be seen that the *speed of the motor is proportional to the applied voltage and inversely proportional to the flux per pole.*

Load and armature current

With increase in load, the armature current will increase, and it will be seen from equation (4.3) that this will necessarily involve a reduction in the back e.m.f. (e). Therefore any increase in load will be followed by a reduction in speed.

If we rewrite Equation (4.3) as:

$$E = e + IR$$

and then multiply throughout by the armature current I we get the power equation for a motor:

$$IE = Ie + I^2 R$$

The watts loss due to the heating of the armature windings is represented by $I^2 R$, whilst Ie gives the watts available for conversion into mechanical power. A small percentage of the latter is expended in overcoming frictional resistance and iron losses, and the remainder in producing the necessary driving torque demanded by the load conditions.

If the radius of a motor armature is *r* cm, the developed torque may be derived from equation (4.2) as follows:

$$T = \frac{Zlr}{10} \cdot Hi \text{ dyne cm}$$

or

$$T = \frac{Zlr}{10} \times \frac{Hi}{1357 \times 10^4} = \frac{Zlr}{1357 \times 10^5} \cdot Hi \text{ ft lb}$$

It can also be shown that

$$\text{Torque } T = \frac{0.1174ZP}{10^8 X} \phi I \text{ ft lb} \tag{4.5}$$

where Z = total number of active armature conductors
   P = number of poles
   X = number of parallel paths through armature winding
   ϕ = flux per pole
   I = armature current.

From these two formulae we see that the torque is directly proportional to the product of the two variables, namely flux and current.

The series motor and its characteristics

D.C. motors, like generators, may be series, shunt, or compound wound, each type having its own particular sphere of application. The series wound motor is noteworthy for its capability to produce a heavy starting torque. It is this characteristic which renders this type of motor particularly suitable as an engine starting motor.

We have seen that at starting the current taken by any motor is great owing to the absence of any induced back e.m.f. As the field winding is in series with the armature winding, the full current taken by the motor also serves to excite the main field. Equation 5 shows that the torque of a motor is directly proportional to the product of the flux and current. Since, with a series motor, these two factors are a maximum at starting, it follows that the torque is then also a maximum.

It has also been explained that the speed of a motor varies inversely as the field strength. With a series wound motor, the effect of additional load is to increase the field strength, which in turn causes a reduction in speed. This type of motor is essentially a variable speed motor and, as the speed-torque curve in Figure 4.2 shows, the speed varies appreciably with load variation. Particularly is this so at light loads, where it will be observed the speed curve rises at a very rapid rate, so that under no-load conditions the motor may attain very dangerous speeds. A series wound motor should therefore never be run without load otherwise the speed may be sufficiently high to set up excessive centrifugal stresses with the liability of very serious damage. The speed variation is not so great at the higher torque values because the flux density in the field system is approaching the saturation point of the iron. A considerable increase in current is then necessary to produce any marked increase in the magnet flux. When heavily loaded, the speed of the series wound motor is substantially constant.

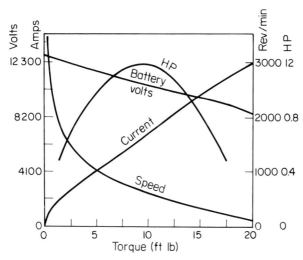

Figure 4.2   Characteristic starting motor curves

With heavy load and current values, the I^2R losses in the windings will increase very rapidly, and there will be a corresponding reduction in the horse-power developed.

Efficiency

The commercial efficiency of an electric motor is given by the formula:

$$\text{Efficiency} = \frac{\text{Mechanical output in watts}}{\text{Electrical input in watts}}$$

$$= \frac{\text{hp} \times 746}{E \times I}$$

where E = voltage at the motor terminals
I = current in amperes taken by the motor

The efficiencies of electric starting motors range between 50 and 70% corresponding to the maximum horse-power developed.

Considerations affecting size of starting motor

In considering the size of a starting motor for a given design of engine, it is necessary to know the torque that must be applied at the crankshaft to start rotation and also the horse-power necessary to maintain rotation of the crankshaft at the starting period. If the starting motor is to operate satisfactorily under all conditions, these values must correspond to the conditions obtaining when the engine is very cold, since the engine resisting torque will then be at a maximum. The power necessary to crank an engine in cold weather is three to four times greater than that required at normal temperatures, or when the engine is warm.

Starting torque values will vary for different engines, being largely dependent upon the piston displacement per cylinder, the compression pressure, and the number of cylinders. The piston displacement per cylinder will decide the peak torque values, and for engines of the same total piston displacement the one with the least number of cylinders will require the greatest starting torque. Thus for a twin-cylinder engine of the same total capacity as a four-cylinder engine, the starting torque will be nearly twice as much as for the four-cylinder engine. Actually in practice the starting torque of two such engines is in the ratio of about 9:5. Similarly, a four-cylinder engine will require a greater starting torque than a six- or eight-cylinder engine of equal total piston displacement.

During the winter months the crankshaft speed for proper vaporisation of the fuel and for the production of a spark at the plug electrodes will, when starting from cold, be higher than when the weather is warm and dry. The carburation and ignition on modern engines is such that even under the adverse conditions of winter a crankshaft speed of 100 rev/min will generally ensure satisfactory starting.

To start an engine, a certain minimum torque is necessary to free the crankshaft, and the starting motor must be capable of rotating the crankshaft at a minimum speed of 100 rev/min in very cold weather. If the starting and driving torques of an engine for adverse climatic conditions are known (and they may be determined by dynamometer tests), it is then a relatively easy matter to decide upon the size of starting motor required. It is, however, also necessary to know the ratio between the starting motor and crankshaft speeds – a ratio largely decided by the engine design and in particular by the size of the flywheel gear – before the size of starting motor can be settled.

By way of illustration, let us assume that a given design of engine, under normal atmospheric conditions, requires a torque of 50 ft lb to free the crankshaft and say 14 ft lb driving torque to maintain an engine speed of 100 rev/min. Allowing for winter conditions, these torque values will then be 200 and 56 ft lb respectively, the latter expressed in horse-power being 1.05 hp. If the design of engine under consideration permits of a gear ratio of 9:1, then the maximum or stalled torque of the starting motor must not be less than 22 ft lb, whilst at 900 rev/min motor speed a driving torque of 6.6 ft lb, equivalent to 1.05 hp will be necessary.

It is difficult to give precise data in regard to starting torques applicable to different types and sizes of engine, owing to the varying factors involved. The following starting motor stalled torques per litre total piston displacement are, however, approximate, assuming a 10:1 gear ratio:

Twin-cylinder engines	9.0 ft lb per litre
Four-cylinder engines	5.3 ft lb per litre
Six-cylinder engines	4.6 ft lb per litre
Eight-cylinder engines	4.2 ft lb per litre
Twelve-cylinder engines	3.8 ft lb per litre

According to whether the gear ratio is higher or lower than 10:1, the stalled torques will be correspondingly lower or higher than the specified values.

Owing to the very heavy discharge current when the starting motor is switched on, there is quite an appreciable drop in the battery voltage. The performance of the starting motor is therefore largely dependent upon the size and condition of the battery employed. A battery of ample capacity is essential, otherwise the heavy discharge current will result in an abnormal voltage drop which will limit the starting motor current and the torque developed.

TYPES OF STARTING MOTOR

The starting motor is similar in construction to the generator with the exception that the windings, brushgear, and terminals are much more substantial in order to cope with the heavy currents involved in starting. A further notable exception is in respect to the armature bearings. It is not general practice to provide ball bearings for the armature, and many modern starting motors have plain bearings either of the lubricated or self-oiling types. In some cases a ball bearing is provided at the driving end and a plain bearing at the commutator end. Since the starting motor is only for occasional use, ball bearings are not absolutely essential, although some makers, particularly British manufacturers on their latest models, fit one ball bearing.

For light and medium cars, motors of 3½ to 5-in yoke diameters and with stalled torques varying from 4 to 20 ft lb are generally employed. Large car and commercial vehicle engines are usually fitted with 5- to 8-in diameter starting motors, where starting torques of 30 to 100 ft lb are necessary. On diesel engines fitted to commercial vehicles, starting torques have greatly increased, necessitating such large power outputs from starting motors that the system voltage had to be increased from 12 to 24 V for the higher powered engines.

In order that the starting motor shall be disengaged from the engine when the latter is running, it is necessary to provide some form of automatic engaging gear.

For this purpose extensive use is made of the inertia-type drive on British and to a lesser extent American types. In fact, the inertia-type form of drive is almost universally employed in this country for cars, but for heavy vehicles, notably the diesel-engined types, the positively engaged starter drive is the more prevalent. The latter form of drive is also used to a large extent for cars in America where both types of drive are employed. Continental practice is to use the positively-engaged drive, particularly for 6-V systems.

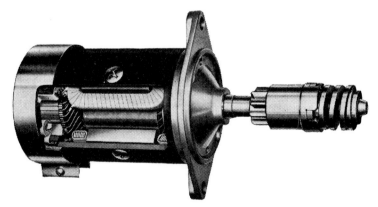

Figure 4.3   Lucas Model M35G starting motor, with SB type drive

Figure 4.3 illustrates the inertia-type drive in which a pinion is loosely mounted on a screwed sleeve, and the shock of engagement is taken up by a coil spring interposed between the motor shaft and the screwed sleeve which is free on the driving shaft. The inertia of the pinion causes it to move along the screwed sleeve into engagement with the flywheel gear when the motor shaft accelerates, and out of engagement when the motor slows down on releasing the starter switch. Contrary to what one might ordinarily expect, the engaging pinion and sleeves should not be lubricated as they operate best when clean and dry. In fact, lubricating with oil or grease is liable to cause the pinion to stick and prevent the free sliding action of the pinion on the screwed sleeve.

Depending on the method of mounting, the pinion may be arranged to move towards the motor to engage the flywheel or away from the motor. The former arrangement is known as the *inboard* type, whilst the latter is termed the *outboard* type of drive. A special housing is necessary for outboard drives in order that an additional bearing can be provided to support the outer end of the shaft.

Whilst the inertia-type of starter drive has proved satisfactory for cars and the lighter commercial vehicles, the high starting torques on modern high compression car engines and on heavy diesel-engined vehicles, has necessitated the use of the alternative pre-engaged form of drive in which the starter pinion is engaged with the flywheel gear under reduced power and full power is only engaged after the gears are fully engaged.

Starting motor mountings are of three main types, namely saddle, flange, and barrel mounting, the last applying particularly to outboard drives. The important dimensions are now standardised to ensure interchangeability, and these dimensions are given in the S.M.M.& T. Standard Specification No. 17. Whatever the form of mounting, it is essential that the starting motor should be installed on the engine or chassis so that the motor shaft is parallel to the crankshaft and the pinion is in correct relationship with the flywheel gear. It is also very necessary to see that the mounting is quite rigid, as the drive is liable to be damaged if the motor is not securely and firmly held in position.

CONSTRUCTION AND PERFORMANCE DETAILS

Lucas starters, Models 2M100 and 3M100, are designs which represent current British practice incorporating the best possible ways of economically obtaining high performance and reliability. These starters are designed for use on petrol engines fitted in cars, commercial vehicles and certain industrial engines. They are not intended for heavy-duty applications as for example where very high vibration levels may exist, such as petrol-engined tractors, portable compressors, earth moving and agricultural machinery.

These starters differ in construction only in respect of their field windings (see under the heading 'Motor'). The resulting two levels of performance enable a wide range of engine sizes and starting conditions to be catered for by a machine of given frame size.

As will be seen from Table 4.1 the engine sizes for which these starters are suitable cover the ranges previously met by models M418G and M45G (37-slot): these later models will be superseded by the 2M100 and 3M100 respectively.

The majority of modern engines using starters of this size require pre-engaged drives to give satisfactory starting and adequate service life. The inertia drive is suitable only for limited applications in temperate climates. Thus, while versions of the 2M100 and 3M100 will be available with inertia drives as replacements for corresponding M418G and M45G (37-slot) starters, it is strongly recommended that new engine applications of 2M100 and 3M100 starters should adopt the pre-engaged drive.

For vehicles with automatic transmission, inertia drive starters are not recommended. Pre-engaged starters are considered to be essential for these applications.

An approximate guide to the petrol engine sizes for which these starters are suitable under temperate and cold climate conditions is as shown in Table 4.1. It must be realised however that many factors can influence starter suitability, e.g. battery, cable, compression ratio, engine moment of inertia, etc. Figures 4.4 and 4.5 illustrate the construction of these starters.

Table 4.1 LUCAS STARTERS FOR TEMPERATE AND COLD CLIMATES

Model	Drive	Temperate climates			Cold climates		
		4 cyl	*6 cyl*	*8 cyl*	*4 cyl*	*6 cyl*	*8 cyl*
3M100	PE	–	4.2 litres	3.5 litres	2.5 litres	4.2 litres	3.5 litres
3M100	Inertia	2 litres	3 litres	–	–	–	–
2M100	PE	2.5 litres	3.3 litres	3 litres	2.0 litres	3.3 litres	3 litres
2M100	Inertia	2 litres	3 litres	–	–	–	–

The motor

The typical starting motor used in cars and some commercial vehicles is a 4-pole 4-brush machine employing a 29-slot armature running in two oil-impregnated sintered bearing bushes. The only constructional difference between the two motors is in their field windings. In 3M100, the nine turns-per-pole series-parallel connected field is of copper. Model 2M100 has a four turns-per-pole series-connected aluminium field. Thus, 3M100 is a higher performance machine than 2M100.

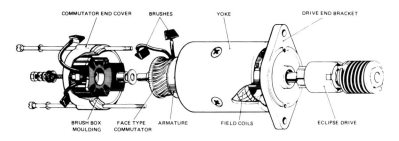

Figure 4.4   Construction of starter with Eclipse type inertia drive

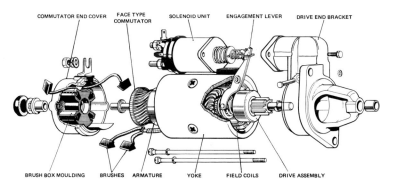

Figure 4.5   Construction of starter with pre-engaged drive

Metric threads are employed, except for pole screws and through bolts which will be changed to metric at a future date.

A face-type commutator is employed, in conjunction with axially-aligned fully-insulated brushgear housed in a phenolic moulding riveted to the commutator end cover. This face-type commutator permits the use of a windowless yoke, giving advantages in external protection. The motor end brackets are secured by through-bolts in conventional manner. Rotation is normally clockwise viewed from the drive end.

Pre-engaged drives

Pre-engaged models incorporate a 19S solenoid and a 7SD roller clutch. Solenoid performance is such that these starters are suitable for use with unrelieved flywheel ring gears.

1. Solenoid model 19S

Pull-in voltage at 20° C (max)	* 7.5 V	
Pull-in current (max)	40	A
Drop-off voltage (max)	4.0 V	
Drop-off current (max)	5.1 A	
Hold-in current (at 8 V)	10	A

* In general, the greater the out-of-mesh clearance between the pinion and flywheel teeth, the lower will be the pull-in voltage required to clear the tooth-to-tooth engagement condition. The voltage quoted is consistent with the minimum out-of-mesh clearance of 0.050 in (1.27 mm).

The operating windings, plunger and moving contact assembly are housed in a pressed body, to which is screw-fixed a moulded base incorporating the fixed contacts and external terminals. The engagement spring is located within the plunger and a lost motion spring ensures that the solenoid contacts open prior to pinion retraction, or if the pinion fails to disengage from the flywheel for any reason when the operating switch is released.

Additional contacts are provided for short-circuiting the resistor used with ballasted ignition. A $^3/_{16}$ in (4.76 mm) Lucar blade is used for this feature, non-interchangeable with the ¼ in (6.35 mm) Lucar solenoid supply terminal. As an option, an M5 shrouded pin for the ballast terminal is available, with plug-in connector on the cable harness. This gives improved sealing and forms a basis for future solenoid design.

2. Roller clutch model 7SD

A sealed five-roller clutch is housed in an extruded body and is driven by an involute spline form rolled on the armature shaft.

Drive engaging mechanism

Drive engagement is effected through a swing-hinged plate on the roller clutch, this arrangement giving reduced noise on over-run. By use of a solenoid plunger clevis which is pierced to a dimension with respect to the solenoid mounting face, the actuating position of the engagement plate in the drive end bracket is pre-set, eliminating the need for pinion setting by eccentric pin adjustment. It is thus essential for each plunger to be kept with its associated solenoid.

Pinion

The pinion fitted to the standard machine is nine tooth 10/12DP (2.54 Module). This provides direct interchangeability with the Eclipse inertia drive machine on an engine which may employ either pre-engaged or inertia drive starters according to the operating conditions.

Where such interchangeability is not required, it is possible to achieve greater advantages offered by pre-engaged starters by using higher gear ratios. A version with ten tooth 12DP pinion is therefore available and additionally this gives lower cranking noise than the standard pinion. Also available is a nine tooth 12/14DP (2.11 Module) pinion with special heat treatment: for quiet operation, this pinion should be used only when accurate centre distance and gear profile can be guaranteed, since it has a marked deterioration of contact ratio with meshing centre increase.

Inertia drives

Model 2M100 is available with either a nine tooth 10/12DP Eclipse drive or a ten tooth 8/10DP S-type drive. Model 3M100 is available with a ten tooth 8/10DP S-type drive, or a run-off helix version which provides drive overload relief. As stated previously, these are primarily intended only as replacements for existing inertia-drive versions of M418G and M45G (37-slot) starters.

In general, the use of 8/10DP pinions is not recommended because of the resulting low gear ratio. The Eclipse drive, while having the advantage of a higher gear ratio than the S-type drive, has a greater tendency to pinion ejection.

INERTIA DRIVE STARTING MOTORS

Figure 4.6 shows a cut-away view of the Bosch starting motor type CB and Figure 4.7 shows a cut-away view of the Bosch starting motor DG.

The operation of these motors is generally as described earlier.

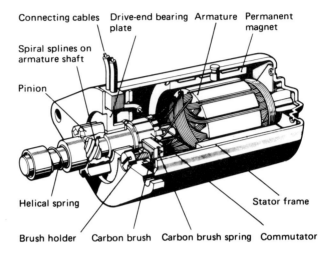

Connecting cables Drive-end bearing Armature Permanent
                   plate   magnet

Spiral splines on
armature shaft

Pinion

Helical spring

Stator frame

Brush holder Carbon brush Carbon brush spring Commutator

Figure 4.6 Cut-away view of starting motor, Type CB

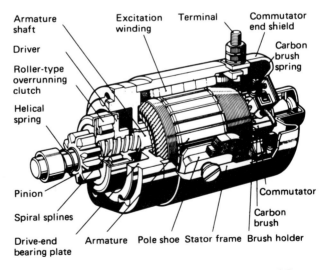

Armature   Excitation Terminal   Commutator
shaft   winding   end shield

Driver   Carbon
                                         brush
Roller-type   spring
overrunning
clutch

Helical
spring

Pinion   Commutator

Spiral splines   Carbon
                                         brush

Drive-end   Armature Pole shoe Stator frame Brush holder
bearing plate

Figure 4.7 Cut-away view of starting motor, Type DG

PRE-ENGAGED DRIVE STARTING MOTORS

The construction and internal wiring of these starting motors are shown in Figures 4.8 to 4.10. Starting motors of this type are switched on by means of a solenoid switch mounted directly to the motor. The end of the solenoid plunger extending out from the coil terminates in a flattened

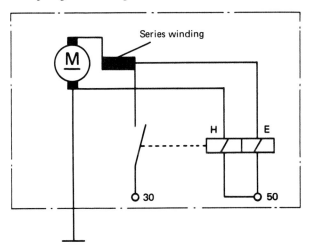

Figure 4.8   Wiring diagram of pre-engaged drive starting motor

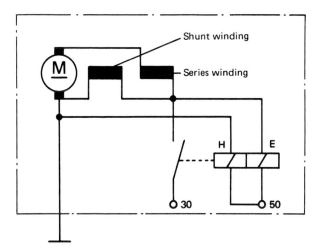

Figure 4.9   Wiring diagram of pre-engaged drive starting motor with shunt winding

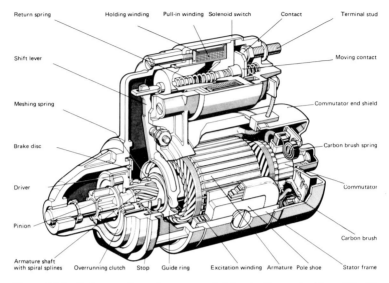

Return spring Holding winding Pull-in winding Solenoid switch Contact Terminal stud

Shift lever

Moving contact

Meshing spring

Commutator end shield

Brake disc

Carbon brush spring

Driver

Commutator

Pinion

Carbon brush

Armature shaft
with spiral splines Overrunning clutch Stop Guide ring Excitation winding Armature Pole shoe Stator frame

*Figure 4.10   Cut-away view of pre-engaged drive starting motor Type EF with
outer-wedge roller-type overrunning clutch*

section with a slot into which the end of the shift lever fits with a certain
amount of play, or plunger clearance. This clearance is needed when the
starting motor is switched off, so that the return spring on the solenoid
switch can move its plunger by the length of the clearance toward the
at-rest position and can thus lift the moving contact from the fixed
contacts. This provision is necessary to allow for the possibility that the
pinion does not disengage from the engine flywheel ring gear. (In older
starting motor models the shift lever is fastened by means of a pin
directly to the solenoid plunger with no play and the starting motor
drive is fitted with a cut-off spring.)

The armature shaft has spiral splined on its pinion end (Figures 4.14
and 4.16) on which is mounted a drive. This drive is coupled to the
pinion by means of a roller-type overrunning clutch. The lead of the
splines is selected so that as the armature shaft rotates, the pinion,
which is held firmly in place and prevented from turning as the shaft
turns, is moved into the flywheel ring gear. Two guide rings or discs
which can shift longitudinally, i.e. parallel to the armature shaft, are
mounted on the drive and the fork-shaped end of the shift lever fits
into them. The so-called meshing spring is mounted as an elastic member
between the guide ring or disc and the drive so that the shift lever
always moves the full distance (only then can the moving contact of
the solenoid switch close) ensuring that the starting motor current is

always switched on even if the pinion and flywheel teeth should butt. In other words, the shift lever pivots and pushes the drive together with the pinion towards the flywheel. The pinion is then pushed all the way into meshed position by the screw effect of the spiral splines. In this way the splines premit a torque to be applied to the vehicle engine only when the teeth of the pinion are fully meshed with the teeth of the flywheel ring gear.

After the meshing of the drive pinion with the ring gear the over-running clutch provides the positive mechanical connection between the starting motor armature and the vehicle flywheel. The connection is broken as soon as the crankshaft speed becomes higher than the speed of the starting motor.

Meshing

In principle, the meshing operation is composed of two successive movements of the pinion — the axial movement and the helical move-ment. For practical purposes the starting motor is actually switched on in one continuous rapid operation.

When the starting switch is turned on, the shift lever is first moved by the solenoid switch against the force of a spring. The excitation and armature winding are not carrying any current at that point, so the armature does not turn. As one end of the shift lever is drawn towards the solenoid, the other end, working through the guide ring and the clutch spring on the pinion end of the armature shaft, pushes the drive and the pinion toward the flywheel ring gear, resulting in the drive and pinion turning due to the screw effect of the spiral splines. If the tooth on the pinion fits directly into a tooth space on the ring gear, the pinion meshes immediately as far as the shift lever can move it, i.e. until the moving contact hits the fixed contacts in the solenoid assembly. This completes the axial movement of the pinion.

If the teeth on the pinion and on the flywheel ring gear should butt when the pinion is pushed towards the flywheel, the pinion is tempor-arily prevented from moving further. The shift lever, acting through the axially-shiftable guide ring on the pinion end of the armature shaft, compresses the meshing spring until the moving contact in the solenoid assembly hits the fixed contacts and the armature begins to turn. The pinion also turns, moving its teeth across the end face of the flywheel ring gear until the pinion teeth drop into mesh in the next spaces between the ring gear teeth. The pinion is forced into the meshed position by the meshing spring pressure, and the screw action of the splines.

At the end of the pinion axial movement, i.e. shortly before the end of the travel, the solenoid switch contacts close in any event and the starting motor current begins to flow through the excitation windings and armature windings. The armature begins to turn and by action of the splines forces the drive pinion, which is firmly held in the ring gear and prevented from turning, further into meshing position inside the ring gear until it reaches the pinion stop at the end of the armature shaft. At this point, further movement of the drive pinion away from the armature is not possible and in that position the pinion is coupled to the armature shaft by a positive mechanical connection through the overrunning clutch and the driver, so the starting motor can now crank the vehicle engine.

Demeshing

As the vehicle engine starts to operate, its rotational speed eventually exceeds the rotational speed of the starting motor. At this instant the overrunning clutch mechanism breaks the positive mechanical connection between the pinion and the armature shaft and thus protects the armature from damage caused by excessive speed.

The pinion remains meshed, however, as long as the shift lever remains in the pulled-in position. Only when the starting motor is switched off do the shift lever, drive and pinion return to their rest positions due to the action of the return spring. This spring also holds the drive pinion firmly in the rest position, despite engine vibration, until the next starting operation.

Overrunning clutch

For protection purposes, screw-push starting motors are fitted with a roller-type overrunning clutch (Figure 4.11). This couples the pinion with the drive as long as the armature shaft is providing driving force to the flywheel, but it breaks the connection when the flywheel tries to turn the pinion faster than the shaft. In order to accomplish this dual function the clutch rollers move in tapered notches in the clutch shell. The notches are so designed that when the engine is being cranked the rollers are clamped tightly, in the narrow part of the notches, between the clutch shell and the cylindrical part of the pinion (pinion collar). When the vehicle engine starts to operate, the excess rotational speed of the pinion spins the rollers back against the force of their springs into the wider part of the notches where they make only loose contact with the clutch shell and the pinion. In the rest position the rollers are pressed back into the narrow part of the notches by means of springs acting either directly or through guide bushings or guide bolts, so that when the starting motor is switched on again the pinion is coupled firmly to the ring gear.

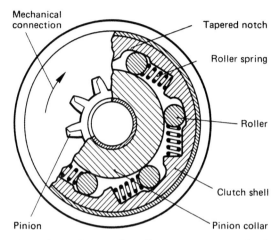

Figure 4.11   Outer-wedge roller-type overrunning clutch

In this outer-wedge type overrunning clutch, the notches are machined into the inside of the clutch shell which turns outside, i.e. around the pinion collar and which is connected to the armature shaft by means of the driver. This arrangement has the advantage that the mass of the pinion can be kept small and that the overhaul torque when the engine speed exceeds that of the pinion is also small. This design therefore contributes to the service life of these parts and of the drive housing or drive-end bearing plate.

Armature brake

In order to bring the starting motor to a stop as rapidly as possible after it has been switched off so that the operator can make another attempt to start the engine if the first attempt has failed, the starting motor is fitted with an armature brake (Figure 4.12). This brake is usually a type of mechanical (disc) brake.

Some higher-power pre-engaged drive starting motors are fitted with a shunt winding (Figure 4.9) to reduce the no-load speed of the

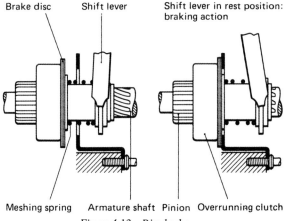

Brake disc   Shift lever   Shift lever in rest position: braking action

Meshing spring   Armature shaft   Pinion   Overrunning clutch

Figure 4.12   Disc brake

armature and to bring it to a stop when switched off. This results in a very short run-down time when the starting motor is switched off.

SLIDING ARMATURE STARTING MOTORS

In these medium-power starting motors designed for more powerful engines, greater forces are applied during the meshing process. Meshing of the pinion with the ring gear teeth must be accomplished gently, however, in order to avoid damaging either the pinion or the ring gear. For this reason, the starting motor is switched on and the pinion is meshed with the ring gear, in two stages.

In this sliding-armature type, the axial movement of the pinion is produced directly by the armature itself which shifts in an axial direction (the sliding armature). The motor, therefore, is fitted with a long commutator (Figure 4.13). The designation 'sliding armature starting motor' stems from the sliding motion of the armature. Figure 4.22 shows the construction of a starting motor of this type.

The sliding armature starting motor has three excitation windings, i.e.

> An auxiliary starting winding,
> A shunt winding (holding winding), and
> A series winding (the main winding).

The auxiliary starting winding and the holding winding are activated during both stages of the meshing process, the series winding is fully activated only during the second stage.

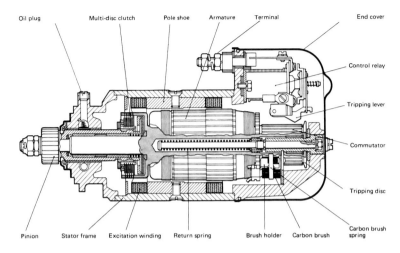

Figure 4.13   Cross-sectional drawing of a Bosch sliding armature starting motor designed for flange mounting

In these starting motors a multi-disc clutch (Figures 4.22 and 4.30) with torque limitation (mechanical overload protection) provides the direct mechanical link between the armature and the pinion.

Meshing

The process involved in meshing and demeshing are shown in Figures 4.14 to 4.19. For purposes of simplicity, the connection between the armature and the pinion has been depicted as rigid, although in practice the multi-disc clutch is located between these parts, the operation of which will be described separately below.

In the rest position (Figure 4.14) the armature and the relay are not under power, i.e. they are de-energised; the armature is stationary and has been moved somewhat out of the magnetic field by spring force, as a result of which the pinion is not engaged with the flywheel ring gear.

When the starting switch is turned on, the first switching stage in the relay becomes activated (Figure 4.15). One of the contacts on the tilting contact bridge switches in the auxiliary starting winding and the holding winding. The armature is drawn into the field of force of these windings where it turns slowly until the pinion meshes smoothly and easily with the teeth of the flywheel ring gear (Figure 4.16); a process which prevents damage to both the pinion and the ring gear. If the pinion and ring gear teeth should butt, the pinion is turned further with only weak torque and limited axial pressure against the side of the flywheel until its teeth reach the next set of spaces between the teeth on the ring gear. The pinion is then driven further forward to drop into mesh with the ring gear.

When the pinion has meshed with the ring gear but shortly before it reaches its fully extended position, the armature has been shifted axially so far that the tripping disc mounted on its commutator raises the tripping lever (Figure 4.17). As a result the tilting contact bridge, which was locked during the first switching stage, is released and can make contact with its other end (second switching stage). This completes the connection between the battery and the series winding.

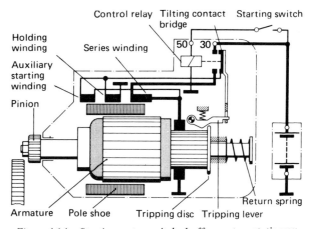

Figure 4.14   Starting motor switched off, armature stationary

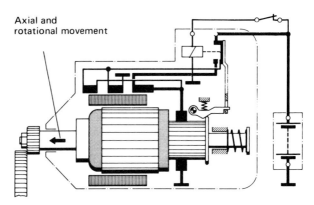

Figure 4.15   Switching stage No 1. Auxiliary starting and holding windings switched on; armature drawn forward and turns slowly

As soon as the series winding is energised, the starting motor develops its full torque and cranks the vehicle engine. When the vehicle engine then increases its rotational speed, the armature is also forced to rotate at a higher speed. The current flowing through the armature and the excitation winding decreases sharply and thus also the magnetic forces exerted on the armature to keep the pinion meshed.

If only individual ignition pulses were to be generated at this point (i.e., before the engine starts to operate independently), the armature would return to its rest position and the pinion would demesh. For this reason, the armature is held in its operating position by the holding winding which is connected in parallel with the armature and whose

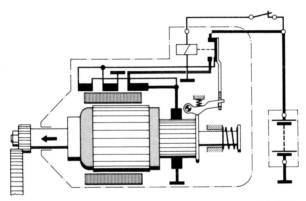

Figure 4.16   Current flows in auxiliary starting and holding winding; pinion meshes with ring gear

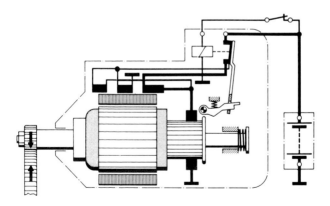

Figure 4.17   Switching stage No. 2. Series winding switched on, complete mechanical connection; vehicle engine is cranked

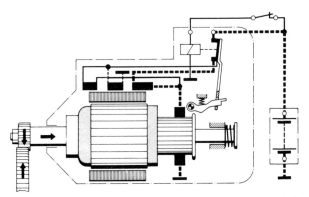

Figure 4.18   Vehicle engine has started, armature speed is increased, current in the series winding decreases, only the holding winding holds the armature engaged

magnetic field is not weakened, until the operator of the vehicle releases the starting switch (Figure 4.18).

Demeshing

When the starting switch is released after the engine starts to operate, the starting motor becomes currentless (Figure 4.19). As a result, the magnetic force exerted on the armature is removed and the pressure between the teeth of the pinion and the ring gear decreases so much that the armature return spring, acting inside the armature shaft,

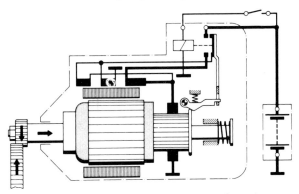

Figure 4.19   Switch-off process. Current through starting motor windings is cut off, pinion demeshes, return spring draws armature back to its rest position

can return the armature to its rest position and thus disengage the pinion.

During the disengaging process the spring-loaded tripping lever is pressed back into the locked position so that the next starting process can again take place in two stages. After the pinion has disengaged, the armature comes to a stop in a few seconds.

Multi-disc clutch

The multi-disc clutch is designed to limit, by means of frictional contact, the torque transmitted to the pinion from the armature shaft to a maximum permissible amount.

The individual discs in the clutch mechanism are mounted so that they can shift axially but cannot turn in the drive flange or on the pressure sleeve (Figure 4.20). These discs are fitted with drives alternately on

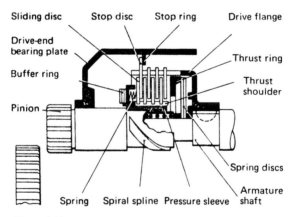

Figure 4.20   Construction of multi-disc clutch with torque limitation and initial stage (rest position)

their outside and inside diameters which are engaged respectively with the drive flange and the pressure sleeve. The drive flange is fixed on the armature shaft, while the pressure sleeve is located on the spiral splines of the drive spindle, which is mounted loosely on the armature shaft; the pressure sleeve can thus be driven forward or backward by the splines as the spindle turns.

A certain initial pressure between the clutch discs provides the degree of mechanical connection which must already be present in the clutch (so that the pressure sleeve can move) before it can lock and provide positive mechanical connection. For this purpose springs press against the first disc (sliding disc) of the pressure sleeve and this disc presses against the larger diameter first disc (stop disc) of the drive flange.

In the rest position the stop disc, under the influence of the armature return spring, is pressed against a stop ring fixed-mounted in the stator frame. Thus the pressure of the springs acts only on the sliding and stop discs — only a slight torque can therefore be transmitted — while the other discs are under no pressure. This slight mechanical connection is designated the initial meshing stage.

The effect of this initial stage is that when the pinion and ring gear teeth butt only the slight torque will be transmitted which corresponds to the pressure between the sliding disc and the stop disc. Although the stop disc is initially lifted from the stop ring when the armature moves forward, if the teeth butt the pressure sleeve moves inward, because of the effect of the spiral splines, until the stop disc again makes contact with the stop ring.

If, as the pinion moved forward, its teeth encounter gaps between the teeth on the ring gear it can immediately drop into mesh and when the pinion has reached its fully extended position the full mechanical connection starts. When the pinion is firmly meshed and the mechanical connection established and, as the armature shaft turns, the pressure sleeve is driven on the spiral splines toward the inside of the starting motor, as a result of which the pressure between the discs is increased. This pressure continues to increase until the friction between the discs is sufficient to transfer the particular starting torque required. At that point the mechanical connection takes the following path: armature shaft — drive flange — outer discs — inner discs — pressure sleeve — pinion.

The torque limitation built into the multi-disc clutch operates in the following way. When the full mechanical connection has been established, a thrust ring transfers the pressure from a spring disc assembly, mounted on the armature shaft and supported on the drive flange, to the clutch disc assembly. The pressure sleeve is fitted with a thrust shoulder on its inner end. The increasing pressure between the discs resulting from the pressure sleeve being 'screwed' inward toward the inside of the starting motor, i.e. the torque which can be transmitted, is limited when the maximum permissible loading is reached by the thrust shoulder pressing the spring discs away from the thrust ring. As a result, the clutch beings to slip.

If the vehicle engine flywheel suddenly begins to accelerate because of ignition pulses or when the engine starts, the flywheel begins to drive the pinion faster than the starting motor has been driving it ('over-running'). As a result, the pressure sleeve is screwed slightly outward toward the pinion and the mechanical connection is broken so that the starting motor armature will not be driven at excessive speeds which could damage it. In such a case the multi-disc clutch acts as an over-running clutch.

SLIDING GEAR STARTING MOTORS

This type of starting motor, designed for large internal combustion engines, operates in two stages in order to protect the pinion and ring gear. The cross-sectional drawings in Figures 4.21 and 4.22 show the construction of Types KB and TB.

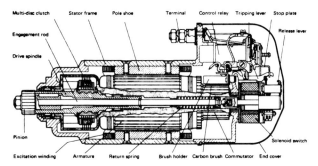

Figure 4.21   Cross-section of the two-stage Bosch sliding gear starting motor, Type KB

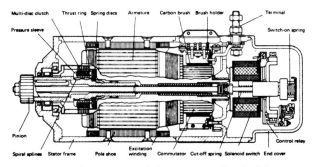

Figure 4.22   Cross-section of the two-stage Bosch sliding gear starting motor, Type TB

The armature is fixed-mounted in the drive-end bearing plate and in the commutator end shield; the armature shaft is hollow and is designed to serve as the housing for a multi-disc clutch on its drive end. This housing is sealed on the end face by a cover on which a rolling or plain bearing is mounted to support the armature in the drive-end bearing plate. On the commutator end the armature is supported in a plain bearing. A solenoid switch for the pinion and a control relay for the two switching stages are flanged to the commutator end shield. In this type of starting motor the pinion is shifted forward (sliding gear) by the solenoid switch acting through an engagement rod leading through the hollow armature shaft (Figure 4.21).

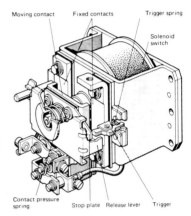

Figure 4.23 Locking device on solenoid switch in starting motors, Types TB and TF

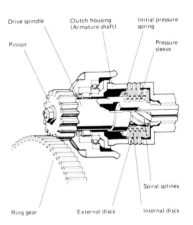

Figure 4.24 Construction of multi-disc clutch (without initial stage)

The resistance, or opposing, winding located in the solenoid switch of starting motor Type TB (Figure 4.22) serves to adjust the starting motor torque when meshing. In order to prevent oil, dirt or dust penetrating into the inside of the starting motor, a radial lip-type seal is fitted on the drive-end bearing plate. The commutator end shield and the solenoid switch are enclosed by an end cover and the carbon brushes are enclosed by either the end cover or a cover band.

The multi-disc clutch is mounted on the spiral splines of the drive spindle which is supported in a rolling bearing in the drive-end bearing plate and in a needle roller bearing in the armature shaft. The pinion is joined in a positive mechanical connection with the spindle by means of a feather key.

The solenoid switch is designed to drive the pinion forward in two stages. For this purpose it is fitted with a locking device (Figures 4.21 and 4.23) with a trigger, a stop plate and a release lever.

This type of starting motor is also fitted with a multi-disc clutch (Figure 4.24) between the armature and pinion, the operation of which was described earlier. However, in this case the clutch does not have an initial meshing stage.

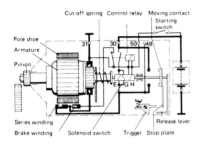

Figure 4.25   Rest position
E. Pull-in winding
G. Resistance (opposing) winding
H. Holding winding

In recent models of T-type starting motors the clutch disc assembly is no longer held under pressure by the helical springs in the pressure sleeve. Instead there are four bolts with helical springs in the pressure sleeve which only pre-stress the pressure sleeve against the shoulder of the drive spindle. This offers the advantage that the overrunning torque is lower and as a result the clutch discs are not loaded so heavily during the overrunning process.

In addition to the series winding, there is a brake winding on the pole shows which becomes active when the starting motor is switched off. Some of the starting motors are fitted with a shunt winding which limits the armature no-load speed. In some cases an additional auxiliary starting winding is fitted which is switched in during the first switching stage and is connected in series with the armature.

Meshing

The following description applies to starting motor Type TB.

When the starting switch is turned on, current flows into Terminal 50, through the holding winding of the solenoid switch and through the control relay winding (Figure 4.26). As a result, the control relay opens the contacts for the brake winding and switches on the pull-in and resistance windings of the solenoid switch. The solenoid plunger acts through the engagement rod to push the drive spindle forward, pushing the pinion against the ring gear. At the same time the main winding receives a relatively low current through the pull-in and resistance windings which together act as a series resistor, as a result of which the motor armature turns slowly (Figure 4.27). In this first switching stage the starting motor does not yet develop its full torque.

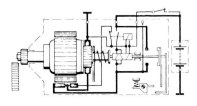

Figure 4.26   1st switching stage. Control relay winding and holding winding on solenoid switch are energised

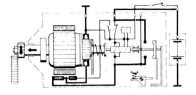

Figure 4.27   Pull-in winding on solenoid switch is energised, pinion is forced forward, armature turns slowly

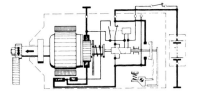

Figure 4.28   Pinion meshes with torque still low

Shortly before the end of the forward pinion travel, movement of the moving contact on the solenoid switch is halted when the stop plate (Figure 4.32) strikes the trigger. The solenoid plunger continues to move, however, and gentle meshing is thus assured by the pinion being moved forward and at the same time rotating slowly (Figure 4.28). If the pinion and the ring gear teeth should butt, the pinion is then turned against the side of the ring gear so that it can drop into mesh easily and gently in the next set of spaces between the teeth on the ring gear.

Although the pinion meshes during the first switching stage, the vehicle engine is not yet cranked because the starting motor torque is not high enough at this point. Immediately before the end of the pinion travel a release lever lifts the trigger, the trigger releases the stop plate and the moving contact is snapped instantaneously against the contact bars because of the effect of the spring (the contact pressure spring, Figure 4.23) which has been stressed during the meshing of the pinion. This contact process is designed to take place as rapidly as possible to prevent self-welding of the contacts (i.e. the contacts welding themselves closed), which could occur if meshing were delayed and thus to extend the service life of the contacts.

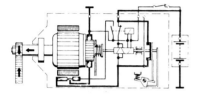

Figure 4.29   2nd switching stage. Trigger is released, series winding is switched in by moving contact, starting motor has full torque, engine is cranked

Even though the pull-in and resistance windings of the solenoid switch are short circuited when the main contacts close, the armature is held in the switched on position by the holding winding through which current is flowing. The starting motor now received full current, develops its full torque and cranks the vehicle engine through the positive mechanical connection now provided by the multi-disc clutch (Figure 4.29).

Demeshing

As the vehicle engine speeds up, it eventually drives the pinion faster than the starting motor has been driving it. This overrunning condition

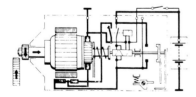

Figure 4.30   Starting motor switched off, pinion demeshes, armature is braked

causes the clutch to release so that the starting motor armature is not driven so fast that it could be damaged. The pinion remains engaged with the ring gear, however, as long as the starting switch remains turned on. Only when the starting switch is released, which switches off the starting motor, does the pinion return to its initial position. When the starting motor is switched off, the winding on the control relay and the holding winding on the solenoid switch are de-energised.

The control relay interrupts the circuit leading to the pull-in winding and switches the brake winding in. A cut-off spring (Figure 4.22) presses the plunger of the solenoid switch back into its rest position, as a result of which the main circuit is interrupted and the drive spindle with the pinion can also be brought back to the rest position through the dorce of its return spring located inside the armature shaft; at this point the pinion demeshes (Figure 4.39). This return spring also has the function of holding the drive spindle in the rest position until the next starting process despite vibrations caused by the operating engine. During the demeshing process, the spring-loaded trigger is pressed back to the locking position so that the next starting process can again take place in two stages.

Armature brake

In T-type starting motors a brake winding (Figure 4.30) is installed so that when the motor is switched off it will come to a stop in the shortest possible time and so that if necessary a fresh attempt to start the engine again can be made as soon as possible. Starting motors designed for operating voltages above 50 V are fitted with a shunt winding instead of a brake winding.

Similar to the series winding, the brake winding is mounted on the pole shoes and has no effect as long as the starting motor is operating. When the starting motor is switched off, however, the brake winding is switched (by one of the control relay contacts) in parallel with the armature, which of course is still turning and thus acts as an electric brake, bringing the armature to a stop in a very short time.

THERMOSWITCHES

In starting systems in which unacceptably long starting processes and constant restarting must be expected (for example, with a low battery voltage, damaged ring gear teeth, or defects in the engine) Type T starting motors are used with two built-in thermoswitches as protection against thermal overload of the starting motor in the first and second switching stages; these thermoswitches are connected in series and are built into carbon brushes or connecting bars.

If the temperature in the solenoid switch windings should exceed certain values in the event of 'blind meshing' (i.e. where pinion and ring gear meet teeth corner against teeth corner and the pinion is blocked from turning further; this situation should not be confused with 'teeth against teeth' which presents no danger) or if temperatures in other current-carrying parts should exceed certain values as the engine is cranked, the thermoswitches interrupt starting motor cable 50 and the starting motor is automatically switched off. After a few minutes have passed, the starting motor will have cooled down sufficiently so that it can be operated again.

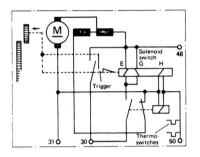

Figure 4.31   Internal circuit of starting motor Type TB designed for 24-V operation with built-in thermoswitches and shunt winding connected to positive brush

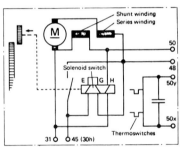

Figure 4.32   Internal circuit of starting motor Type TB designed for 40–110V operation with built-in thermoswitches and shunt winding connected to terminal 50

The internal circuitry of these starting motors designed for an operating voltage of 24 V is somewhat different from the design of motors operating at higher voltages (Figures 4.40 and 4.41). In the starting motors operating at higher voltages, a capacitor is installed in parallel with the thermoswitches for purposes of spark suppression.

STARTING MOTORS WITH INTERMEDIATE TRANSMISSION

Large diesel engines with relatively large diameter ring gear are fitted with Type T starting motors having an intermediate transmission (Figure 4.33). These starting motors have an eccentric drive system (intermediate transmission) in which the pinion is displaced from the armature.

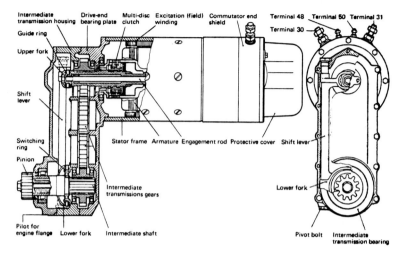

Figure 4.33   Type TF starting motor designed for 24V operation, with inter-mediate transmission

The intermediate shaft with the pinion is supported in the inter-mediate transmission bearing where it can turn and move axially. A shift lever located in the intermediate housing transfers the thrust movement of the engagement rod to the intermediate shaft and pinion.

THE DYNASTART

The Dynastart combined starter motor/dynamo has been designed to meet the demand for combined electric starting and generating equip-ment in a single compact unit for fitment to existing engine designs, without major modification. Drive is usually by a V-belt mounted on the crankshaft extension or by toothed belt or chain.

Two units are produced with nominal generator output of 8 A and 5 A at 12 V continuously rated. The units are of four-pole design, two poles carrying the series windings for starting and two the shunt windings for generation. The construction and flange mounting arrangements are similar to the conventional d.c. generator.

Starting is effected by a key switch or push-button switch which operates the starting solenoid and causes the Dynastart to operate as a compound motor with predominantly series characteristics. When the engine starts, the starter switch is released and the Dynastart is then driven by the engine, as a self-excited shunt wound generator. As soon as the rated voltage is reached, a separate regulator begins to maintain the rate of charge to suit the state of the battery.

CHAPTER 5

THE BATTERY

The battery serves to store electrical energy when there is little or no demand for current, and acts as a secondary source of electrical energy when the engine is not running, or the generator speed is insufficient to meet the full requirements of the load on the system. A wide range of batteries, varying from 5- to 200Ah capacity is necessary to meet the requirements of the motor industry.

Automobile batteries must necessarily be of robust construction, not only to withstand severe vibration but also the high charging rates and heavy discharge currents associated with automobile service. As the battery is not always in an accessible position on the vehicle, it is sometimes not given the attention it requires, and for this reason it is often regarded as the weak link of the system. The modern battery, if properly cared for, is capable of giving good service for several years. The periodic attention required in service is quite small, but this attention must be given if the battery is to be maintained in an efficient and healthy working condition. It is well to remember that the life of any battery is, in very large measure, dependent upon the attention it receives in service.

There are two main types of battery employed on automobiles: the lead-acid and the nickel-alkaline types. The former is by far the most extensively used, although during recent years the nickel-alkaline battery has been increasingly used on motor cycles and commercial vehicles.

THE LEAD-ACID BATTERY

In the lead-acid battery, the positive and negative electrodes are immersed in dilute sulphuric acid and when fully charged the active material on the positive plate is lead peroxide and that on the negative plate spongy lead. During discharge, chemical reaction converts the active materials of both plates to lead sulphate, water being liberated in the process. This results in the lowering of the specific gravity of the acid or electrolyte. When charging, this chemical action is reversed, with a consequent rise in the specific gravity of the electrolyte. The specific gravity is therefore a good indication of the state of charge of the battery.

The following reversible equation will serve to show the constituent changes as a result of *discharging*, if read from left to right, and *charging* if read in the reverse direction

	Charged			*Discharged*	
Positive Plate	*Electrolyte*	*Negative Plate*	*Positive Plate*	*Electrolyte*	*Negative Plate*
PbO_2 +	$2H_2SO_4$ +	Pb =	$PbSO_4$ +	$2H_2O$ +	$PbSO_4$
(Lead peroxide)	(Sulphuric acid)	(Spongy lead)	(Lead sulphate)	(water)	(Lead sulphate)

It is presumed in this equation that discharge is carried to the point of complete exhaustion. In practice, however, the battery should not be discharged beyond a certain point, for two main reasons. One is that lead sulphate ($PbSO_4$) occupies a greater volume than lead peroxide (PbO_2); consequently excessive sulphation will set up mechanical stresses in the positive plate, causing shedding of the active material and cracking of the plate. The other reason is that if sulphation is carried to excess, the sulphate becomes stubborn and difficult to re-convert to the proper active material during subsequent charging.

Principles of operation

The principles underlying the operation of an accumulator or battery are those of electrolysis. It is well known that water, to which a trace of sulphuric acid is added to render it a conductor, may readily be split up into its constituent gases, hydrogen and oxygen, by the passage of an electric current through it.

In the process of electrolysis the constituents, whether elements or compounds, are called *ions* and are liberated only at the electrodes by which the current enters and leaves the electrolyte. The positive electrode, that is the one at which the current enters the electrolyte, is termed the *anode*, whilst the negative electrode is called the *cathode*. According to which electrode the ions are urged, they are called *anions* or *cations*. These ions are really carriers of electric charges, the anions being negatively charged whilst the cations are electropositive ions. The ions of hydrogen and metals are cations and those of oxygen, chlorine, iodine, and the acid radicals such as sulphion (SO_4) are electronegative and form anions.

Under the influence of an electric pressure or voltage, the ions are urged through the electrolyte, the cations moving in the direction of the current from anode to cathode and the anions in the reverse direction. Since the anions are *negative* charges moving in a *negative* direction, it can be assumed that they are *positive* in effect.

Upon reaching the electrodes, the ions give up their charges, cease to be ions and depending upon their chemical constitution, may enter into chemical union with the electrode. The absorption of the negative and positive charges by the anode and cathode respectively sets up a back or polarisation e.m.f. in opposition to the applied voltage. It will be evident then, that in the case of the battery the charging current will gradually diminish as charging proceeds, assuming, of course, that the applied voltage remains constant. It will also be apparent that upon interrupting the charging current and connecting the battery electrodes to an external circuit, the back e.m.f. set up will cause current to flow through the electrolyte in the reverse direction to that of the charging current. As the discharge continues the back e.m.f. will diminish, depending upon the rate and time of discharge.

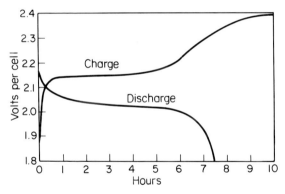

Figure 5.1   Voltage curves for a lead-acid battery

Typical voltage curves for a lead-acid battery are given in Figure 5.1. It will be noted that, on charge, the voltage per cell rises rapidly at first and then increases gradually until towards the end of the charge, when the voltage rise becomes more rapid. On open circuit, the cell voltage drops almost immediately to about 2.2 V and for the initial period of discharge rapidly falls to approximately 2.0 V, remaining fairly constant until towards the end of the discharge, when a sudden fall in voltage occurs. It is at this stage that further discharge should be promptly discontinued and the battery recharged, otherwise excessive sulphation will occur, which (as stated earlier) is liable to cause permanent injury to the plates. From the foregoing we see that each cell of a lead-acid battery gives a nominal back e.m.f. of 2.0 V, so that three and six cells will be required for 6- and 12-V batteries respectively.

Chemistry of charge and discharge

By electrolysis we are able to restore the power of a battery to generate an electric current. When a battery is in a discharged condition, both the positive and negative plates are coated with lead sulphate ($PbSO_4$), which presents a whitish appearance. The application of a voltage to the battery, so that the positive and negative plates are the anodes and cathodes respectively, produces an ionic stream through the electrolyte, causing dissociation of the constituents of the electrolyte and electrodes. Lead (Pb) is dissolved from the anodes, the released lead cations being urged through the electrolyte in the direction of the current, to be deposited on the negative plates or cathodes. At the same time the water of the electrolyte is decomposed into its constituent gases. The hydrogen cations liberated at the cathodes unite with the sulphion (SO_4) anions to form sulphuric acid, leaving the lead (Pb) of the lead sulphate ($PbSO_4$) on the negative plates. Expressed as a chemical equation, the changes at the negative plates may be given as follows

$$PbSO_4 + H_2 = Pb + H_2SO_4$$

The free oxygen anions react on the lead sulphate of the positive plates and on the water of the electrolyte to form a brown coating of lead peroxide, and to increase the sulphuric acid content of the electrolyte, as represented by the equation

$$PbSO_4 + H_2O + O = PbO_2 + H_2SO_4$$

Authorities differ as to the precise nature of the chemical reaction, but the foregoing will suffice to show that on charge the plates are de-sulphated, the acid radical SO_4 entering the electrolyte. So long as the plates are being de-sulphated the hydrogen and oxygen will enter into chemical combination with the other constituents, but when this chemical action ceases, *gassing* will occur as evidence of the liberation of hydrogen and oxygen at the negative and positive plates respectively. When these gases are freely evolved at the electrodes it is an indication that the battery is fully charged. Further charging then only results in the decomposition and reduction of water of the electrolyte. The periodical examination of the battery, at least once each month, is strongly recommended so that any loss of electrolyte can be replaced. Since it is only water that is driven off, distilled water should be used for replenishment.

In a battery on discharge, positively charged hydrogen ions travel to the positive plate and combine with the oxygen of the lead peroxide to form water, whilst the released sulphion ions (SO_4) combine with the lead of both positive and negative plates to produce lead sulphate. The

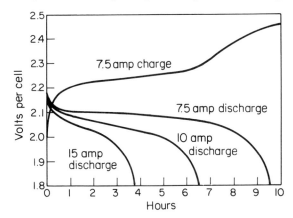

Figure 5.2   Battery discharge voltage curves

formation of water in the vicinity of the positive plates, and the abstraction of the SO_4 radical from the acid, lowers the specific gravity of the electrolyte.

Capacity and rate of discharge

The period of discharge before the voltage per cell begins to fall rapidly to 1.9 V or less will, of course, depend upon the rate of discharge; but in general the lower the rate of discharge the greater will be the ampere-hour output or capacity of the battery. This will be noted from the discharge voltage curves in Figure 5.2 for an automobile battery rated at 75 Ah on a 10-hour discharge rate. The battery was discharged at 7.5, 10, and 15 A rates, after previous charges at 7.5 A for 10 hours. From the curves it will be seen that the ampere-hour outputs, before the voltage per cell dropped to 1.8 V, were 72, 65, and 56.5 Ah respectively, thus showing that rate of discharge has an important bearing upon the capacity of the battery.

In referring to the capacity of a battery, it is therefore necessary to state also the rate of discharge. Present practice in regard to automobile batteries is to specify the capacity of the battery on the basis of a 10-hour rate of discharge. For example, a 75 Ah battery on this rating will give, when fully charged, 7.5 A for a period of 10 hours, without the voltage falling below 1.8 V per cell. Some manufacturers also specify their batteries on a 20-hour rate of discharge, in which case the capacity is approximately 20 per cent higher than for the 10-hour rated capacity. Thus a battery specified on the 10-hour discharge rate as 75 Ah capacity would have a 90 Ah capacity on a 20-hour rate of discharge.

The capacity of a battery is dependent upon the surface area of the positive and negative plates, and in well-formed plates the porosity of the active materials greatly increases the effective area exposed to the acid. When a battery is on discharge there is a localised weakening of the electrolyte at the electrodes and in the pores of the active materials, which is only strengthened by diffusion with acid in the main body of the electrolyte. The fact that at high discharge rates this weakening effect is more rapid than the diffusion is the explanation of the more rapid fall in voltage and lower capacity. Should the battery be given a period of rest it will be found that the voltage will rise slightly owing to the subsequent diffusion of the acid into the pores of the active materials.

EFFICIENCY OF A BATTERY

The efficiency of a battery may be determined either as the quantity or the energy efficiency. In the former case it is given by the ratio of ampere-hours input and output and in the latter case by the ratio of the watt-hours. Thus

$$\text{Quantity efficiency } \% = \frac{\text{Ampere-hours output}}{\text{Ampere-hours input}} \times 100$$

$$\text{Energy efficiency } \% = \frac{\text{Watt-hours output}}{\text{Watt-hours input}} \times 100$$

As already pointed out, the output of a battery is dependent upon the rate of discharge and, as the curves in Figure 5.2 show, the quantity efficiency may vary from 96 to 75% and it may even be as low as 50% if the discharge rate is very high. In computing the watt-hour efficiency, the average voltage during charge and discharge is taken into consideration. Therefore, owing to the difference in the cell voltage on charge and discharge, the energy efficiency is much lower than the quantity efficiency. The characteristics of the 75-Ah battery given in Figure 5.2 indicate a watt-hour efficiency of 86% for normal discharge and 66% for the higher discharge rate. Here again, with heavy discharges, the energy efficiency may be much lower than stated above.

BATTERY TESTING

In comparing the performance of different batteries, it is generally advisable to determine the quantity and energy efficiencies for various discharge rates such as normally obtain in service. To ensure truly comparable results, each battery should be normally charged and discharged at least twice before taking test readings.

New batteries should be given the first or conditioning charge to the makers' instructions, and then given two or more cycles of charge and discharge. The specific gravity of the electrolyte should also be checked at frequent intervals during charge and discharge to ensure that the conditions in each cell are in accordance with the specified requirements of the manufacturers.

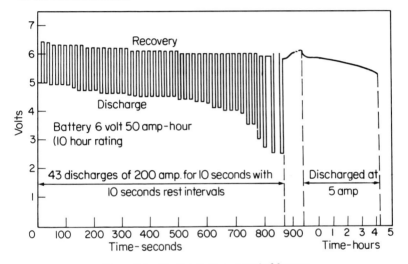

Figure 5.3 Starting tests on a typical battery

With starter batteries it is not unusual, as a further test, to subject the battery to heavy current discharges comparable with cold engine starting conditions. Figure 5.3 graphically illustrates such a test and shows the remarkable performance of a small 6-V battery discharged for short periods in rapid succession at 200 A. After a period of 14 hours rest, the battery was discharged at its normal 10-hour rate and gave an output of half its rated capacity. For the purpose of testing under load when required, many modern battery testers have provision for measuring the battery voltage under suitable loads.

The cadmium test

This test can be made only where access to individual cell plates is possible. Whilst voltage per cell and the electrolyte specific gravity readings are generally a good indication of the condition of a battery, they do not necessarily show the relative chemical condition of the positive and negative plates. For this purpose it is necessary to apply the *cadmium test*, which consists of determining the voltage between a cadmium testing strip immersed in the electrolyte and the positive or negative plate. The cadmium strip is generally enclosed within a perforated ebonite tube, one extremity of the strip being connected by a rubber insulated conductor to the negative terminal of a cell-testing voltmeter. The other terminal of the voltmeter is connected to the positive or negative plate as desired. The cell voltage is equal to the algebraic difference of the cadmium to positive and cadmium to negative voltage readings.

For a healthy battery nearing completion of its charge at normal rate, the cadmium to positive voltage will be 2.5 and the cadmium to negative −0.14, giving a cell voltage of 2.5−(−0.14) or 2.64 V. In the case of a battery discharged at normal rate (one-tenth of its rated capacity on the 10-hour rating) for 10 hours, the cadmium to positive reading should not be less than 2.0 V and the cadmium to negative voltage not greater than +0.2 V, giving a cell voltage of 1.8 V. Should the cadmium to negative reading approach +0.3 V it would indicate a poor chemical condition of the negative group of plates, whilst if the cadmium to positive voltage is appreciably less than 2.0 V the chemical condition of the positive group is below standard.

Essential test conditions

A high-grade high resistance voltmeter or a modern 'DVM' is essential for cadmium tests. Traditional dial type voltmeters will be either 3-V centre zero or scaled to give a negative reading of 0.2 V, and a positive reading of 3 V. It is also necessary when making cadmium tests that the battery should be either on charge or discharge. Open-circuit voltage readings, whether for cadmium or cell-voltage tests, are no indication of the state of a battery. For example, a battery may give a cell voltage reading of 2.05 V on open circuit and upon being discharged at normal rate would only show 1.8 V per cell. On the other hand, this same battery may have a cell voltage of 1.98 V when on normal discharge. In the former case the battery would be in a discharged condition and in the latter would only be partially discharged, although in both cases the open circuit voltage was the same.

BATTERY CONSTRUCTION

A battery normally consists of several cells connected with each other in series, each cell containing 1 group of positive plates and 1 group of

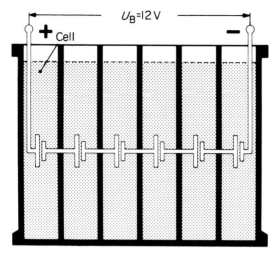

Figure 5.4 The nominal voltage of a storage battery depends on the number of individual cells connected in series

where V_B = $n V_Z$
  n = *number of cells*
  V_B = *nominal voltage of battery*
  V_Z = *nominal voltage of cell*

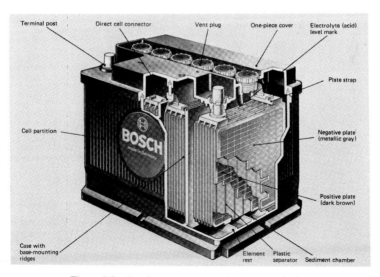

Figure 5.5 Bosch storage battery for motor vehicles

negative plates. The purpose of connecting the cells in series is to obtain a higher battery voltage, for example, a 12-V storage battery consists of 6 cells, each of 2 V (Figure 5.4).

An operational storage battery designed for motor vehicles consists basically of the following parts:

Case with partitions to separate the individual cells.
One-piece cover with cell openings and vent plugs.
Positive plates with plate straps.
Negative plates with plate straps.
Separators (between the individual plates).
Cell connectors.
Terminal posts (positive and negative terminals).

In addition, the storage battery contains the electrolyte (dilute sulphuric acid).

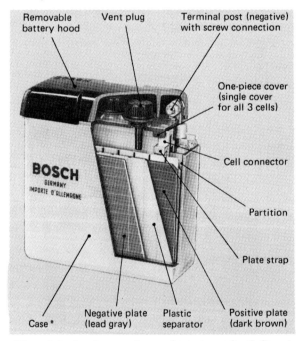

Figure 5.6   Bosch storage battery for motor cycles. Sediment chamber with element rests not visible

For comparison, Figure 5.6 shows a 6-V motorcycle storage battery. The most important parts of a storage battery will be described in greater detail below.

THE BATTERY CASE

Modern battery cases are constructed of insulating, acid-resistant material (hard rubber or plastic) and have ridges around the bottom of the case on the outside for mounting purposes (Figure 5.5). Inside, element rests run along the full length of the case floor and the feet on the bottom edges of the plates sit on these rests. The spaces between the element rests, i.e. below the bottom edges of the plates, form what is known as the sediment chamber, which accumulates the particles of solid material which are sloughed off from the plates during operation and fall to the bottom of the case. This sediment, which contains lead and is electrically conductive, can thus accumulate in these spaces without contacting the lower edges of the plates where it would cause short-circuits.

The battery case is sub-divided by partitions into individual cells, for example, there are 6 cells in 12-V batteries and 3 cells in 6-V batteries. These cells are the basic assembly in the construction of the storage battery. In them are located the elements (cell packs), i.e. the assemblies of positive and negative plates with separators.

One-piece cover

In modern batteries (Figure 5.5) the cells, together with the elements, are all covered and sealed by a one-piece cover which may have one opening for each cell so that the cells can be filled with electrolyte, or may be of the low-maintenance or 'sealed for life' type. Openings are closed by vent plugs which are screwed into place, each plug having a small hole serving as a vent.

Elements

The elements consist of positive and negative plates grouped together with separators between each pair of plates (Figure 5.7). The number and size of these plates are two factors which determine the nominal ampere-hour capacity of the cells.

The plates are constructed of lead grids (the supporting framework for the 'active material'), and of the active material itself which is coated or 'pasted' onto the grids. The active material of the positively-charged plate contains lead peroxide (PbO_2 dark brown in colour) while the active material of the negatively-charged plate contains pure lead in the form of sponge lead (Pb, metallic grey in colour). This porous active material provides a large effective surface area.

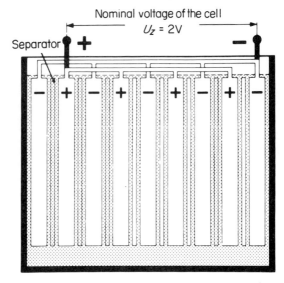

Figure 5.7 Element (with positive and negative plates and separators fitted between them). The nominal ampere-hour capacity of the cell is dependent, among other factors, on the number and size of these plates

Each group of positive plates and each group of negative plates is held together by its own plate strap to which the plates are welded. Each element usually has one more negative plate than positive plates, so the two outside plates are normally negative (Figure 5.7).

Separators

Since motor vehicle-batteries must be built to take up as little space and to be as light as possible, the plates must be very close to each other, but they must not touch (which might happen if a plate were bent or particles crumble off their surface) because this would cause a short-circuit which could destroy the battery. For this reason, separators are installed between the individual plates when an element is assembled.

These separators ensure that plates with opposite polarity are separated sufficiently from each other and are electrically (i.e. with regard to the flow of electrons) insulated from each other. The separators must present no notable resistance to the movement of ions in the electrolyte and moreover must consist of an acid-resistant but permeable (micro-porous) material so that the electrolyte can pass through them. A microporous structure is necessary so that fine lead fibres cannot grow through the separators, this would result in a short-circuit.

Cell connectors

The individual cells in the battery are connected in series, and these connections are made by the cell connectors (Figure 5.7).

In order to decrease internal resistance and weight, so-called direct cell connectors are used in modern batteries. Here, the plate straps of the individual battery cells are connected by the shortest path, directly through the cell partition.

Terminal posts and terminals

The plate strap joining the positive plates in the first cell is connected with the positive (plus) terminal post of the battery and likewise the plate strap joining the negative plates in the last cell is connected with the negative (minus) terminal post of the battery. The maximum voltage thus exists between the two terminal posts, 6 or 12 volts.

The battery cables are attached by means of special cable terminals. In order to be sure that the cables are not connected to the wrong terminal posts, the cable terminals are marked with their polarity and some batteries have different terminal post opening diameters to match the respective terminal posts. The positive post is thicker than the negative post.

A now well established British design is the Lucas 'Pacemaker'. This design has been evolved after many years of intensive development to meet the demands of the automotive industry for a more powerful battery, but smaller and lighter. This was achieved by the use of a container and lid moulded in polypropylene. This material allowed 'through-the-wall' connections to be made between adjacent cells. By drastically shortening the current path and so reducing the internal resistance, higher cranking speeds and more reliable engine starting was achieved. Comparison between the Pacemaker and an earlier battery of similar volume shows a 100% increase in starting power.
power.

Designed to provide a battery life comparable with the expected vehicle life, the Oldham type PG battery, the constructional features of which represent a notable departure from the conventional battery plate construction. The distinctive feature of this design is the use of p.v.c. tubes in conjunction with glass fibre sleeving to retain the positive active material. The positive plate consists of about fifteen thin perforated p.v.c. tubes, each of which is lined with a woven glass fibre sleeve and

centred on a feathered 'X' metal spline of which alloy the Oldham
grids are made. The active material is closely packed into the space
between the spline and the glass sleeve. In use, the glass sleeve becomes
securely embedded in the active material and so effectively retained.
The thinness of the p.v.c. and fibre glass wall contributes to greater
capacity and it is also claimed that, compared with the more orthodox
design of battery construction, appreciable saving in weight is effected.
A noteworthy advantage of the double-sleeve multi-tube construction
is its high resistance to bursting and the shedding effects of vibration.

Battery electrolyte

The electrolyte in automobile batteries is normally of higher density
than that used in batteries for other purposes. There are two reasons
for this, the first being that the volume of electrolyte is limited by
consideration of battery size and weight; and secondly, starter batteries
are frequently called upon to give very heavy discharge currents. The
curves in Figure 5.8 show that whilst the voltage of a lead-acid cell is
not seriously affected by the electrolyte density, except for extremely
weak solutions, there is a limited density range so far as the specific

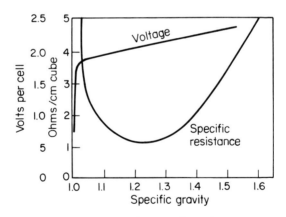

Figure 5.8   Electrolyte specific gravity curves

resistance is concerned. If the internal resistance of the cell is to be
kept low, then the density is limited to a specific gravity range of
1.15 to 1.30. The minimum specific resistance corresponds to a density
of 1.22 and this is a very suitable working density for batteries operating
at normal rates.

In batteries discharging at very high rates the localised dilution of the electrolyte at the plate surfaces is very rapid, and in order that the rate of dilution shall not be greater than that of diffusion of the acid a higher density is necessary. If, however, the density is too high, the plates, and in some cases the separators, will be attacked by the acid and damaged in consequence. This consideration fixes an upper limit of 1.30 for the density. In general, the specific gravity of the electrolyte should not exceed 1.28 for a fully-charged automobile battery or be less than 1.15 when the battery is discharging.

In tropical climates the battery is required to operate at high normal temperatures, and for this reason the specific gravity of the electrolyte of a fully-charged battery should not be greatly in excess of 1.22. The lower working density slightly reduces the capacity, but on the other hand the higher operating temperature increased it and so compensates for the effect of the lower density electrolyte.

The electrolyte is prepared by mixing commercial concentrated sulphuric acid of 1.840 s. g. with distilled water and for a density of 1.28 approximately 3 parts of acid to 8 parts of water are necessary. It is very important to remember when mixing sulphuric acid and water that the *acid must be added very slowly to the water*. To add water to concentrated acid is very dangerous, owing to the violent chemical action that occurs when this is done. The water used for batteries must be quite free from impurities such as chlorine and iron, which will cause injurious local action. Although the water supply in certain districts is satisfactory for use in batteries, the safest course to adopt is to use only distilled water for mixing or replenishment.

In specifying the density of the electrolyte it is assumed that this is measured at the normal temperature of 15° C (60° F), because the specific gravity will vary with temperature. Owing to the expansion of the electrolyte with rising temperature, the specific gravity will decrease although the acid content remains constant. Similarly, at lower temperatures the normal contraction of the electrolyte results in higher specific gravity readings. It is therefore necessary to correct for these variations in the density readings at temperatures higher or lower than 15° C. For every 0° C (5° F) above normal, 0.002 should be added to the hydrometer reading and correspondingly deducted if the temperature is lower. Thus electrolyte of 1.250 s. g. at 15° C would give a hydrometer reading of 1.242 at 26° C (80° F) and of 1.258 at 4° C (40° F).

THE BATTERY IN SERVICE

The battery manufacturers' instructions in respect to the charging of the battery prior to installation on the vehicle should be most rigidly followed. This will ensure the proper conditioning of the plates and the battery will, in consequence, give better and longer service. When first installed, it is also advisable to work the battery lightly for a week or so and to examine it frequently to see that the level of the electrolyte is maintained ¼ to ½ in (6–12 mm) above the top of the plates.

Primarily for overseas markets, batteries may now be supplied 'dry-charged' and sealed, a procedure which obviates lengthy initial charging and also ensures that there is no deterioration during the storage period, which can be for an indefinite time in all climates. When required for service, all that is necessary is to fill each cell with electrolyte of the correct specific gravity.

A battery in a discharged or even partially discharged state should never be allowed to stand for long periods in this condition, otherwise the lead sulphate will become stubborn and difficult to reconvert to lead peroxide and lead. There is also the further risk of buckling the plates and shedding the active material if the battery is not promptly recharged. When a battery is out of commission or standing for a period of weeks, a full charge should be given immediately after being taken out of service and the battery given a freshening charge every two to four weeks at a little over half the normal charging rate until 'gassing' freely occurs. Another way of maintaining an idle battery in a healthy condition is to pass a 'trickle' charge of about 2 to 5 per cent of the normal charging current for about two days every month. Continuous trickle charging is not recommended by the battery makers.

To obviate corrosion, the terminals and the extremities of the cables connected to the battery should be regularly wiped free of any acid with a cloth moistened in liquid ammonia and the terminals smeared with Vaseline or proprietary compound now marketed for this purpose. The top of the battery should also be kept clean and free from any accumulation of dirt, dust and electrolyte.

STEEL-ALKALINE BATTERY

By virtue of its construction, this type of battery has great mechanical strength which renders it suitable for certain classes of service. There are two distinct types of steel-alkaline battery: the tubular positive nickel-iron type and the flat-plate nickel-cadmium type. It is the latter type only which is suitable for automobile service and the following reference relates only to the flat-plate nickel-cadmium battery.

Owing to high initial cost and a relatively high internal resistance, the steel-alkaline battery has not found widespread application on automobiles. During recent years, however, notable advances have been made in design and particularly in the improvement of the electrical characteristics. In consequence, there has been a very substantial increase in the usage of this type of battery, more especially on public service and commercial vehicles. In this class of service, the advantage of long life of the steel-slkaline battery, which is 10—15 years, more than offsets the higher initial cost as compared with the lead-acid battery.

Operation

The positive and negative active materials are nickel hydroxide and cadmium oxide enclosed in pockets made from finely perforated steel ribbon. The electrolyte is an alkaline solution of potassium hydrate and distilled water of approximately 1.2 sp. gr. Unlike the lead-acid battery, the electrolyte does not enter into permanent chemical union with the active materials of the positive and negative plates, but merely acts as a liquid conductor. Consequently the specific gravity of the electrolyte remains substantially constant during charge and discharge and therefore is no indication of the state of charge.

The precise chemical action of the nickel-alkaline battery is very complex and not yet definitely established, but in the process of charging the positive material is oxidised, whilst the negative is de-oxidised, and reduced from cadmium oxide to pure spongy cadmium. On discharge the reverse action takes place, the positive active material being de-oxidised to nickel hydrate and the negative re-oxidised to cadmium oxide. The operation of the battery depends therefore upon the electrolytic carrying of oxygen from the negative to positive plates during charge and in the reverse direction when discharging. The passage of current through the electrolyte splits it up into its constituents, which are immediately neutralised by secondary reactions occurring between the dissociated potassium and the excess water. Since the active materials are insoluble in the electrolyte, there is practically no self-discharge and therefore the battery will stand for long periods in any state of charge without loss of capacity or fear of permanent injury to the plates.

Construction

In the CAV-type battery, the electrochemical active materials are contained within channel receptacles, formed from finely perforated nickelled-steel strip, mounted in steel frames to form the plates. Thus, even under the most severe jolting, the plates do not lose any active material and no sediment is produced.

In this cell, the active material of the positive plate consists of nickel hydroxide mixed with other ingredients to improve conductivity. The negative active material consists mainly of cadmium oxide. Nickelled sheet steel is employed for the cell containers, the seams being welded and the containers specially treated to prevent rusting. The cells are assembled in hardwood insulating crates, being firmly held in position by suspension bosses located in tough rubber insulators.

The range of CAV alkaline batteries for road vehicles include types for starting and trolley-bus manoeuvring and for lighting purposes only. The former, owing to their ability to give a very high percentage of their capacity at a high discharge rate, are also very suitable for oil-engine starting.

Attention in service

It is one of the special advantages of the nickel-cadmium cell that it requires a minimum of attention in service and it will adapt itself without damage to widely varying operating conditions. Routine topping up (with distilled water only) is necessary, as for lead-acid batteries, and provided the cells and crates are kept clean and dry no other attention is required. Routine gravity readings are unnecessary but once or twice during the life of the battery the solution should be changed completely; this is usually done when the specific gravity falls to about 1.160.

Advantages

This type of battery has, owing to its construction, distinct advantages, chief among which are

1. Long life.
2. Saving in maintenance and labour costs; no special battery staff or battery maintenance sheds are required.
3. Saving in weight; this alone is sometimes sufficient to offset the higher initial cost of the alkaline battery.
4. Ability to withstand widely varying operating conditions, e.g. change from winter to summer schedules, without damage.
5. The life of the battery approximates to the life of the vehicle and so can be depreciated as a capital item without further provisions for spares or replacements.

Now that the initial resistance and voltage characteristics have been brought to approximation with lead-acid characteristics, it may be said that virtually the only drawback of the nickel-cadmium battery is its high initial cost, but on many types of service this is more than counter-balanced by the extra life and the advantages enumerated above.

SILVER-ZINC ACCUMULATOR

A battery development of outstanding interest is the silver-zinc accumulator manufactured in this country by Venner Accumulators Ltd. The advantages claimed for the accumulator are such as to render it particularly suitable where weight and volume are important considerations. Whilst not a dry accumulator, the electrolyte is in an absorbed condition and therefore it is unspillable. There are no plates and separators, and resistance to shock is limited only by the strength of the case. No corrosive or poisonous fumes are released, either during charge or discharge, and high charging and discharge rates are permissible without fear of damage to the accumulator.

It can be stored in any state of charge without deleterious effect and in the half-charged condition any tendency to self-discharge is negligible.

In regard to weight and volume, it is claimed that the silver-zinc accumulator is one-third that of the normal lead-acid type and one-fifth that of the nickel-iron design, whilst the volume is reduced by one-half and two-thirds respectively.

The life of this accumulator is stated to be comparable with that of other types of accumulator and owing to its simplicity of construction, the price is competitive, despite the use of silver as an electrode.

The principle of this accumulator is based on that of the silver-zinc electrochemical couple, well known for its use as a primary cell.

In external appearance the silver-zinc accumulator resembles the lead-acid type. Apart from the constructional features already mentioned, the active materials are retained in two sacks, which are folded and contained with the electrolyte in the absorbed state, within the accumulator casing. The nominal cell voltage is 1.5 V. On charge the volts per cell rise to 2.0 V and, except for high discharge rates, the cell voltage is maintained between 1.7 and 1.3 V for the major period of discharge.

THE ZINC-AIR BATTERY

To meet the requirements of the battery-electric vehicle and particularly the electric car, a novel form of zinc-air battery has been developed which shows considerable advantages over existing lead-acid types, and will operate down to temperatures as low as $-40°$ C.

The basic bi-cell consists of a porous zinc anode in the centre with two thin air cathodes on either side. These cathodes are in the form of a conductive mesh and a layer of catalyst. The electrolyte is in the form of aqueous potassium hydroxide, and the porous anode is impregnated with this material.

The present system of recharging is by replacing discharged anodes with new ones, though an electrically rechargeable battery is being developed.

This design of battery shows such great weight/output advantage over all other commercial designs that it could have a very big bearing on the future development of the battery-electric car.

BATTERY CHARGING

Storage batteries installed in motor vehicles are charged by the generator (either the d.c. generator or the alternator with a built-in rectifier) whilst the vehicle is being driven. Under normal conditions, the electrical energy produced by the generator system is sufficient to supply all electrical loads with power and to adequately charge the battery.

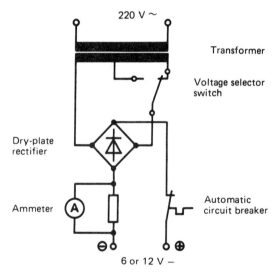

Figure 5.9   Circuit diagram of a home battery charger

If, on the other hand, the battery has been discharged for any reason (for example, the vehicle has been out of service for a long time) its state of charge should be checked and, if necessary, it should be charged with a battery charger.

For the charging of batteries a d.c. supply is essential. This is obtained in the most widely used type of battery chargers by some form of rectifier. The circuit arrangement of such a charger is shown in Figure

5.9, from which it will be seen that the unit is energised from normal 200/250 V a.c. mains. A transformer steps this voltage down to a considerably lower value which is then applied via a selector switch or variable resistance to a full wave metal (dry plate) rectifier. The d.c. output from this rectifier is then connected to the battery or batteries to be charged. It is usual practice to fit an ammeter in the negative output lead and an automatic circuit breaker or fuse in the positive output lead.

By series or parallel-series arrangement of the rectifier elements, battery chargers can be supplied for a wide range of outputs. These vary from small trickle chargers for home use to large industrial type units for commercial vehicle fleet operators or for industrial battery-electric truck users.

In the simplest designs the voltage developed by the charger is kept constant and the charging current is determined only by the internal resistance of the battery itself and no special control gear is required. As the level of charge increases so the charging current falls and the battery becomes fully charged in 8–12 hours. A charger to give a much more rapid rate of charge in which a battery can be charged in about half an hour must be capable of maintaining a constant charging current which can be much higher than the normal charging current. This is obtained by the use of much more complex control circuitry.

When selecting a charger it is necessary to note how many storage batteries and of what capacity will usually have to be charged with it at any one time.

LOCATION AND INSTALLATION OF THE BATTERY IN A VEHICLE

The following basic points should be observed when installing a battery in a vehicle:

(a) Easy accessibility.
(b) Protection against excessive heating or cooling.
(c) Location giving required weight distribution of the various items installed on the vehicle.
(d) Protection against moisture and dirty water splash from the road.
(e) Protection against mechanical damage such as excessive vibration.
(f) Protection against oil, fuels, solvents or any corrosive fluids.

The requirement for easy accessibility is of the utmost importance to ensure regular care of the battery, particularly as regards 'topping-up' the cells with distilled water at regular intervals. The battery should also be located where it is naturally well ventilated, as it is impossible to prevent acid fumes escaping through the vent holes in the cell caps, which would cause corrosion on any iron parts near the battery even though these parts may be plated or painted.

Location

The battery should be installed in a location where there can be no build-up of heat such as could occur near the exhaust system, as it should not be exposed to temperatures higher than $+60°$ C for any length of time. This would reduce the service life through high self-discharge. These requirements sometimes necessitate the thermal insulation of the battery and even its ventilation and cooling by a motor driven fan and an air cooling system controlled by a thermostat adjacent to the battery.

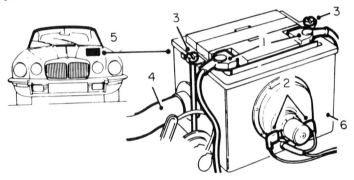

Figure 5.10   Showing under bonnet battery installation where auxiliary cooling is essential. 1. Battery terminal connectors secured by pinch bolts. 2. Motor driven cooling fan fitted with snap connectors. 3. Battery retaining bolts. 4. Outlet pipe for cooling air. 5. Under bonnet location. 6. Battery cooling jacket

Figure 5.10 shows an example of an under bonnet battery installation in which the battery is air cooled and the cooling fan motor automatically switched by a thermostat.

From a cooling point of view and also in the interests of weight distribution it is advantageous to fit the battery in the boot, but in many vehicles it is located on the cooler side of the engine compartment in order to keep the cable runs as short as possible to minimise voltage losses between the battery and the starting motor and alternator or generator. The heavy cables required for connection to the battery should always be so positioned and long enough so that in no circum-stances is a pull exerted on the battery terminal posts.

As the temperature drops, capacity and discharge voltage of a battery decrease. The capacity of a motor vehicle battery, therefore, must not be too low because of the necessity to operate under winter conditions. The starting ability of a storage battery at low temperatures can be gauged by the magnitude of the discharge voltage and the discharge time when the battery is loaded with the cold-discharge test current.

Batteries should be protected from low temperatures (below freezing), especially when discharged, if at all possible. A good state of charge is the best protection against freezing. (The electrolyte in a discharged battery freezes at about $-11°$ C, but in a charged battery at about $-69°$ C.)

Mounting

The battery must be properly mounted in the vehicle so that it does not move out of place under road vibration. It should be securely clamped on to a flat supporting structure by means of a holding down frame or bracket secured by clamping screws. The construction of such mounting fixtures should be such that battery maintenance and regular service is not impeded in any way.

Cable connections

When installing a battery always connect the positive (+) cable first and then the negative (−) cable (assuming that the negative cable is connected to 'earth' on the vehicle). When removing the battery from the vehicle detach the cables in the reverse order.

The cable connections must not become loose even under maximum vibration conditions because if they are loose, contact resistances can develop which result in considerable power losses at high current levels. In addition, under certain conditions the danger of explosion could arise through the ignition of the hydrogen-oxygen mixture released during gassing of the battery.

After the cable terminals have been attached to the battery terminal posts, they should be lightly greased with an acid-free and acid-resistant grease or petroleum jelly.

Installation of motorcycle batteries

Motorcycle batteries are subjected to especially high mechanical loads. They should be installed as close as possible to the centre of gravity of the motorcycle where vibrations and movements are least. A motorcycle battery fitted with a covering hood is best mounted by means of a clamping band over the hood. A piece of elastic material (e.g. foam rubber) should be placed between hood and clamping band.

CHAPTER 6

AUTOMOBILE LIGHTING AND SIGNALLING

Automobile lighting is a subject of the utmost importance both as regards the performance of the vehicle and its safety. Sudden failure of the lighting whilst travelling at high speed at night could result in a fatal accident. The possibility of headlamp failure due to a bulb failing or a short circuit in the headlamp wiring is regarded so seriously on public transport vehicles that a headlamp safety change-over circuit breaker is fitted to provide alternative lighting automatically and instantaneously. Details of the requisite switching arrangement are described in Chapter 11. Whilst the primary object of headlamp design is to provide good illumination for a considerable distance ahead of the vehicle, it should also take into consideration the effects such illumination produces on the vision of other road users.

Innumerable devices have been designed over many years in an endeavour to produce ideal illumination by headlights and adequate safety for the driver without interfering with the vision of approaching persons, whether cyclists, pedestrians or other drivers. So diametrically opposed are some of the requirements for good vision in the two cases, that it is doubtful whether the solution can ever be more than a compromise when all the factors involved are considered.

HEADLAMPS

Light sources

Light sources are of two kinds, those that emit light and those that reflect light. In the automobile headlamp we have a combination of both, the incandescent filament of the electric lamp being the primary source of light, whilst the reflector is the secondary source. The important characteristics of any light source are its intensity, colour and distribution, the latter including intensity as well as direction.

Automobile filament lamps

The tungsten-wire filament lamp bulb now in general use utilises a tungsten wire filament heated to incandescence by the electrical current and in the case of evacuated or vacuum type bulbs, the filament attains a working temperature approaching 2300° C. Although the melting point of tungsten is 3500° C it is not advisable to operate the filament in a vacuum at a temperature higher than 2300° C owing to the rapid volatilisation of the metal which occurs when this temperature is exceeded. On the other hand, the vacuum serves to prevent the conduction of heat from the filament and further, the absence of any combustible gas, such as oxygen, prevents the burning away of the filament.

From the foregoing it will be understood that the voltage at which automobile lamp bulbs are operated is quite critical and if a bulb is operated at a voltage appreciably higher than that for which it is designed, the life of the filament will be greatly reduced. The consequent evaporation of the filament at the higher working temperature will also cause blackening of the bulb.

Gas-filled lamps

Headlamp bulbs of the gas-filled type are now in very wide use. In this type of lamp the filament is operated at a much higher temperature resulting in a much whiter light. This higher efficiency, of the order of 0.7 watt/candle power compared with about 1.0 watt/candle power for a vacuum lamp, is achieved by operating the filament under pressure by filling the bulb with an inert gas such as argon which does not support burning of the filament. The difficulties due to convection causing cooling of the filament are also greatly minimised by winding the filament in the form of a close spiral.

Tungsten halogen lamps

Tungsten halogen bulbs sometimes called iodine bulbs are a comparatively recent development now in general use. A major advantage of such a bulb is that about 30% greater lumens output can be obtained for equivalent wattage, rating and life.

In a normal gas filled incandescent lamp, approximately 10% of the filament metal is deposited onto the bulb wall during its life and because of the internal convection currents this gives a black patch above the filament as shown in Figure 6.1(a). The bulb must therefore

Automobile lighting and signalling

be relatively large to allow the tungsten to spread over a large area and so prevent undue obscuration. Even so, the light output gradually diminishes and at half life is about 90% of the initial value.

If a halogen, such as iodine or bromine, is added to the gas filling of a tungsten filament lamp it will combine with evaporated tungsten on its way to the bulb wall. As this tungsten halogen combination (or

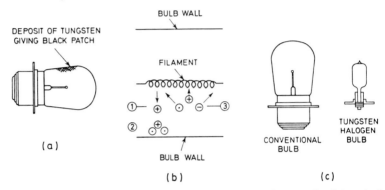

Figure 6.1 (a) Conventional tungsten bulb showing where patch of deposited tungsten forms. (b) The cycle in stages of combination and dissociation of the tungsten and halogen atoms. (c) Comparison of sizes of a conventional bulb and a tungsten halogen bulb of the same light output

molecule) approaches the hot filament, it breaks up, returning the tungsten and releasing the halogen to take part in another cycle of events.

Considering this cycle in stages with reference to Figure 6.1(b):

1. Tungsten ⊕ is evaporated from the hot filament.
2. On its way to the bulb wall, a tungsten atom is intercepted by two or more halogen atoms ⊙. This molecule cannot settle onto the bulb because it is too hot.
3. As the tungsten halide molecule diffuses into the region of the incandescent filament, it splits up, returning tungsten and releasing halogen. This continuous regenerative process pevents the bulb from blackening and the light output therefore remains constant throughout its life.

Because the bulb does not become blackened by collecting tungsten it can be made very much smaller than a conventional type as indicated in Figure 6.1(c).

This small envelope is usually fused silicon and is very strong. Such lamps can be gas filled to a pressure of several atmospheres and this leads to much brighter filaments for the same life, or longer life for the same brightness. These high efficiency filaments are also smaller than conventional lamp filaments which allows more precise focusing or optical control to be obtained.

The normal distribution of light from an incandescent filament is in itself quite unsuitable for automobile road lighting where the light is required to be projected in the form of a beam in order to obtain illumination some distance ahead of the vehicle. It is the purpose of the reflector to modify the light distribution from the primary source and concentrate it in the desired direction. A reflector in no way increases light, but merely re-directs it to a plane where illumination is required.

With all forms of reflector, there is some loss of light due to absorption, but for a reflector of good design and finish, this loss is quite small. Modern headlamp reflectors are very efficient due to the accuracy of the paraboloid shape which is maintained to 5/100 mm and also the extreme care taken in manufacture to produce the most highly polished surface. This is achieved in many instances by a deposition process, in which a very efficient and durable reflecting surface is obtained by the deposition of aluminium on the surface under vacuum. The construction of most modern lamps is such that the reflector is not accessible for cleaning purposes, as any attempt at cleaning with an abrasive material such as metal polish would greatly impair the reflector surface.

Intensity of a light source and its measurement

By the intensity of a light or luminous intensity, we mean its power to produce illumination at a distance. For a long time it was customary to express this quality as *candle-power*, the unit then adopted being the *international candle* and equal to one-tenth of the Vernon Harcourt Pentane standard lamp, burning under specified conditions of atmospheric pressure and humidity. It is now general practice to accept the international unit, the *candela* (cd) as the unit of luminous intensity. (See 'Units and standards of light', *Jour. I.E.E.,* Vol. 7, No. 78, p. 379.)

The method of measuring the luminous intensity of a light source is a comparative one and is based on the law of inverse squares; that is, that the illumination on a surface is inversely proportional to the square of the distance from the light source. If, as shown in Figure 6.2, a light source is centrally located within a sphere of one foot radius, all rays of light emitted will be distributed over a surface area of πD^2 or 4π ft². Supposing now the radius of the sphere is increased to 2 ft, the surface area over which the rays will be then distributed will be 16π ft² or four times greater than before. Therefore the intensity of illumination in the latter case will only be one-quarter of that for the sphere of half the radius.

The simplest apparatus for measuring luminous intensity is a Bunsen photometer, consisting of a white paper screen, in the centre of which is a grease spot. When illuminated from the front, the grease spot appears darker than the surrounding surface, and when illuminated from behind it appears lighter. Assuming the luminous intensity to be equal on both

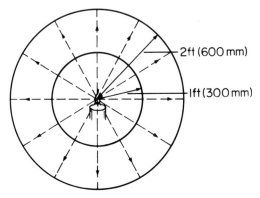

Figure 6.2   Intensity of a light source

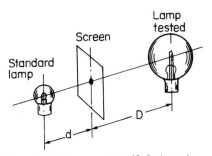

Figure 6.3   Measurement of light intensity

sides of the screen, then the spot will be invisible. Figure 6.3 illustrates diagrammatically the principle of apparatus of this kind and its application in the measurement of luminous intensity. Assuming that for equal intensity on both sides of the screen the positions of the standard and tested lamps relative to the screen are given by distances d and D respectively, then the luminous intensity of the tested lamp can be computed from

$$cd_T = cd_s \times \frac{D^2}{d^2}$$

where cd_T = intensity of the tested lamp
cd_s = intensity of the standard lamp.

Various types of photometer, both stationary and portable, are employed in the measurement of luminous intensity. They operate on the simple principle illustrated, whereby the illumination from any desired light source is compared with, and adjusted to be equal in intensity to, the illumination produced by a known standard source of light.

Unit of luminous flux

By luminous flux we mean the light rays emitted by or radiating from any concentrated light source. The unit is the *lumen* and is the flux of light contained within unit solid angle from a uniform light source emitting one candle in all directions. Referring again to Figure 6.2 unit solid angle would embrace one square foot of surface, of the one-foot radius sphere, so that the light flux incident upon this area would be equal to one lumen if the light source is one candela. Since the area of this sphere is 4π ft^2, the total lumens emitted by a uniform light source of one candela, is 4π or 12.57.

Thus: Luminous flux (F) in lumens $= 12.57 \times$ no. of candelas.

Illumination intensity

The illumination intensity on a surface is defined as the luminous flux reaching it per unit area. Consequently, latest practice is to make use of terminology more representative of luminous flux density than the hitherto standard unit, the foot-candle. The unit is termed the lumen per square foot (lm/ft^2) and is equal to the foot-candle (fc). It represents the illumination obtaining on a surface everywhere one foot distant from a light source emitting one candela equally in all directions towards the surface. It has already been illustrated by Figure 6.2 that the intensity of illumination for a point source of light, varies inversely as the square of the distance of the surface from the light source.

Thus: $E = I/D^2$ lumen/ft^2 (6.1)

where E = illumination intensity in lumen/ft^2
I = light source intensity in candelas
D = distance of illuminated surface from the light source in feet.

When the light rays fall obliquely on a surface, as is the case with automobile road illumination, the illumination intensity will obviously be less than for a plane at right angles to the rays, owing to the larger area illuminated. In this case, the illumination intensity is given by the formula:

$$E = \frac{I}{D^2} \cos \theta \text{ lumen/ft}^2 \qquad (6.2)$$

where θ is the angle of incidence or the angle between the light rays and the normal to the surface, at the point of incidence. By Lambert's Law, expressed mathematically as:

$$E = \frac{I}{h^2} \cos^3 \theta \text{ lumen/ft}^2 \qquad (6.3)$$

The illumination intensity at any point can be computed if the height, angle of incidence, and intensity are known.

The international unit of illumination intensity is the lumen per square metre, termed the *lux* (lx) and is equal to 0.093 lumen/ft².

Brightness or luminance

Brightness is sometimes confused with illumination. When travelling at night along a road, the colour and surface of which vary, the brightness of the road will also vary, but not the illumination, although on light surfaces, it will appear to be more intense. Illumination depends solely upon the light flux incident upon the surface. Brightness, however, not only depends upon the light incident upon the surface, but also upon the percentage of light the surface reflects and may be defined as the light emitted or reflected per unit area in a direction at right angles to the surface. The definition of brightness applies equally to light sources as well as illuminated surfaces.

The brightness or luminance of a surface is expressed by its luminous intensity per unit projected area, the plane of projection being perpendicular to the direction of view. The unit is, therefore, the candela per unit area. In the British system of units, the units most commonly employed are the candela per square inch (cd/in²) for primary light sources where the luminance is high or candela per square foot (cd/ft²) for secondary surfaces reflecting light of a relatively low order. In the metric system, the units are candela per square centimetre, termed *stilb* (sb) or candela per square metre, termed *nit*.

Flux of light method

The rapid progress since about 1920 in the art of illumination has led to the almost universal use of the *flux of light method* of determining illumination intensity. Since 1 lumen per square foot is produced by 1 lumen uniformly distributed over 1 square foot, it is a relatively easy matter to compute the resultant intensity of illumination on a surface, if the quantity of light incident on the surface and the area are known. The incident light flux will, in the case of headlamps, be the total lumens emitted by the electric lamps multiplied by the reflector efficiency.

Thus $E = \dfrac{F}{A} \epsilon$ (6.4)

where E = illumination intensity in lm/ft^2
F = light flux of light source in lumens
A = area of illuminated surface in ft^2
ϵ = reflector efficiency.

As an example two car headlamps with 36-W gas-filled bulbs illuminating a strip of road 20 ft wide at the car and 30 ft wide at a distance of 250 ft, the approximate lumens per watt for gas-filled bulbs is 16, so that the two 36-W bulbs will give a total of 1160 lumens of light flux. Assuming a reflector efficiency of 85 per cent, and computing the area of the road surface illuminated as 6250 ft^2, the average intensity will be

$$E = \frac{1160}{6250} \times 0.85 = 0.158 \text{ lm/ft}^2$$

If the intensity is given in terms of lux units, it is necessary to express the area in square metres.

Thus $E = \dfrac{1160}{582} \times 0.85 = 1.7 \text{ lux}.$

In order to give some idea of the illumination intensity represented by the above values it may be mentioned that it is equal to the illumination of a well lighted public thoroughfare. As the light flux extending to the more distant areas is contained within a much smaller solid angle than the light flux illuminating the area near the car, the amount of light incident on the distant areas will be proportionally less. Therefore the illumination will be most intense immediately in front of the car and will diminish as the distance from the headlamps increases in a manner typified by the curve in Figure 6.4.

Beam intensity

The beam intensity of a correctly focused headlamp will depend upon the light flux emitted by the filament, the reflector efficiency and the design and size of the reflector. It may be measured either

1. by moving a photometer across the beam and measuring the intensity at definite angles either in the horizontal or vertical planes, the headlamp remaining stationary, or
2. by keeping the photometer stationary and swivelling the headlamp, on either its horizontal or vertical axis as desired, until the beam has traversed the photometer screen.

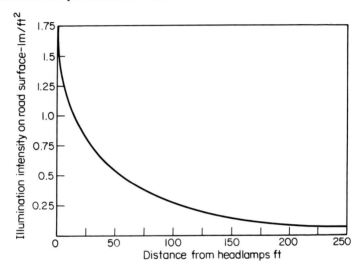

Figure 6.4   Headlamp beam characteristics

In this way the characteristics of the beam can be ascertained, the beam intensities for the various angular displacements from the central horizontal axis of the lamp being determined from the formula

$$cd_B = ED^2$$

where E = illumination intensity in lm/ft^2 as measured by the photometer

D = distance in feet, of the photometer from the headlamp.

American practice in regard to headlight tests is to make photometric measurements of this kind, and the Society of Automotive Engineers' (SAE) standard specifications, include limitations for the beam intensity above and below the horizontal axis. Above the horizontal the beam intensity must not exceed 3000 candelas, whilst the maximum permissible intensity should not be greater than 50 000 candelas and this beam intensity should fall within an angle of $1\frac{1}{2}°$ to $2\frac{1}{2}°$ below the horizontal axis.

Reflector theory

Light may be projected in the form of a beam by means of lenses or reflectors. For automobile headlights, reflectors of the parabolic type are almost universally employed. Lenses are also used as headlamp cover glasses, their purpose, however, being to increase side illumination and also to re-direct the light rays in a downwards direction (see Figure 6.5) rather than to concentrate the light rays in a central beam.

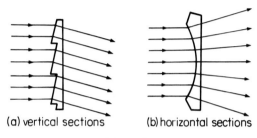

(a) vertical sections (b) horizontal sections

Figure 6.5   Reflector patterns

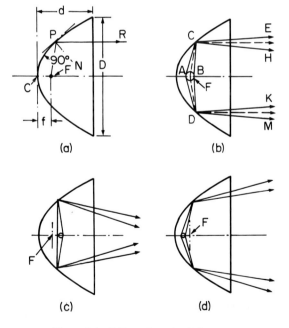

(a)   (b)

(c)   (d)

Figure 6.6   Effect of moving light source

The section of a parabolic reflector is shown in Figure 6.6a, where *c* is the central point of the reflector, *D* and *d* the diameter and depth respectively, *f* the focal length, and *F* the focal point. The focal length of a paraboloid may be determined from the formula

$$f = \frac{(D/2)^2}{4d}$$

If we take any point P on the surface of the reflector and draw a line PN at right angles to the tangent at this point, we shall find that the line PN bisects the angle formed by the lines joining the focal points with point P. One focal point is at F whilst the other is at an infinite distance from it. Therefore, the line joining P and the second focal point will be a line along PR, and will be parallel to the axis of the reflector. A line drawn at right angles to the tangent *at any other point* on the reflector will also bisect the angle defined by the focal points and the point on the reflector.

The law of reflection states that the angle of reflection is equal to the angle of incidence, so that if we assume FP to be a ray of light from an *absolute point* source of light, it will be reflected along PR. Under these conditions the beam would consist of reflected rays parallel to the reflector axis. Since it is physically impossible to have an absolute point source of light, a parallel beam is only theoretically possible.

Light formation

In Figure 6.6b it is shown that the effect of size of the light source is to produce cones of light. Each point on the reflector will emit a cone of light and the resultant beam will be a cone. If the centre of the light source is at the focal point of the reflector, the spread of the projected beam will be dependent upon the size of the light source. So far as the side and downward rays are concerned, this beam spread is a desirable feature, as it greatly increases the area of road surface illuminated. On the other hand it is not advisable to have too wide a beam spread, otherwise the intensity of the beam will be so much diminished that distant illumination will be poor. It can be stated that the filaments of present-day headlamp bulbs are of such dimensions as to permit of a satisfactory spread of beam without undue loss of intensity of the central portion of the beam, providing the filament is symmetrically disposed relative to the focal point of the reflector.

The movement of the light source relative to the focal point affects, in a very marked degree, the distribution of the light rays from the reflector, as illustrated in the diagrams Figures 6.6c and d. When the filament is forward of the focal point, the angle of incidence of the light rays on the reflector is reduced, and since the angles of incidence and reflection are equal, the reflected rays converge. As the angle of incidence is increased when the filament is behind the focal point, the resultant rays projected from the reflector will be divergent.

THE IMPORTANCE OF ACCURATE FOCUSING AND AIMING

Focusing

The accurate focusing of the bulb filament relative to the reflector is of the utmost importance. A correctly focused headlamp is essential to project an intense far reaching beam and also provide near illumination over a wide angle to show clearly the sides of the road and at the same time will not produce upward rays likely to cause extreme dazzling effect on other road users.

Bulb manufacturers take great care in the design and manufacture of modern pre-focused bulbs so that manufacturing discrepancies in the positioning of the bulb filament in relation to the cap and reflector are kept to an absolute minimum. Provision is also made for the filament to be asymmetrical to cater for either 'right-hand' or 'left-hand' driving. This is very important in the 'dipped' position.

Aiming of headlights

Regulations regarding the aiming of headlights are given in BS AU156. Some units aim on main beam and some on dip and some by mechanical means referred to later. BS AU162a 1976 specifies the requirements for optical devices for aiming headlights.

Problem of headlight dazzle – Department of Transport Regulations

So far as Great Britain is concerned, it is essential that vehicle lighting shall comply with the Road Vehicle Lighting Regulations SI 1971 No. 694 which requires the lighting system to be arranged so that it will provide road illumination which is 'incapable of dazzling any person standing on the same horizontal plane as the vehicle at a greater distance than 25 feet from the lamp whose eye level is not less than 3 ft 6 in above the plane'.

Many factors are involved when the problem of headlight dazzle is considered one of which is that wonderful human organ, the eye. It is the final judge of both illumination and dazzle and whilst an intensely bright illumination is necessary for good driving vision, the clear vision of all road users in general must not be sacrificed to the requirements of the individual.

Some characteristics of the eye

Probably the most striking characteristic of the eye is the extensive range of illumination intensity over which it will operate. From bright sunlight to a photographer's darkroom conditions represents a reduction in illumination intensity from 10 000 lm/ft^2 to less than 0.01 lm/ft^2. A reduction ratio greater than 1 000 000 to 1, yet the eye will adapt itself to see under both conditions.

In an extreme case such as this, some considerable time must necessarily elapse before the eye responds to the changed conditions and this will explain another characteristic, namely that the eye is unable to see clearly at the same time two objects illuminated at widely different intensities, nor can it view a bright object and then immediately see a dark one. If then, satisfactory vision is to be provided by headlights, one essential feature of the illumination is that it shall be uniformly distributed. Further, the dazzle effect of headlights will be far less pronounced if the road surface is uniformly illuminated, for as will be pointed out later, the contrast between intensely bright and dark areas is the chief contributory case of dazzle. The continuous adjustment of the eye to suit varying intensities also results in eye fatigue, which is another sound reason for a uniform distribution of light.

The area illuminated also has an important bearing upon the capability of the eye to see objects clearly. A much lower illumination intensity is permissible to enable an object to be seen distinctly if the illumination is widely distributed, than if the illuminated area is small and surrounded by a darker area. Illumination of the sides of the road as well as of the road surface traversed is therefore essential for good vision from the driving seat, and by reducing contrast such light also minimises the dazzling effect on approaching road users.

DAZZLE – ITS CAUSES AND PREVENTION

Dazzle may be defined as brightness in the field of vision that will cause interference with vision and is dependent upon three factors, namely, contrast, brightness and the angle which the bright area subtends at the eye. Dazzle may to some extent be prevented with these causes in mind by:

1. Reducing the contrast between the headlights and the surrounding area by a wide distribution of light of uniform or lower intensity.
2. Reducing the brightness of the headlights by methods such as dimming. This is not satisfactory.
3. Stopping light of high intensity from entering the eyes of the oncoming persons. This is done by projecting the headlight beams so that the light flux above the horizontal level of say 1 to 1½ metres is of very low intensity.

An intensive study of this problem was made as early as 1952 by Dr. J. H. Nelson of Joseph Lucas, Ltd., and many points raised then are still relevant. (See *Dazzle – an examination of a world wide problem*, by Dr. J. H. Nelson, Road International, August, 1952).

Anti-dazzle devices

As mentioned earlier, these have been numerous over the years and it is only possible here to illustrate the principles of anti-dazzle by a few typical examples.

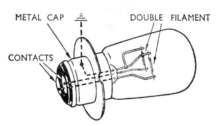

Figure 6.7   The double-filament headlamp bulb. One filament is so adjusted in position in relation to the reflector that it gives the main forward beam, and the other filament gives the dipped beam when the dipping switch is actuated

In some early designs the offside headlamp was switched off, the nearside headlamp was mechanically dipped and later the nearside reflector and bulb holder mechanically dipped. The system at present in most general use is to re-direct or dip the light rays by means of lenses, reflectors and the use of two separate bulbs, one for main beam and one for dipped beam, or the use of bifocal or twin filament bulbs.

These systems are controlled by the driver, either by a foot operated switch or by a switch mounted on the steering column.

The principles of re-direction of light rays by lenses formed in the glass front cover of headlamps, is used in a variety of designs to reduce the beam intensity above the horizontal, whilst the intensity of the beam below the horizontal is increased.

Bifocal or double-filament bulbs

The most extensively used anti-dazzle arrangement is that afforded by bifocal bulbs. These bulbs have two filaments in which one filament is positioned in relation to the reflector to give the main forward beam, whilst the other filament gives the dipped beam. The 'main beam' filament is usually of higher wattage than the 'dipped' filament, Figure 6.7.

Recent developments in European international regulations have led to the general use of bifocal bulbs giving an asymmetrical beam pattern for the dipped beam, with emphasis on near-side kerb illumination.

Setting of headlamps

Very practical instructions for the setting of headlamps are given in the manufacturers' handbooks where it stated that to comply with these regulations the lamps must be set as shown in Figure 6.8. To effect the

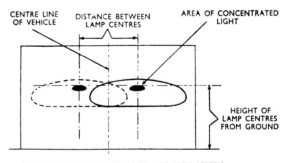

(A) FRONT OF VEHICLE TO BE SQUARE WITH SCREEN

(B) VEHICLE TO BE LOADED AND STANDING ON LEVEL GROUND

(C) RECOMMENDED DISTANCE FOR SETTING IS AT LEAST 25 FT.

(D) FOR EASE OF SETTING ONE HEADLAMP SHOULD BE COVERED

*Figure 6.8   Beam setting for main beams of
a two-headlamp car*

necessary setting or checking, the car should be placed on level ground with the headlights facing a blank wall. The headlamps must be set so that the light beams are parallel with the ground and with each other. This will ensure that when the 'dip' switch is operated the lamps will give a non-dazzling effect. Each lamp should be covered in turn whilst resetting the other.

HEADLIGHT CONSTRUCTION

There are two forms of light unit construction in general use. These are the 'Sealed beam' unit in use in the USA, UK, South America, Canada and Japan, and the 'European headlamp' which is permitted in several different constructions but is obligatory in the principal continental countries and is equally recognised with the sealed beam in the UK, South America, Canada and Japan.

The sealed beam unit

This unit, suitable for mounting in a modern flush fitting headlamp, is essentially an all glass unit performing the functions of light source and optical unit at the same time. The unit comprises two tungsten filaments, accurately located in an aluminised glass reflector, fused to a front lens suitably internally fluted to give the required beam pattern. The US standard has fixed dimensions (initially two 7-in diameter light units per vehicle, then later four 5¾ in diameter units). Three distinct advantages of this system are:

1. Perfect sealing against ingress of air, dust, moisture or rain, so it will therefore maintain its optical efficiency indefinitely, particularly as the filaments are also designed to give a longer life than the standard bulb.
2. Interchangeability.
3. Low cost due to mass production.

On the other hand there are the inherent disadvantages:

1. The difficulty of manufacturing accurately a device which must perform two functions.
2. The necessity of replacing the entire unit each time a filament fails which is inherently expensive.

The construction of the all-glass sealed beam unit is illustrated in Figure 6.9, which shows a sectional drawing of the Lucas 7 in unit. Figure 6.10 shows comparative light distribution diagrams for sealed beam and the Lucas F700 headlamp with separate (pre-focus) bulb.

The European headlamp

The primary characteristics of this headlamp are that it comprises two discrete components:

1. A removeable bulb forming the light source.
2. An optical unit (reflector lens assembly).

Advantages of the European arrangement are:

1. The possibility of future improvements as a result of new techiques.
2. The precision possible in methods of manufacture of two separate components.
3. The ease of bulb replacement.

The two main disadvantages are:

1. The large number of variants.
2. The problem of sealing.

It will be appreciated that the possibility of variants led to the development of rectangular shapes which enabled a larger surface area

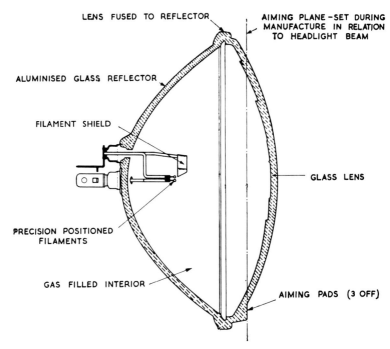

Figure 6.9   Sealed-beam light unit

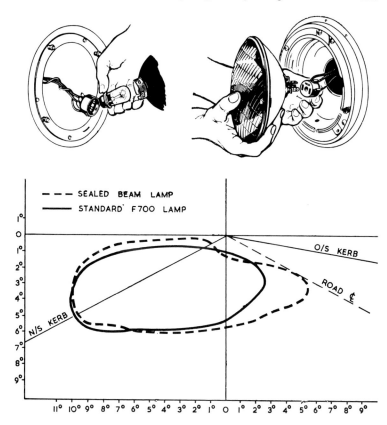

Figure 6.10 Relative light distributions of the sealed beam and pre-focus bulb headlamps. Details of the two light fittings are shown above

to be used giving improved main beam illumination and also a greater horizontal dimension giving improved performance beneath the cut-off on the dip beam.

Figure 6.11 shows the general arrangement of the Lucas flush fitting headlamp with separate double filament bulb and indicates how the bulb is accurately located by its flange and it is also held firmly in position in the reflector by a bayonet fitting cap with spring loaded contacts. The method of adjustment of the headlamp beam in both vertical and horizontal planes already referred to under the heading 'Setting of headlamps' is also shown in Figure 6.8. The Lucas rectangular headlamp of this design is shown in Figure 6.12.

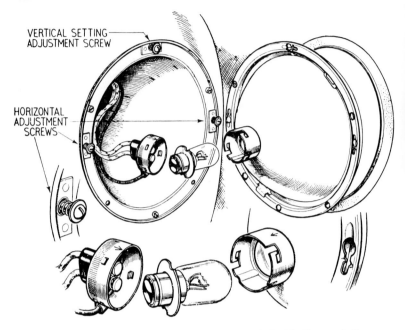

VERTICAL SETTING
ADJUSTMENT SCREW

HORIZONTAL
ADJUSTMENT
SCREWS

Figure 6.11   Lucas flush-fitting headlamp, with double filament bulb

Figure 6.12   Lucas rectangular headlamp

The two-bulb two-reflector construction

This highly efficient form of headlamp construction gives exceptional road illumination. At the same time it conforms with the European regulations, particularly regarding dazzle and bulb replacement, by the use of two reflectors which permit the filaments of each of the two bulbs fitted to be focused with great precision.

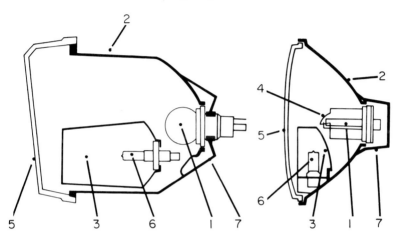

Figure 6.13   Marchal headlamps
1. *Dipped/main beam bulb*
2. *Dipped reflector*
3. *Main beam reflector*
4. *Deflector for dipped headlights. This eliminates the stray upward light rays*
5. *Glass reflector*
6. *Main beam bulb*
7. *Watertight rubber cap*

This arrangement is exemplified in a design from a range of head-lamps produced by S.E.V. Marchal. Their Ampilux design is made in round or rectangular form as shown in Figure 6.13. Both orthodox (classic) and iodine bulbs are used and the lamps are also available with yellow or white beams.

A single reflector headlamp produced by S.E.V. Marchal using the latest iodine bulb, known as H4, is shown in Figure 6.14. In this design special precautions against dazzle are taken by the provision of an occulteur in front of the bulb to eliminate stray rays of the 'dipped' beam. The H4 headlamp is now being further developed in favour of the 'Ampilux' design.

Figure 6.15(a) and (b) show the road illumination for the 'dipped' and 'main' beams respectively for the 'Code European' headlamps. Figure 6.16 shows a comparison of the road illumination as seen from

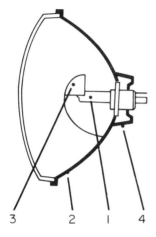

Figure 6.14   Marchal round headlamp
1. *Bulb*
2. *Reflector*
3. *Deflector eliminating stray upward rays*
4. *Watertight rubber cap*

Figure 6.15   Effect of main beam and dipped headlamps

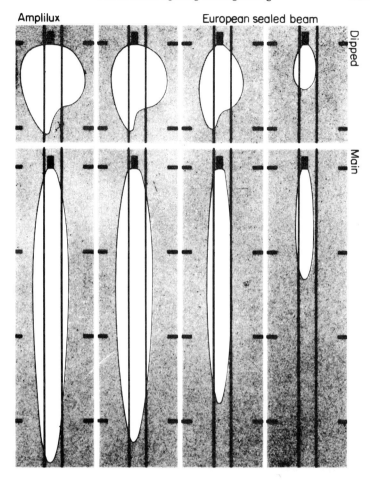

Figure 6.16 Characteristics of main beam and dipped headlamps

above for the functions Dipped and Main for four types of headlamps referred to.

Four-headlamp system

However optically accurate and efficient a bifocal lamp may be, and the sealed beam light unit is certainly outstanding in both of these respects, it is nevertheless something of a compromise.

This has led to the development of the four headlamp system which gives improved standards of vehicle lighting. This system was originally introduced in this country for heavy transport vehicles but is now used on a considerable number of makes of private cars, particularly on high performance models.

It is a fact that the filament-wattage of the four-headlamp system is some 25% greater than that for the two-headlamp system and there is little doubt that the uniformity of light distribution of the four-headlamp system notably increases the range and clarity of vision which ensure greater safety of night driving.

Automatic headlamp dipping and levelling

Various forms of headlamp levelling device have been developed for automatically correcting the headlamp aim whatever the condition of the road or the loading on the vehicle. Such devices can be mechanical, hydraulic, pneumatic, electrical and electronic. The systems are more readily adapted to four-headlamp arrangements and automatically re-direct light towards the road irrespective of the angle of the vehicle.

A French development deals with the problem of headlamp beams pointing skywards when the rear of a vehicle is lower than the front due to vehicle loading or road undulations, by arranging for two of the four headlamps to be mounted on swivels linked to the vehicle's anti-roll bars thus automatically re-directing the light beams towards the road whatever the angle of the vehicle. European Economic Community Directive 76/756/EEC Paragraph 4.2.6. states the requirements for alignment of the dipped beam with changing conditions of vehicle load.

An American system of automatically dipping headlamps when another vehicle is approaching is triggered by a photo-electric cell device fitted to the front of the car and connected to the dipping circuit via a relay. When light from an approaching vehicle falls on the photo-electric cell a voltage is set up in a pick-up circuit which is then amplified by a further transistorised circuit which in turn provides current to energise a relay to switch the car's headlamps to the dip position.

AUXILIARY LAMPS

A wide range of auxiliary lamps are in use to complement and improve the illumination produced by the main headlamps with which vehicles are equipped originally. These lamps are similar to the main headlamps but differ in the design of their optical systems.

Auxiliary lamps for use in fog give a wide flat beam (Figure 6.17) concentrated over a short distance, but across the road on to the kerbs near to the car and gives a sharp horizontal cut-off thus preventing any light becoming reflected back from the wall of fog.

Spot or driving lamps have a powerful narrowly focused beam to give a sharply defined spot of light for a long distance in front of the vehicle used as a supplement to the main headlamps for high speed vehicles (Figure 6.17).

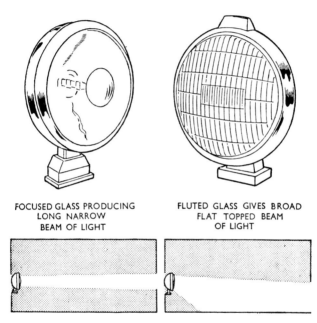

FOCUSED GLASS PRODUCING LONG NARROW BEAM OF LIGHT

FLUTED GLASS GIVES BROAD FLAT TOPPED BEAM OF LIGHT

Figure 6.17   Auxiliary driving lamps. The glass lens in front of the lamp is shaped to give the required beam. (left) Long-range driving lamp; (right) fog lamp

It should be noted that auxiliary lamps are not required by law (except in certain cases) but if they are fitted they must comply with the relevant SAE or IEC regulations.

Side lamps

Forward-facing side (or parking) lamps are mandatory and fitted to all vehicles and in some countries their location and illumination levels are precisely specified.

In some cases side-lamps are incorporated within the headlamp shell and project through a window in the main reflector. In the UK 6-W bulbs were generally used for sidelamps but these have been largely replaced by 5-W capless bulbs. This type of bulb has no metal cap with

PROJECTING LOOP OF WIRE BENT
OVER EACH SIDE OF LAMP BASE
TO FORM CONTACTS

Figure 6.18   The capless sidelamp bulb

soldered contacts but simply projecting wire bent over to form contacts (Figure 6.18).

Rear lamps

Tail lamps are mandatory in all countries. Their size, shape, colour, wattage and visibility angles, visibility distances, mounting positions and light output and permissible colour variations are specified in various national regulations as well as BS, SAE and IEC regulations.

It is generally required that there are two rear lamps showing red together with two reflectors as defined in S.I. 1971 No. 694 and it is general practice to use 6-W bulbs (maximum) to prevent dazzle to drivers approaching from the rear.

Stop lamps coloured red and direction indicator (flasher) lamps coloured amber are usually incorporated with the mandatory rear lamps and reflectors in a single light unit covered by a moulded glass or plastic lens/cover. Tail and stop lamps are incorporated in a double filament bulb in many systems using a 6-W 'tail' filament and a 21-watt 'stop' filament, the latter giving a much brighter red light to the rear when the brakes are applied.

Rear red fog lights have also become very popular in the UK. Two of these light units are fitted to the rear of a vehicle additional to the normal rear lamp/flasher unit in the interests of road safety in fog. They are fitted with a 21-W bulb and a red lens/cover having a large surface area (of the order of 112 cm^2), producing a light intensity of the foglight some hundred times stronger than ordinary red rear lights.

Rear number plate illumination must meet statutory regulations and it is general practice to use one or two 6-W bulbs, simultaneously switched with the side and tail lamps.

Reversing lamps are now generally incorporated in the rear lamp cluster or within the overall styling of the rear and signal lamps. This governs the overall size and shape and mounting position; 21-W rating is widely used and as a white light is being shown to the rear of a vehicle, switching is usually arranged to be automatic by means of a specially designed gearbox mounted switch which closes automatically when reverse gear is engaged, the battery supply coming from the switched side lamps feed. As mentioned under the heading Lighting Regulations, in some countries it is illegal to have a separate switch for the reversing lamps, since such a switch could be inadvertently left 'on' thus causing the white reversing light to be 'on' when the vehicle is being driven ahead causing confusion to following traffic. A separate switch may be fitted if there is a clear warning or signal lamp in the switch knob or on the instrument facia connected to the same circuit.

Trailer lighting

There are strict regulations regarding the lighting of any trailers being towed by commercial vehicles. Side marker lamps and under certain conditions, front corner marker lamps must be fitted.

Caravans must also be fitted with rear lamps and direction indicator lamps which conform to the appropriate international regulations. Suitably reliable weatherproof wiring coupler units should also be used to connect the caravan's electrical loads to the wiring of the towing vehicle.

Boot light

A boot or trunk light is provided in many cars which lights automatically when the lid is opened. A 6-W bulb is usually employed and this is connected in the side and rear lamp circuit.

The action of opening and closing the lid operates a simple push or gravity operated switch near to the hinge of the lid.

INTERIOR LIGHTING

Interior lights are 6-W bulbs usually fitted in ornamental housings having plastic diffuser type front covers or lenses. The light is usually fitted in the roof and serves a dual purpose as a courtesy lamp which is switched on by a very simple push switch on both front doors, or an interior lamp when it can be switched on from the lamp housing or facia board to provide interior illumination for map reading. On some vehicles two lamps are employed, one on either side which may also be automatically switched on by push switches near the door hinges whenever the rear doors are opened.

All these interior lights are wired so that they can operate at any time whether the engine is running or not and are therefore not connected through the ignition switch.

FLUORESCENT LIGHTING

For certain applications of interior lighting such as on public service and commercial vehicles, fluorescent lighting has distinct advantages particularly where the lighting load is high.

In its most popular form, the fluorescent lamp consists of a 5-ft long 1½ in diameter sealed glass tube containing a gas at low pressure and a minute quantity of mercury. The inside surface of the tube is coated with fluorescent powder and at each end of the tune is an electrode or cathode.

The conductor for the current, instead of being a wire filament, is mercury vapour and when the two electrodes are energised and current flows, short-wave radiation occurs. This is mainly invisible and within the ultra-violet wavelength band. The function of the fluorescent powder on the surface of the tube is to convert the invisible ultra-violet energy into visible radiation or light. The ultra-violet radiation, by excitation of the fluorescent powder electrons, causes fluorescence and the emission of light.

One advantage of the fluorescent lamp is that instead of having a light source, the surface area of which is but a minute fraction of a square inch as in the case of the filament lamp, the surface area of a comparable fluorescent lamp is several thousand times greater. Consequently the light flux emitted per square inch of surface area is considerably higher for the filament than for the fluorescent lamp. Therefore the latter provides lighting quite free from glare and eyestrain and is eminently suitable for interior lighting of passenger vehicles, where it is difficult to keep the lamps outside the normal range of vision of thepassengers.

The luminous efficiency of the fluorescent lamp is notably higher than that of the filament lamp, the light output for the same wattage being approximately three times greater. Even allowing for losses in the auxiliaries necessary with fluorescent lighting, the light output for a given battery load is nearly twice that of the filament lamp. The higher efficiency of the fluorescent lamp is due to the fact that less energy is dissipated in heat than is the case with the filament lamp.

By modification of the fluorescent powder coating, the colour of the light emitted can be varied over a wide range without any serious loss of luminous efficiency. This feature may be used for decorative effect in some interior applications.

The dimensions of the fluorescent lamp are decided by three main considerations. First, the electrode losses are a constant waste of power; second, for a given brightness the discharge voltage is roughly proportional to the length and inversely proportional to the diameter; and third, the light output in lumens for a given voltage varies as the square of the length. Of these considerations, the first and last clearly indicate the desirability of length for efficient operation, whilst limitations in the supply voltage will have a controlling influence on the ratio of length to diameter.

Another consideration peculiar to transport lighting is that of available mounting length which limits the maximum size of lamp that can be used and may in some circumstances necessitate series operation of lamps.

ANCILLARY EQUIPMENT

Owing to the fact that fluorescent lamps have a negative resistance, in consequence of which the discharge path resistance decreases with increase in current, it is necessary to have, in series with the lamp, some current limiting device. In the case of a.c. circuits this device is in the

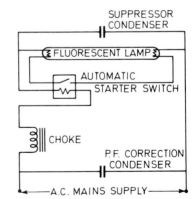

Figure 6.19   Typical fluorescent lamp circuit

form of an iron-cored choke, whilst for d.c. supplies a resistance or tungsten filament lamp ballast is employed.

For starting, a voltage higher than the operating voltage is necessary and this is obtained through the medium of some form of starting device or switch. For industrial and domestic lighting as shown by the circuit in Figure 6.19 an automatic starter switch is employed which short-circuits the lamp discharge path and connects the cathodes in series with the choke and mains supply. By this means the cathodes are quickly heated to operating temperature and commence the emission of electrons. The starter switch is then, by thermostatic or other means, opened automatically and a voltage surge, due to the interruption of the highly inductive circuit, initiates the discharge between the cathodes. Unless the supply voltage falls appreciably, the discharge once started will be maintained. The circuit in Figure 6.19 also includes a condenser across the mains to effect power factor correction and one across the lamp which absorbs high frequency impulses generated in the lamp and reduces radio interference from this source.

APPLICATION TO TRANSPORT VEHICLES

To provide the necessary a.c. voltage energised from the vehicle's battery, motor/alternator and also vibrator inverters were originally used, but it is now general practice to use a transistorised inverter.

An example of such a d.c./a.c. inverter is a CAV fluorescent lighting power unit in which two transistors are provided for the inverter. These are connected to the transformer so that current flows through thr transformer primary in one direction when one transistor is conducting and in the reverse direction when the other transistor conducts. Control windings on the same transformer switch the transistors so that they conduct in alternate half-cycles, giving rise to an alternating magnetic flux in the transformer core. This in turn induces in the transformer secondary winding a suitable a.c. voltage for the fluorescent lighting.

The secondary voltage is applied to the fluorescent tube via a choke ballast which limits the lighting tube current to the required maximum value. Power units are available for 12- and 24-V systems for supplying current for one 20-W 2-ft tube in the case of the 12-V unit and for two 20-W 2-ft tubes and one 40-W 4-ft tube on a 24-V system. Owing to the fact that the inverter operates at the high frequency of 5.5 kHz, the components can be small in size and weight.

SIGNALLING SYSTEMS

Direction indicators

The earliest type of direction indicators took the form of an illuminated oscillating semaphore type of arm. Two arms were provided, one fitted each side of the vehicle, operated electromagnetically by a two-way switch usually mounted on the steering wheel hub.

Flashing light direction indicators/hazard warning systems

The flashing light direction indicator was legalised in the UK from January 1954 and comprises a four-lamp system, two amber lights showing forwards and two amber lights showing rearwards and on some vehicles one on either side, the bulbs having 21-W filaments. These operate in conjunction with a flasher device and a panel warning light.

Some vehicles are fitted with two panel indicator lamps, one to indicate movement of the vehicle to the left and one to the right. The lamps are connected so that the operation of the direction indicator switch causes either the offside or nearside pair of lamps to flash in unison, according to the intended direction of turn.

As a hazard warning to other drivers when a vehicle is stationary in a dangerous situation, all four amber lamps are arranged to flash simultaneously by suitable switching.

The Lucas hot wire type flasher unit

Figure 6.20 shows the circuit arrangement of the Lucas flasher unit. It depends for its operation on the linear expansion of a straight length

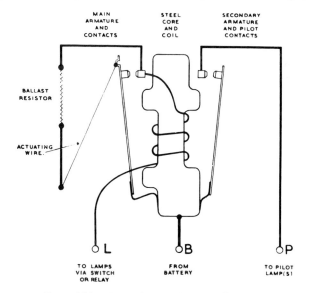

Figure 6.20   Circuit diagram for Lucas flasher unit

of wire in tension, which becomes heated by the passage of electric current through it. When the direction-indicator switch is turned to either left or right, current flows through this actuating wire and a current-limiting or ballast resistance to earth via the filaments of the indicating lamps. This current is insufficient to light the indicator lamps but suffices to produce the requisite linear expansion of the actuating wire for the flasher lamp contacts to close and energise the indicating lamps. In doing this, as will be seen from Figure 6.20, the actuating wire and ballast circuit is short-circuited.

The consequent cooling of the actuating wire results in the flasher lamp contacts being opened, thereby extinguishing the indicator lamps and causing current to flow again through the actuating wire. This sequence of operations continues so long as the direction-indicator switch is closed and the indicator lamps flash on and off about 80 times per minute.

It will also be noted from Figure 6.20 that when the flasher lamp contacts are closed, the indicator lamp current flows through the winding of a solenoid actuating the pilot or warning light switch contacts and in consequence the warning light flashes in unison with the indicator lamps. The solenoid is adjusted so that more than the current of one indicator lamp filament is required to actuate and close the pilot light switch contacts. Thus, should one bulb be faulty, the warning light will not flash, thereby providing instant warning of any bulb failure.

A criticism of the flashing light indicator system is the brightness of the rear signalling lamps at night-time. On the other hand, experience has shown that for the signals to be really effective under all daylight conditions, even higher illumination intensities than those in general use are essential.

The Lucas vane type flasher unit (Model 8FL)

This type of flasher unit has superseded the hot wire type. It is of vane type construction giving a snap action, and is resistant to damage by mechanical shock and electrical overload. With this flasher unit, the indicator (signal) lamp filaments are illuminated as soon as the direction indicator switch is operated, giving immediate indication of the driver's intention – an important safety factor when driving in high density traffic conditions.

The flasher unit consists of a moulded base carrying a snap action metal vane held in tension by a metal ribbon and a pair of normally-closed contacts. An aluminium pressed cover is gimped to the moulded base. The supply and output terminals are ¼ in (6.35 mm) 'Lucar' blades. The input terminal B is connected to supply and the output terminal L to the direction indicator switch.

The unit may be mounted in any attitude either by plugging into a moulded socket, or by means of clip fixing. Connections are made either by the socket or by individual 'Lucar'.

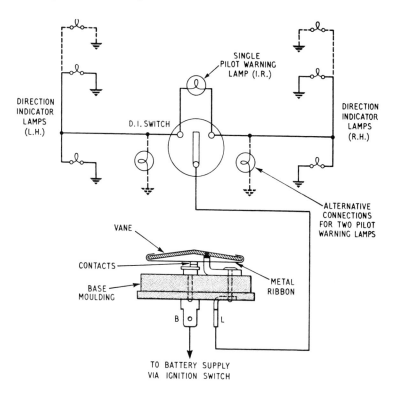

Figure 6.21   Circuit diagram of Lucas Type 8FL flasher unit

The unit operates such that when the direction indicator switch is turned to left or right (Figure 6.21) the appropriate signal lamp bulbs are immediately illuminated, the current flowing via flasher unit terminal B, the normally-closed contacts, the metal ribbon, the metal vane, and terminal L. This current, flowing through the metal ribbon, causes the latter to heat and expand, allowing the vane to relax and so to open the contacts. The signal lamp bulbs are now extinguished, the ribbon cools and re-tensions the vane, closing the contacts for the cycle to be repeated.

The snap action of the vane provides audible indication of flasher unit operation, while a pilot lamp connected as shown gives visual indication of operation. Normally, if one signal lamp bulb (out of two, three or four) fails, audible warning ceases while the pilot lamp and remaining signal lamp(s) remain on *but do not flash*. An occasional unit may however continue to operate, but at a significantly slower rate, immediately obvious to the driver.

The performance limits over a voltage range of 11–15 V for fixed-load vane-type flashers as specified by SAE (to which SMMT proposals also conform) are as follows:

Flashing rate	60–120 per min
Percentage 'on' time (i.e. percentage of each flash cycle for which signal lamps are on)	30–75%

Figure 6.22 illustrates the typical performance of Lucas Type 8FL flasher units over the specified voltage range.

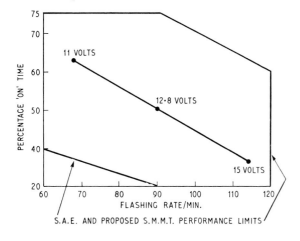

Figure 6.22   Performance curves for Lucas flasher unit

From the instant of switching on, the flasher unit contacts open within 1 second for two-bulb units and 1.25 sec for three or four-bulb units. Frequency variation with increase of voltage over the range 11–15 V is up to +12 flashes per minute/volt, and average frequency variation with increase of load wattage is +4.5 flashes per minute/watt.

The unit should be positioned such that it will be:

1. Protected from water splash.
2. Not subjected to excess vibration.
3. Not subjected to ambient temperatures exceeding 52° C (125° F).
4. Capable of being heard in operation by the driver.

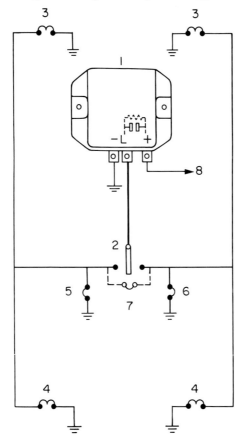

*Figure 6.23 Direction indicator circuit
incorporating a Model 15FL electronic flasher
unit*

1. *Flasher unit model 15FL*
2. *Direction indicator switch*
3. *Front direction indicator bulbs*
4. *Read direction indicator bulbs*

5. *L.H. pilot bulb*
6. *R.H. pilot bulb*
7. *Alternative single pilot bulb*
8. *To battery positive via ignition switch*

The varying bulb loadings of vehicle signalling systems necessitate a number of differently rated versions of Model 8FL, the current rating being marked on the cover. It is extremely important that a unit be used only with the bulb loading for which it is designed, to ensure that its performance conforms to the limits required by law.

Model 8FL is not directly interchangeable with the Lucas FL5 flasher widely used in recent years and described above.

Recent forms of electronic flasher units (with no moving contacts and immediate response) are exemplified by the Lucas electronic flasher units. Three models (15FL, 16FL and 19FL) of electronic flasher units have been designed for use in direction indicator systems or combined direction indicator/hazard warning systems.

Model 15FL flasher unit (Figure 6.23) is intended for use on rigid vehicles having direction indicator systems with fixed two-bulb loadings. Model 16FL is intended for vehicles which either have direction indicator systems with fixed three-bulb loadings, or are required periodically to tow trailers (including articulated vehicles) where direction indicator bulb loading on the flasher unit varies.

These units (in negative earth form only) flash the following bulb loads in direction indicator applications:

Model	System voltage	Bulb load
15FL	12 V	2 × 21 W + 2.2 W pilot
15FL	24 V	2 × 21 W + 2.8 W pilot
16FL	12 V	(2 + 1) × 21 W + 2 × 2.2 W pilots
19FL	12V	2 x 21W + 2.2W pilot or
		2 x 21W + 5W repeater
		+ 2.2W pilot

These models will flash at double their direction indicator bulb rates on failure of one 21W bulb. This serves as an indication to the driver that one flasher bulb has failed, (see bulb failure indication).

These electronic flasher units comprise resistors, capacitors, diodes and transistors in discrete component form and a relay mounted on and soldered to a printed circuit board (Figure 6.24). The components are protected by a black plastics cover, but must nevertheless not be mounted where they are subjected to water splash or excessive vibration. Two holes are provided in the covers of the 15 and 16FL units. Some installations of the 19FL have a clip for mounting on a panel of up to 2 mm thickness. Alternatively the unit may be mounted on a one piece terminal mounting block.

Performance

The above models of flasher unit have been designed to operate in either direction indicator systems, or combined direction indicator/hazard warning systems. For hazard warning purposes, all flasher units will flash double their rated direction indicator bulb loading.

Normal flash rate: 60—120 flashes per minute.

In systems employing these flasher units, the direction indicator bulb filaments are illuminated as soon as the direction indicator switch is operated. In addition to visual indication given by the pilot bulb(s), the action of the relay armature provides audible indication of flasher unit operation.

Bulb failure indication

Failure of a main bulb in direction indicator systems incorporating Model 15FL and 19FL units (Figures 6.23 and 6.24) is indicated by the flashing rate of the remaining bulb and pilot bulb increasing from the normal 60—120 f.p.m. to approximately 200 f.p.m.

In systems incorporating Model 16FL units (Figure 6.26), when three main direction indicator bulbs are operating, the conventional pilot bulb and the second pilot bulb (9, Figure 6.26) flash in unison with the main bulbs at the normal 60—120 f.p.m. When only two main direction indicator bulbs are operating, due to the failure of the third bulb, then only the conventional pilot bulb flashes in unison with the main bulbs at the normal rate. However with cold direction indicator bulb filaments, the second pilot bulb will give one flash when initially switching on. The driver thereby obtains indication as to whether the second pilot or a main bulb has failed.

Similarly, when a trailer is not being towed and only two operable direction indicator bulbs are connected to the flasher unit, only the conventional pilot bulb flashes in unison with the main bulbs at the normal rate. With only one main bulb operational, the conventional pilot bulb flashes at an increased rate of approximately 200 f.p.m.

Operation

Wiring diagrams of direction indicator systems incorporating model

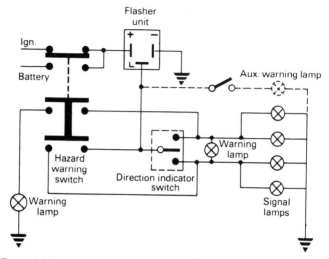

Figure 6.24   Typical wiring diagram for 19FL showing direction indicators controlled by ignition switch.

Figure 6.25   Lucas Model 19FL flasher unit showing electrical components and relay.

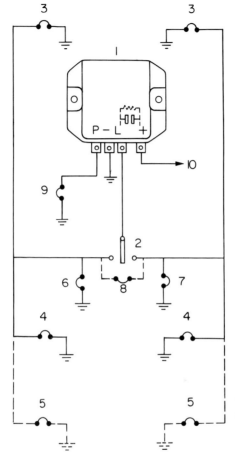

*Figure 6.26 Direction indicator incorporating
a Model 16FL electronic flasher unit*

1. Flasher unit
2. Direction indicator switch
3. Front direction indicator bulbs
4. Rear direction indicator bulbs of towing
    vehicle
5. Rear direction indicator bulbs of trailer
    (when attached)
6. Conventional l.h. pilot bulb
7. Conventional r.h. pilot bulb
8. Alternative conventional single bulb
9. Second pilot bulb (indicating three main
    direction indicator bulbs operating)
10. To battery positive via ignition switch

15FL, 16FL and 19FL are shown in Figures 6.23, 6.24 and 6.26 respectively.

Flasher unit operation is basically dependent on the action of an electrolytic capacitor which is alternately charged and discharged between two voltage levels. At the lower voltage level, the relay incorporated in the flasher unit is energised and its normally open contacts, which are in series with the selected direction indicator bulbs and the supply (see Figure 6.23) are closed. The bulbs are thus illuminated. At the upper voltage level, the relay is de-energised and the direction indicator bulbs extinguished. The bulbs are therefore lit during the charge period of the capacitor and extinguished during its discharge period. The procedure is repeated 60–120 times per minute to give the required flashing rate.

In order to provide an indication of the number of operable or failed direction indicator bulbs, a relatively high value resistor (shown in broken line in Figures 6.23 and 6.26) is connected across the relay contacts and in series with the bulbs. With the direction indicator switch operated and the relay contacts open, most of the supply voltage is dropped across this resistor leaving only a small voltage across the bulbs. While the magnitude of this voltage across the bulbs is insufficient to light them, it is dependent on the number of bulbs that are operable. Consequently by monitoring this voltage any change in bulb load can be detected and the appropriate indication given.

In direction indicator systems incorporating model 15 FL units, an increase in this monitored voltage above the normal for two bulbs is indicative of a failed bulb. When such a change occurs, the lower voltage level to which the capacitor discharges and at which the relay becomes energised, is raised. This reduces the time taken for the capacitor to charge and discharge between the two voltage levels and there is an increase in bulb flash rate. The percentage 'on' time of the remaining operating bulb remains nominally the same.

In systems incorporating model 16FL units, however, an increase in the monitored voltage above the normal for three bulbs can be due to either the failure of a bulb, or the fact that a trailer is not being towed when only two bulbs are connected to the system. In either case the change in monitored bulb voltage is utilised to inhibit operation of the second pilot bulb. Failure of one in two direction indicator bulbs on vehicles not towing trailers produces a further increase in the monitored voltage above that obtained when two bulbs are operable. This increase is used to effect an increase in flash rate as described for model 15FL units.

Installation

The flasher unit should be sited where it is not subject to water splash, excessive vibration or engine heat. A suitable surface inside the passenger compartment is suggested.

Whilst the unit is not appreciably attitude conscious, it is recommended that it be mounted with its terminals lowermost.

The Simms transistorised flasher unit

The Simms type ELT flasher has no moving parts and a power switching transistor energised by pulses from a multi-vibrator circuit provides the interruption.

The power switching transistor is conservatively rated for normal use so that when trailer indicator lamps are connected in parallel with those on the towing vehicle there is little effect on the flashing rate. The clock pulse circuit is designed so that there is negligible change in the flashing rate by comparatively large changes in voltage supply.

Two types of unit are available, ELT12/1 (12 V) and ELT11/2B (24 V), both basically similar in design and operation. The components are mounted on a printed circuit secured to a base plate by two screws which are also used for clamping the transistors in position. The base plate is of dish form to facilitate the dissipation of heat.

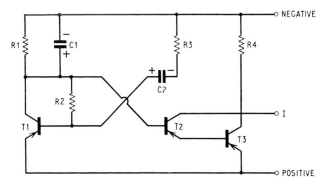

Figure 6.27   Circuit diagram of Simms transistorised flasher unit

A steel cover is secured to the base plate by detachable rubber grommets fitted with metal bushings for the fixing bolts. The cover is designed to allow ample air flow over the printed circuit. The inter-rupter unit should be mounted vertically with the terminals at the bottom. The positive terminal (+) is a female blade connector; the negative terminal (−) and the panel lamp terminal (1) are male blade connectors.

The principle of operation is based on a free running astable multi-vibrator circuit. Referring to the circuit diagram (Figure 6.27), the two transistors T1 and T2 generate the clock pulse which is fed to the power switching transistor T3.

The lamp current passes through transistor T3 which operates in cascade with T2. Lamp load is taken across R4 thus generating volts

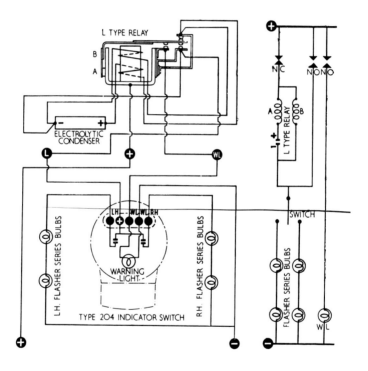

Figure 6.28   Wiring and technical diagrams for CAV flasher unit

for the panel lamp. Diminished lamp load is indicated by diminished panel lamp brilliance.

The ratio of light to dark period is determined by resistor R4 and capacitor C2. Adjustment of frequency is carried out during manufacture by variation of resistor R2 which is specially selected for each unit. The capacitor C1 prevents any incoming pulse from upsetting the flashing rate.

Among other types of flasher control unit are two which are specially designed by CAV for heavy duty in commercial vehicle applications. One is the condenser-relay flasher unit, type CRF, wiring and technical diagrams for which are given in Figure 6.28, in which the necessary delayed action intermittent switching is controlled by a relay switch working in conjunction with a hermetically-sealed electrolytic condenser. The other, known as type MF is motor driven with worm reduction gearing to a three-lobe nylon cam which operates the flasher contacts.

The CAV 489 flasher unit has been developed to meet the future requirements of the Statutory Instrument No 1970/49 for dual intensity flashers, as well as complying with existing UK legal requirements.

Apart from the relay used to switch the flasher lamp load, the 489 unit is fully transistorised. All components, including the blade terminals are contained on one printed circuit board, the whole being fitted into a tough moulded housing of flame retardant polypropylene.

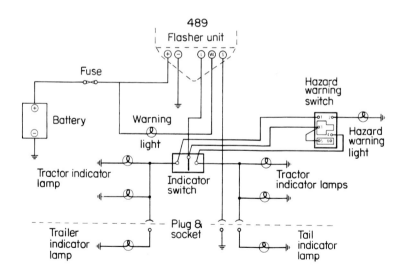

Figure 6.29   Circuit diagram for tractor and trailer

The 489 unit has been designed for a life of 2000 hours and will operate satisfactorily over a temperature range of $-30°$ C to $+70°$ C and over a voltage range of 10–15 V for a nominal 12-volt system and 20–30 V for a nominal 24-V system.

With the addition of a suitable double pole switch and warning light the 489 can be used in a hazard warning system.

The dual-level system is achieved by the use of a dimmer-relay, which is automatically energised when the vehicle side and tail lights are switched on and inserts a series resistance in the circuit of the rear facing flasher and stop lights so that a signal of reduced intensity is given when these lamps are used. In order that the operation of certain types of flasher unit which are current sensitive should not be affected, a second resistance, in parallel with the flasher lamp bulb, is also introduced so that the total current passing through the flasher unit is approximately the same for both day and night conditions. The resistors are an integral part of the dimmer-relay. The values used are the subject of international agreement and are as follows:

	12-V systems	*24-V systems*
Series resistor	3.3 ohms	11.5 ohms
Parallel resistor	30.0 ohms	140.0 ohms

These resistance values must be strictly adhered to.

The warning light normally flashes in anti-phase to the flasher lamps but in some applications, where r.h. and l.h. warning lights are specified the warning lights are connected in parallel with the flasher lamps and therefore flash in phase. In the event of a bulb failure the relay contacts stay closed so that the remaining flasher lamps are illuminated continuously, but the warning light will go out with the normal system and will remain on where twin warning lights are used. Figure 6.29 shows a typical circuit diagram.

SPECIAL PURPOSE BEACONS

This type of light unit for roof mounting is produced in various forms. Mechanical systems comprise a rotating beacon in which a reflector unit is rotated by a small permanent magnet motor and a flashing beacon in which a central electronic flasher unit is fitted similar to that used for direction indicators. Both designs are fitted with either yellow, blue or red covers. Both beacons are fitted with iodine bulbs to produce a powerful light from the smallest possible size of bulb. The more recently introduced flashing xenon lamp is, however, tending to replace rotating beacons. It should be noted that there are EEC and SAE regulations for all signalling lamps.

LIGHTING AND SIGNALLING REGULATIONS – BRITISH STANDARDS

BS 941 covers most forms of filament lamps used for head, side, rear and warning lights. BS AU40 covers motor vehicle lighting and signalling equipment and has been prepared at the request of the Department of Transport and under the authority of the Automobile Industry Standards Committee of the B.S.I.

These regulations specify methods of dipping headlights to prevent dazzle and obligatory requirements such as the size, colour, mounting and use of head, fog, side and rear lamps.

Obligatory rear lamps and reflectors

The character and positioning of obligatory rear lamps and reflectors are also precisely defined in SI 1971 No. 694.

Regulations governing reversing lamps on vehicles

These regulations quoted below are very clearly stated in the interests of road safety, applicable to reversing lamps:

No vehicle shall carry more than two reversing lamps.

Every reversing lamp shall comply with the following conditions:

(a) It shall be illuminated by electricity.
(b) It shall be so constructed that it cannot be switched on otherwise than either:
(i) automatically by the selection of the reverse gear of the vehicle; or
(ii) by the operation of a switch by the driver of the vehicle, being a switch which, except in the case of a mechanically propelled vehicle first used before July 1954, serves no other purpose.

(c) The rated wattage of the elctric bulb or the total rated wattage of all such bulbs with which it is fitted or the rated wattage of the sealed beam lamp with which it is fitted shall not exceed 24 watts; and

(d) It shall be so constructed, fitted and maintained that the light emitted thereby is at all times incapable of dazzling any person who is standing on the same horizontal plane as the vehicle at a greater distance than 25 feet from the lamp and whose eye-level is not less than 3 ft 6 in above that plane.

Except in the case of a mechanically propelled vehicle first used before 1st July 1954, where a reversing lamp is so constructed and fitted that it can be switched on by the operation of a switch by the driver of the vehicle, the vehicle shall be equipped with a device so fitted as to be readily visible to the driver at all times when in his seat and so designed as to indicate when the reversing lamp is illuminated.

No reversing lamp constructed and fitted as provided in the last preceding Regulation shall be illuminated except in so far as is necessary for the purpose of reversing the vehicle.

(See also *Lighting Regulations* 1971 by Wilkinson, p. 750, App. I, Pt. 5).

Other relevant regulations and standards have been prepared by EEC, BS, SAE and FMVSS. These deal with important details such as beam and lens patterns; permissible wattages and light outputs; standard sizes, switching regulations, aiming and adjustments, two and four headlamp systems.

CHAPTER 7

ELECTRICAL IGNITION SYSTEMS

On studying the history of the electric spark, we find that Sir Humphrey Davey was the first man to note its characteristics as early as 1820. He found that on breaking a circuit comprising a battery of voltaic cells, it was possible to produce a spark between the two metallic parts of the circuit when they were slowly drawn apart. No such spark occurred when the same parts were brought together again and the circuit thus closed. By substituting carbon rods for the metal points, the spark became more brilliant and a carbon arc was obtained – the forerunner of the electric arc later to become a special form of lighting and which now has many industrial welding applications.

By 1831 Faraday had produced a spark by electromagnetic means and at the same time, as mentioned earlier, laid the foundation of all electric ignition apparatus now used. A Frenchman, Ruhmkorff, produced induction coils on a commercial basis by 1851, developed from the principles laid down by Faraday which still apply to present day ignition coils. Lenoir, also mentioned in connection with his work on sparking plugs, devised a system of electric ignition using a Ruhmkorff coil and a battery in 1860 with many features still in use.

Since these early developments, the electrical ignition system has passed through four distinct stages of evolution:

1. Hot wire ignition.
2. Break spark ignition.
3. Trembler coil ignition.
4. Jump spark ignition, of which there are two kinds, (a) High tension magneto ignition and (b) Battery-coil ignition.

The first three of the systems listed are now only of historical interest and magneto applications 4(a) are also now very limited.

Battery-coil systems 4(b) are in world-wide use and their main advantage is that they are simple and reliable and can be produced at relatively low cost.

Their main disadvantages are that the spark voltage decreases with higher spark rates because limited by the current load of the contacts, the coil current cannot reach its final value in the short intervals between pulses. Spark duration is limited and the system is sensitive to high tension leakage or side tracking, contact wear is also considerable resulting in short maintenance intervals. Electronic systems, which are described later in this chapter, overcome these disadvantages particularly for higher speed applications and are therefore increasing in popularity.

The fundamental principles involved in the functioning of all systems of spark ignition are outlined below and also the practical constructions of the various systems in current use.

Primary function of the electrical ignition device

The main function of the spark generator is to produce a spark at the plug electrodes inside the cylinder at the appropriate time to ignite the air-fuel mixture. The spark timing must be automatically controlled so that it is correct for engine starting and also for all conditions of speed and load over the complete speed range.

In the so called Otto engines, the spark is timed to occur when the piston is nearing the end of its compression stoke. In the case of a four-cycle (or four-stroke) engine, a spark must occur in each cylinder every two revolutions of the crankshaft whilst with a two-cycle (or two stroke) engine, it is necessary to produce a spark twice as rapidly; that is a spark in each cylinder every revolution of the crankshaft.

For the spark to initiate ignition satisfactorily, the air-fuel mixture must be readily igniteable whether it be supplied from a carburetter or fuel injection system and the air-fuel mixture must have easy access to the ignition spark. The position of the spark gap in the combustion chamber, the duration of the spark and the movement of the mixture are decisive for reliable ignition.

The process of ignition

At the moment when the spark is initiated the gaseous mixture in the cylinder is compressed to approx. $80-120$ lbf/in^2 and the piston is still on the up stroke. The initiation of the explosion occurs simultaneously with the beginning of the spark discharge and at the same instant a pressure wave arising from the explosion is propagated throughout the gaseous medium, the rate of propagation being accelerated by the turbulent state of the mixture.

The voltage applied to the sparking plug by the ignition generator increases until the so-called firing or sparking voltage is reached. This voltage may vary between 5000 and 20 000 V depending on the compression ratio of the engine, geometry of the cylinder head and conditions of temperature and gap length of the spark plug.

The air-fuel mixture around the plug gap at high pressure offers an enormous resistance to the passage of an electric current through it. But when the gap has become ionised as a result of the electric stress set up by the voltage applied, the gap is broken down and current flows across the gap. The energy generated appears as light and heat in the form of the spark.

The voltage at the plug causes the gap between the spark plug electrodes to become electrically conductive so that the ignition spark can jump across the gap. This spark is at a temperature of several thousand degrees centigrade, sufficient to ignite the air-fuel mixture, which then continues to burn by itself.

The ignition spark must have a certain minimum energy which is converted to heat. If the spark jumps across the gap but does not have sufficient energy, ignition will not take place. The rate of dissipation of the spark energy is probably the most important factor affecting ignition of the air-fuel mixture.

Improved fuel economy has been achieved by widening spark plug gaps. To meet these conditions, higher sparking voltages are required, which in turn result in higher spark energy. There are of course limits to which the sparking voltage can be increased and these are determined not only by the maximum voltage required for cold starting, which may be limited by the design of the ignition generator, but also by the increased plug erosion. Starting conditions however require high spark energy because very rich mixtures may obtain at the same time as excessive plug leakage and due to cold plug conditions.

The energy per spark under normal mixture and temperature conditions may be around 0.1 mJ, whereas for cold starting the energy required will be many times greater and a spark duration of only a few microseconds.

In order to ensure satisfactory ignition under all conditions, modern ignition coils and other ignition systems are designed to provide sparking voltages of the order of 20 kV and spark energy well in excess of starting requirements. At the same time spark energy must not be so great that it causes rapid or excessive rotor arm and spark plug electrode erosion.

BATTERY COIL IGNITION (INDUCTIVE IGNITION SYSTEM)

The essential components of this system can be itemised as follows:

1. An engine driven generator for charging the battery.
2. A battery which is also used for many other purposes, lighting, starting etc.

3. An induction or ignition coil which stores the ignition energy and delivers it to the distributor via the h.t. ignition cables. This coil comprises a primary and secondary winding on a common axial laminated iron core.

4. A contact breaker which opens and closes the primary circuit of the ignition coil for the purpose of energy storage and voltage conversion.

5. A capacitor which suppresses most of the arcing between the contact points and also assists in the precise timing of the ignition spark.

6. A distributor which distributes the sparking voltage at the appropriate time to the spark plugs in the firing order required.

7. A centrifugal advance mechanism which automatically adjusts the spark timing in accordance with engine speed.

8. A vacuum advance mechanism which automatically adjusts the spark timing in accordance with engine load. This is achieved by means of a diaphragm operated unit which produces movement on the contact breaker base related to induction manifold depression.

9. Spark plugs which carry the spark electrodes across which the spark passes to initiate combustion.

Items 4, 5, 6, 7 and 8 are usually incorporated in a single engine driven unit referred to as the Ignition Distributor.

The ignition coil – contact breaker operation

The growth of current in the primary circuit and its effects will now be considered. When the ignition switch is closed (by turning the ignition key) the primary winding of the ignition coil will be connected directly across the battery during the 'closed' period of the contact breaker (see Figure 7.1). During the 'open' period, the primary winding will be isolated and no current can flow. The rate of growth of current in the primary winding during the 'closed' period is a vital factor because this determines the value of the primary current broken when the contacts are separated; that is, the 'break' current I_B.

Knowing the electrical constants of the system it is an easy matter to calculate the growth of current in the primary winding by applying the well-known exponential law

$$i = \frac{V}{R} \left(1 - e^{\frac{-Rt}{L}}\right) \tag{7.1}$$

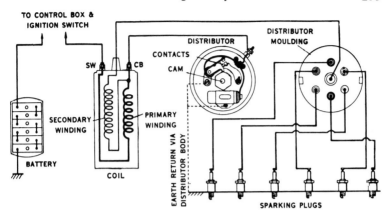

Figure 7.1   Ignition wiring diagram

where i = instantaneous primary current (A)
$R = (r + r_1)$ = total resistance in primary circuit (Ω)
r = primary resistance (Ω)
r_1 = ballast resistance (Ω)
L = self-induction of primary (henrys)
V = battery voltage
t = time in seconds measured from the instant when the contacts close
e = base of Napierian logarithms.

The application of this formula can be exemplified by considering the case of a known design of coil incorporating a ballast resistance which had the following characteristics

r = 1.3 Ω when hot
r_1 = 0.25 Ω when cold
r_1 = 0.6 Ω when hot
L = 0.00514 henry
V = 6 V

The maximum current that can flow in the primary winding will depend on the temperature of the ballast resistance, which is made of iron wire having a high temperature coefficient, so that its resistance when hot may be from two to three times the resistance when cold. We therefore get

Maximum current when ballast resistance is cold

$$= \frac{V}{R} = \frac{6}{1.55} = 3.87 \text{ A}$$

Maximum current when ballast resistance is hot

$$= \frac{V}{R} = \frac{6}{1.9} = 3.17 \text{ A}$$

Two curves, determined from equation (7.1), are plotted in Figure 7.2 to show the growth of current in the primary circuit, first, when the ballast resistance is cold, and second, when this resistance is hot. At high speeds the primary current will grow in conformity with curve 1, whilst at very low speeds it will tend to follow more closely curve 2. For intermediate speeds, the growth curve will lie somewhere between these two limiting curves.

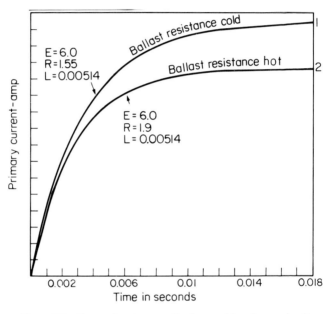

Figure 7.2 Curves showing growth of current in primary circuit

We can easily determine from the 'growth' curves the current at 'break' provided we know the time during which the contacts are closed. With this particular design of coil and contact breaker the total time period during which the contacts are closed and opened — corresponding to one complete spark cycle — is divided as follows

Total time period   = T
Period of closed-circuit = $0.66\,T$
Period of open-circuit   = $0.33\,T$

If the ignition unit is mounted on a 6-cylinder engine, then at a speed of 4000 rev/min the number of sparks generated per minute would be

$$\frac{4000}{2} \times 6 = 1200$$

that is a spark frequency of 200 per second. We then get

$$T = \frac{1}{200} = 0.005 \text{ sec}$$

period of closed circuit = 0.66 × 0.005 = 0.0033 sec

Referring to curve 1, in Figure 7.2, we note that after a time interval of 0.003 sec the primary current has attained the value 2.45 A as indicated on this curve. This would be the value of the 'break' current I_B at this particular speed.

Taking the other extreme case, a very low speed of 100 rev/min, we find the spark frequency to be

$$\frac{100}{2} \times 6 = 300 \text{ per minute or 5 per second.}$$

We then get

$$T = \frac{1}{5} = 0.2 \text{ sec}$$

Period of closed-circuit = 0.66 × 0.2 = 0.133 sec

Referring to curve 2, in Figure 7.2, we find that at this speed the steady current is reached in about 0.015 sec, so that this condition is

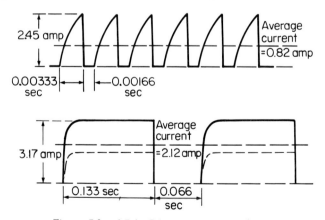

Figures 7.3 and 7.4   Primary current waveforms

Electrical ignition systems

established long before the contacts separate. The value of I_B is, therefore, the value of the steady current, namely, 3.17 A.

These calculations enable the primary current waves to be plotted for each speed, and this is done in Figures 7.3 and 7.4. The *average*

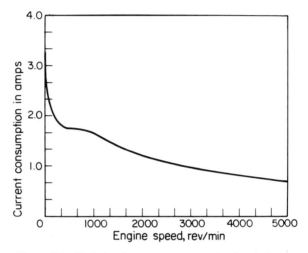

Figure 7.5   Variation in current consumption in relation to engine speed

current for each speed is also indicated on each diagram, this being the value of the current which is measured by the ammeter in the primary circuit. Summarising, we get

Speed of 4000 rev/min

    'Break' current I_B = 2.45 A
    Average current   = 0.82 A
    Energy taken by coil = 0.82 × 6 = 4.92 W

Speed of 100 rev/min

    'Break' current I_B = 3.17 A
    Average current   = 2.12 A
    Energy taken by coil = 2.12 × 6 = 12.72 W

These calculations are checked by Figure 7.5, which shows the variation in the current consumption of the particular design of coil over a speed range of from 0 to 5000 rev/min, as determined by actual experiment.

Secondary voltage and spark energy

At the moment when the contacts begin to separate, the voltage rises very suddenly in the secondary winding due to the collapse of the magnetic field built up by the growth of current in the primary winding. The design of ignition coil to which the figures used in the previous calculations relate had 165 turns on the primary winding and 10 500 turns on the secondary winding. This gives a ratio between turns of 64, and these figures can be taken as representative of other types of ignition coil designed for operation from a 6-V battery. The same coil designed for a 12-V battery is provided with the same secondary winding, but with a primary winding having 244 turns of a smaller size of wire, giving a ratio between turns of 43.

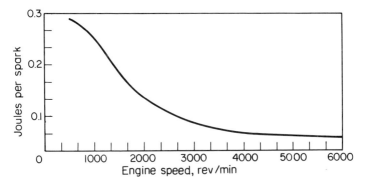

Figure 7.6   Showing Joules per spark in relation to engine speed

The energy stored in the magnetic field at the moment of 'break' is $\frac{1}{2} LI_B^2$. This energy is precipitated very suddenly into the secondary winding, causing a rapid rise of secondary voltage which in turn initiates a spark at the plug electrodes. There is, therefore, a very sudden conversion of energy in the electromagnetic form to heat energy represented by a spark discharge. The amount of energy in each spark discharge is very little less than $\frac{1}{2} LI_B^2$ at low speeds, but at high speeds owing to increased losses, the spark energy determined by experiment is about one-half the energy calculated by this formula.

In Figure 7.6 the measured spark energy in joules per spark is plotted for the particular design of coil under consideration, against a speed range of 500 to 6000 rev/min. At low speeds the spark energy is about 0.03 J per spark, but this falls away rapidly as the speed is increased to 2000 rev/min and more slowly thereafter, reaching a figure of about 0.006 J per spark at a speed of 6000 rev/min.

Construction of the ignition coil

The construction and essential parts of the ignition coil are illustrated in Figure 7.7, depicting the components of the Lucas ignition coil and Figure 7.8, showing a cut-away view of the assembled unit. On a laminated iron core is first wound the secondary winding of about 20 000 turns of 44 SWG enamel-covered wire, the layers being insulated from each other by thin paper strip. Over the secondary winding and insulated from it by varnished paper, is wound the primary winding which for a 12-V system comprises some 360 turns of 25½ SWG enamel-covered wire in three layers, with varnished-paper interlayer insulation. A slotted iron sheath is wrapped around the windings and the components inserted within an aluminium container, together with a porcelain base support, which serves to insulate the core from the container. The inner end of the secondary winding is connected to the laminated iron core which, together with a compression spring and terminal screw, constitutes the high tension connection. The other end of the secondary winding is connected to one end of the primary winding.

On complete assembly of the components, shown in Figure 7.7, the upper edge of the container is spun over the flange of the terminal moulding to secure the components firmly against vibration. The primary leads are also connected to the two low tension terminals.

For many years it was customary to wax impregnate the windings prior to insertion in the casing and run pitch into the container to retain the windings and core assembly firmly. Failures due to air inclusions with this method led to what has become well-known as the oil-filled coil. This practice of insulating the windings by immersing them in oil in a sealed container follows the established practice with large power transformers. With this method of insulation, mechanical means for holding the winding assembly firmly are necessary. Winding the primary on the outside of the core assembly enables the heat losses of this winding to be more rapidly dissipated and heat flow to the casing is still further facilitated by using oil instead of wax and pitch insulation. The higher resistance of an outside primary also obviates the necessity of a ballast resistance, which was a feature of ignition coils with the primary wound next to the core.

Another example of the adoption of oil as the insulating and cooling medium is the Runbaken Oilcoil shown in Figure 7.9. The primary is wound on the outside of the secondary winding and it is claimed that rapid heat dissipation is obtained through oil circulation and that the windings are not damaged even when the coil is inadvertently left on for long periods. The glass container is hermetically sealed as renewal of the oil is never necessary. The maximum operating voltage is stated to be 30 kV and the recommended plug gap setting for use with this

Figure 7.7   Lucas ignition coil

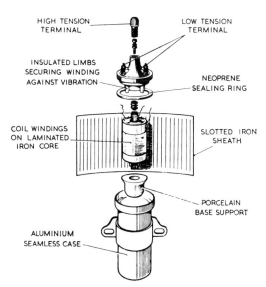

HIGH TENSION
TERMINAL

LOW TENSION
TERMINAL

INSULATED LIMBS
SECURING WINDING
AGAINST VIBRATION

NEOPRENE
SEALING RING

COIL WINDINGS
ON LAMINATED
IRON CORE

SLOTTED IRON
SHEATH

PORCELAIN
BASE SUPPORT

ALUMINIUM
SEAMLESS CASE

*Figure 7.8   Cut-away view of Lucas Model LA12
ignition coil*

ignition coil is 0.032 in, with which setting good starting and general performance is obtained. The primary current c.r. oscillograph traces given in Figure 7.10 indicate that the current at 'break' is well maintained at high operating speeds, the upper trace corresponding to 2000 rev/min on a 6-cylinder engine.

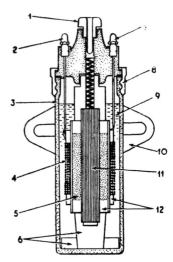

Figure 7.9   The Runbraken Oilcoil

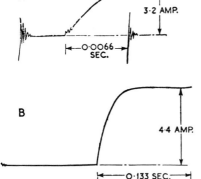

Figure 7.10 I_B values for the 6-V Oilcoil. Oscillograph traces showing primary current waveforms produced (A) at 100 sparks per sec (9 kV ball gaps) and (B) at 5 sparks per sec (9 kV ball gaps)

Figure 7.11 shows a schematic cut-away view of a Bosch ignition coil.

As mentioned earlier, some ignition coils are provided with an external ballast or series resistor which limits the primary current and also the thermal loading of the coil. In some systems this ballast resistance can

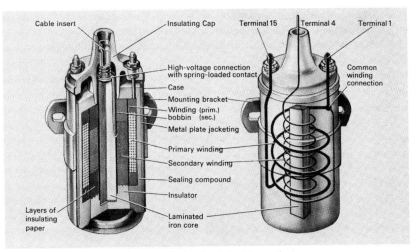

Figure 7.11   Bosch ignition coil

be short circuited during engine starting to give higher ignition energy to compensate for the temporary reduction in battery voltage during starting. Such a resistance has a value of 1–2 ohms. This reduces the thermal loading on the coil because the heat generated in the resistor can be more easily dissipated to the surrounding air than from within the coil.

The ignition distributor

The complete distributor carries out the function of distributing the ignition pulses to the individual spark plugs via the h.t. cables in the correct firing sequence and at the correct instant of time. To enable the timing function to be carried out efficiently the distributor unit houses the contact breaker capacitor and centrifugally-operated spark advance mechanism and also the vacuum control device for the contact breaker.

An example of the general construction of a distributor unit is shown in Figure 7.12. This illustrates the Lucas distributor model 45D which is produced in 4- and 6-cylinder versions for use on engines having maximum speeds up to 9000 rev/min (4-cyl.) or 8000 rev/min (6-cyl.).

The drive for this type of distributor is via an offset driving dog pinned to the shaft which engages with a corresponding slot in the end of a drive shaft gear driven from the engine crankshaft. The capacitor is internally mounted on the contact breaker base.

Figure 7.13 illustrates a Bosch design of distributor. This unit is gear driven and the ignition capacitor is fitted externally.

The electrical processes which take place in the distributor (arcs, glow discharges, corona effects) produce nitric oxide and ozone gases, which are poisonous and corrosive. In order to prevent the formation

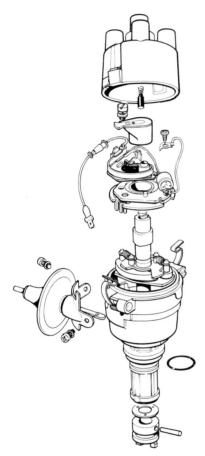

Figure 7.12   Lucas distributor, Model 45D

of acids resulting from accumulation and concentration of these gases, the distributor cap is provided with ventilation holes. If the ventilation is not adequate, however, the surfaces of the metal parts corrode, which results in increased wear of the breaker lever rubbing block and thus in changes in the ignition timing and dwell angle. The surfaces of the insulating parts are also attacked and the insulating capabilities of these parts are reduced, resulting in leakage currents. A shield made of plastic

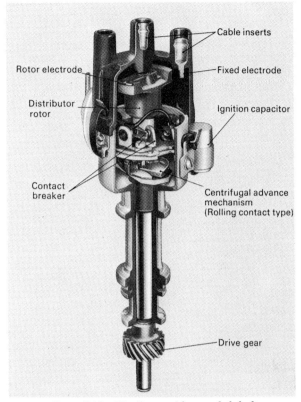

Figure 7.13   Distributor with extended shaft

or metal separates the distributor compartment from the compartment which houses the contact breaker and the spark advance mechanisms. This shield prevents larger quantities of dirt and carbon dust from reaching the contacts and also prevents moisture from entering the distributor section.

Contact breaker

The contact breaker is essentially a cam-operated switch which closes and opens the primary circuit of the ignition coil in synchronism with the engine rotation and thus triggers an ignition pulse every time the contacts open. Thus the cam has as many lobes as the engine has cylinders.

Dwell angle

As described earlier, the current in the coil primary winding does not rise from zero to its maximum value instantaneously when the contact breaker closes, but requires a short time interval to overcome the inductive effect of the coil. The contacts must therefore be closed long enough for the current to build up.

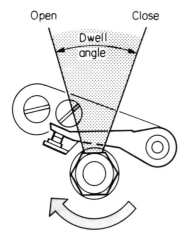

Figure 7.14   Cam dwell angle

The angle through which the cam rotates between the instant when the points close and re-open is called the 'Dwell' or 'Cam' angle (see Figure 7.14). To ensure smooth running of an engine, the dwell angle for each cylinder must be as nearly equal as is practicable, so for this reason the cam profile has to be machined to close tolerances.

As engine speed increases the dwell angle remains virtually constant but the revolution of the cam becomes more rapid and the time during which the contacts are closed is reduced. The primary current may not have time to build up to its maximum value, resulting in reduced magnetic field intensity and reduced secondary voltage.

To meet the requirements of 8- and 12-cylinder engines regarding open and closed contact periods, it is general practice to use double lever contact breakers. This means a four-lobed cam with two contact levers is used on an 8 cylinder engine and a six lobed cam with two contact levers on a 12-cylinder engine. This arrangement permits the contacts to remain closed for sufficient time to enable the ignition coil to built up its magnetic energy to meet the high sparking speeds required by multi-cylinder engines.

When the contacts open to break the l.t. circuit, the back e.m.f. induced in the primary tends to send a current surging back across the gap, causing arcing and burning at the points. This effect would prevent the efficient cut-off of the primary current and the rapid collapse of the magnetic field. A condenser is therefore connected across the points to act as a 'by-pass'. The voltage charges up the condenser instead of arcing across the gap. When the plug fires, the condenser discharges through the primary circuit.

The ignition capacitor

The capacitor or condenser is connected across the contacts to perform two main functions:

1. To assist the collapse of the magnetic field by passing current rapidly through the primary winding in the reverse direction to normal flow.
2. To minimise contact arcing so prolonging service life.

To appreciate how the capacitor acts to reduce sparking, we must first consider an electrical property peculiar to this component. When a voltage change occurs across a capacitor, a current will flow between the plates and the size of this current is directly proportional to the rate of change of the voltage. That is to say the faster the change in voltage the less resistance to the flow of current offered by the capacitor and with no change in voltage, the capacitor will appear an open circuit.

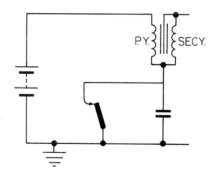

Figure 7.15   Coil-ignition circuit

If we consider the circuit shown in Figure 7.15 when the contacts are closed and the circuit conditions are stable, no voltage changes are happening and the capacitor will appear as an open circuit. When the contacts are opened, the inductance of the coil will generate a large rapidly increasing voltage, which if the capacitor was not there, would be dissipated in the form of a spark across the contacts. However, due to the property described earlier, the rapid change in voltage will cause current to flow into the capacitor, charging it up. This stops a build up of voltage at the points and stops sparking across the gap.

Since the capacitor is connected to a circuit which is earthed through the battery, it will discharge through the primary winding and ignition circuit, following the path of least resistance. This 'back voltage' occurs very rapidly effecting a complete collapse of the coil's magnetic field. The speed, at which the charge and discharge takes place, is proportional to the capacitor size and the resistance of the circuit.

To have the maximum effect, the resistance and inductance of the connecting leads should be kept to a minimum and therefore the capacitor should be mounted as close to the contacts as possible.

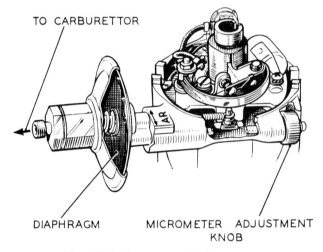

TO CARBURETTOR

DIAPHRAGM   MICROMETER ADJUSTMENT
KNOB

Figure 7.16   Lucas vacuum ignition control

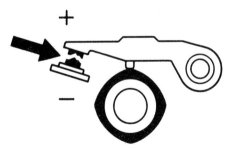

Figure 7.17   Wear of contacts. As a result of transfer of metal, a crater develops on the breaker lever contact, while contact material accumulates on the fixed contact

Most modern capacitors are made from metallised paper, as this material allows reasonably compact construction and also has the property of being self-healing after a minor breakdown. The repair is the result of the heat generated at the failure point, vaporising the metal and hence clearing the fault. The average value of these components is in the order of 0.2 mf.

Various forms of contact levers are in use, but it is general practice for the lever 'heel' or 'rubbing block' which is in contact with the hardened steel cam lobes to be made from synthetic bonded fabric or glass filled nylon material (Figure 7.16). This is retained in contact with the cam surface by means of a stainless steel spring. The lever construction is such that there is sufficient flexibility in the lever assembly to produce a 'wipe' effect when the contacts open and close. Experience has shown that this effect is beneficial in keeping the contacts clean and minimises contact pitting tendencies.

Normal wear on the lever 'heel' and at the contacts can result in variations in ignition timing and contact point sparking can occur, which can result in a transfer of material from the positive contact on the breaker lever to the stationary negative contact point as shown in Figure 7.17. Part of the metal is vaporised and an oxide coating is formed on the contact point surfaces which slightly increases the electrical resistance in the primary ignition circuit. Wear on the rubbing block and on the contact points have opposing influences, but eventually the wear on the rubbing block assumes more importance and this means that the point gap is slowly reduced during the course of time, the dwell angle is slightly lowered and the arcing across the contacts increases.

SPARK ADVANCE MECHANISMS

The purpose of the spark advance mechanism is to assure that under every condition of engine operation, ignition takes place at the most favourable instant in time, i.e. most favourable from a standpoint of engine power, fuel economy and decontamination of the exhaust gases. By means of this mechanism the advance angle is accurately set so that ignition occurs before the top dead-centre point of the piston.

The engine speed and the engine load are the control quantities required for the automatic adjustment of the ignition timing. Depending on these quantities the ignition timing adjustment mechanism adjusts the instant of ignition in accordance with characteristic curves developed by the engine manufacturer, showing the required degrees of advance before TDC plotted against speed in rev/min.

It is general practice to provide automatic timing control (which varies the ignition timing in relation to speed) combined with some form of vacuum ignition control, to be referred to later. This varies the timing in relation to engine load, thus rendering the ignition control fully automatic.

The centrifugal advance mechanism

The ignition timing for full load operation is incorporated below the contact-breaker mechanism (see Figure 7.12) and forms what is virtually a flexible coupling connecting the *driving* shaft, geared to the engine, to the *driven* shaft which carries the contact breaker cam and distributor

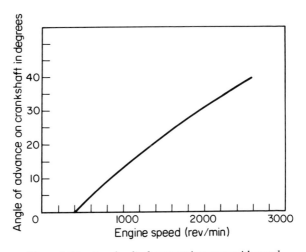

Figure 7.18   Angular displacement increases with speed

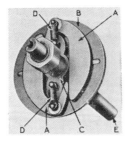

Figure 7.19   Automatic timing unit

rotor arm. In all constructions there is a system of weights which are constrained to move under the influence of centrifugal force and this controlled movement produces an angular displacement between the driving and driven shafts which increases with speed (see Figure 7.18).

The general form of construction established over many years is illustrated in Figure 7.19 which shows a Delco Remy design. The two weights A are pivoted on a plate B rigidly secured to the driving shaft. Near the pivoted end of each weight there is a projection which embraces the end of the camplate C, rigidly fixed to the base of the cam which is free to rotate on the extremity of the driving shaft E. When everything is at rest the various parts are in the relative positions shown in Figure 7.19 but the weights A move outwards once the whole unit is rotated, the springs D limiting the movement to bring about a steady angular displacement of the cam with increase in speed.

Vacuum ignition timing control

As described earlier, the centrifugal advance mechanism compensates for an increase in engine speed to allow sufficient time for the air/fuel mixture to burn. However, when a vehicle is cruising under light load the intake of air and fuel is reduced and the mixture is at a lower pressure when the spark occurs. Under these conditions the mixture burns more slowly and it is again necessary to advance the spark position to ensure that maximum impact is delivered to the system.

The vacuum at the carburetter is proportional to engine load; on light load the vacuum is relatively high, on heavy load the vacuum is low. We can therefore use a vacuum operated mechanism to advance the timing on light load.

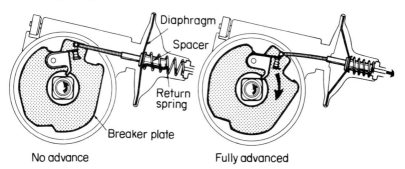

No advance   Fully advanced

Figure 7.20   Operation of contact-breaker plate

Since the alternative to increasing the combustion time by advancing the timing would be to increase the air/fuel charge, the vacuum advance mechanism may be regarded as an economy device.

Figure 7.20 illustrates the vacuum advance mechanism in a typical distributor. It comprises a spring-loaded diaphragm, one side of which is open to atmosphere, while the other side is connected, via a copper pipe, to the carburettor near the choke and hence is subjected to the inlet vacuum.

Conditions affecting vacuum advance are:

1. At idle speed, although the inlet depression is large, the throttle butterfly is not opened wide enough to expose the vacuum tapping. Due to this, there will be no pressure differential across the diaphragm and it will not deflect.

2. On light load, the inlet depression is still large and the butterfly is opened wide enough to allow the depression to be applied to the diaphragm. This will now move against the pressure of the return spring and rotate the breaker plate, causing the ignition to be advanced.

3. On heavy load, when the throttle is opened wide, the inlet depression is reduced and the action of the spring will return the diaphragm and breaker plate to their original positions. Ignition timing is then dependent solely on centrifugal advance.

It can be seen that the vacuum advance characteristic will depend on the spring rate and the length of the spacer limiting the movement of the diaphragm.

Manual adjustment

Initially, the distributor is set by the manufacturer to suit the engine and anticipated operating conditions. However, on occasion, it may be necessary to slightly adjust the ignition timing to compensate for a change in condition, for example the use of fuel having a different octane rating.

To facilitate this, some distributors have a vernier or micrometer adjustment control which permits small adjustments to be made without disturbing the position of the distributor body, as shown in Figure 7.16.

To meet the exacting requirements of exhaust emission control laid down in various countries, the vacuum systems may be combined in one unit containing a dual-diaphragm system. Such a system is exemplified in the Bosch design illustrated in Figure 7.21. Here the vacuum timing control is also utilised for exhaust emission control, a process which requires retarding the timing back past TDC to the true retarded position. For this purpose the first vacuum unit used for the partial load adjustment in the advance direction (the 'early' unit) is supplemented by a second unit (the 'late' unit) which moves the breaker plate in the direction of rotation when there is a vacuum. Two vacuum systems are thus combined into one compartment.

The 'late' unit is also connected to the intake manifold of the engine but through a second, i.e. a separate vacuum line; this line enters the manifold at a point behind the throttle valve where a vacuum exists when the engine is operating in neutral or when the vehicle is over-running the engine. When a vacuum exists here, the ring shaped

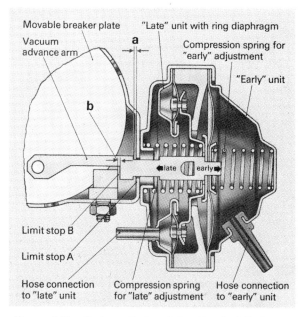

Figure 7.21 Timing adjustment by the dual-diaphragm vacuum advance mechanism to either advance position or true retarded position. (a) Adjustment in advanced position as far as limit stop A. (b) Adjustment in true retarded position as far as limit stop B

diaphragm in the 'late' unit draws the limit stop for the zero position of the 'early' unit diaphragm a certain distance away (b). This movement is transmitted through the linkage system to the moveable breaker plate moving this plate to the 'late' position. The essential feature of the dual diaphragm vacuum advance mechanism is that both systems are de-coupled; the adjustment of the ignition timing point in the 'advance' direction has priority (a). This means that simultaneous vacuum in both chambers results in the correct partial load adjustment in the advance direction.

ELECTRONIC IGNITION

In the conventional form of ignition distributor described previously, the contact breaker operates as a high-speed switch in the primary circuit of the ignition coil. Whilst developments have been made in improving reliability, assisting cold starting and in accuracy of timing control to improve engine performance, there has been no major change in design to the basic form of energy transfer. This means that a sparking rate of about 400 sparks per second is the maximum which can be reached. This sparking rate is equivalent to an 8-cylinder engine at 6000 rev/min.

Present day racing engines have been designed to run at speeds up to 12000 rev/min. Since conventional battery/coil ignition systems would not meet these speeds, electronic ignition systems have been developed which are capable of providing a sparking rate of 1000 sparks per second, thus meeting the requirements of an eight cylinder engine running at 15 000 rev/min.

There are both mechanical and electrical reasons why engine performance is restricted by the limitations of the conventional coil ignition system. Mechanically it is difficult to manufacture a cam, breaker heel and spring mechanism which will provide and maintain the required timing accuracy; and electrically, contact-breaker points are a limiting item.

It is therefore quite natural that solid-state or electronic switching has been used to meet these requirements, which has been made possible by the development of reliable high voltage transistors and silicon-controlled rectifiers.

To achieve the high sparking rates required it has also been necessary to break away from the traditional energy storage principles of the ignition coil and to take energy for the spark directly from the battery in the form of a high amplitude current pulse of short duration.

The first development in this field retained the mechanical contact breaker to trigger a transistor circuit in which the breaker contacts are able, with the aid of a transistor, to switch on a high current in the coil primary without carrying a heavy load themselves.

An example of this arrangement is the Lucas TAC system described below.

THE LUCAS (TAC) TRANSISTOR ASSISTED CONTACTS SYSTEM

This system operates like the conventional coil ignition system except that instead of interrupting the primary current by the breaker contacts, a high voltage transistor performs this duty. Thus contact wear resulting from the extremely rapid making and breaking of a highly inductive circuit is eliminated because the existing contact breaker now serves to make and break only a small non-inductive control current in the base circuit of a driver transistor. Consequently, contact duty is greatly eased.

Since there is no arcing at the contacts, low speed performance is improved. This is beneficial, particularly for low temperature starts. High speed performance is improved because the high-voltage transistor, which can handle higher values of primary current, permits an ignition coil of lower primary inductance to be employed and used in series with an external ballast resistor.

Less periodic maintenance is required and after adjusting the contact gap at 500 miles to compensate for the initial bedding in of the breaker heel in the contact breaker, no further contact gap adjustment is required for at least 25 000 miles and then only to compensate for possible further wear of the heel.

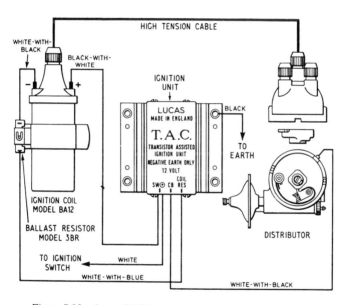

Figure 7.22   Lucas TAC (transistor-assisted contacts) ignition

The ignition unit comprises four main components: an aluminium heat sink; printed circuit board; aluminium cover and steel baseplate; the arrangement of these components is shown schematically in Figure 7.22. The heat sink is robust and is generously flanged for effective heat dissipation; it carries the high voltage power transistor and two capacitors. The printed circuit board carries the driver transistor and associated base-current-limiting and collector load resistors. Shrouded blade terminals conforming to BS AU17 are employed. An extra 'CB' terminal is provided as a tapping point for engine-speed sensing equipment, when used.

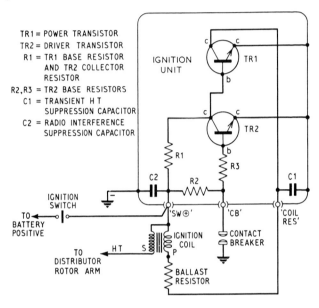

Figure 7.23   Circuit diagram for Lucas TAC ignition system

To follow through the operating cycle of the TAC4 electronic circuit on negative-earth vehicles, reference should be made to the circuit diagram (Figure 7.23) and assume the ignition switch to be closed, the engine slowly turning and initially, the contact-breaker to be open. Under these conditions, a small current from the battery flows through resistors R2 and R3 and the base-emitter portion of transistor TR2 and to earth via the ignition unit mounting studs. Transistor TR2 is therefore switched on with resistor R1 acting as its collector load. This short-circuits the base-emitter portion of transistor TR1, which is therefore switched off. Thus, at this stage, no current can flow through the ignition coil primary winding. As the engine turns, however, the contact breaker closes and switches off transistor TR2. This allows transistor TR1 to conduct and the ignition coil to be magnetically energised, with resistor R1 now acting as a base current limiter. When the contact breaker re-opens, the original circuit conditions are restored with transistor TR2 switched on and TR1 switched off. The rapid switching off of the ignition primary current by transistor TR1 causes a collapse of magnetic flux within the ignition coil which in turn, results in a high tension secondary voltage being induced in the normal way for distribution to the sparking plugs.

Capacitor C1 protects transistor TR1 from any freak high tension voltages of a transient nature to which it might possibly be subjected in service. Capacitor C2 serves to suppress interference with radio equipment.

It will be observed that the conventional ignition capacitor normally connected across the contacts is not now required. Radio interference suppression is taken care of within the ignition unit.

The next step in the development of electronic ignition systems was the replacement of the mechanical contact-breaker by an electronic means of triggering the onset of the spark and the switching of the coil primary circuit. This has been achieved in the breakerless inductive semiconductor system.

The breakerless transistorised coil ignition system

An example of this system is that using the Lucas Models 43DE and 45DE (4- and 6-cyl.) and 47DE4 (4-cyl.) distributors which show superior performance over the contact-breaker system in three important directions; namely, stability of ignition timing without the need for periodic adjustment, a factor of increasing importance in the light of present and possible future legislation concerning the control of pollution due to vehicle exhaust emissions.

Consistent sparking over the entire speed range because the rate at which the primary current is interrupted does not vary with distributor speed. Coil output is therefore constant down to the lowest engine speeds contributing to easier starting. Higher sparking rate; this is considerably beyond the limit of 400 sparks/second imposed by the mechanical contact breaker.

Figure 7.24 illustrates the construction of distributor Model 45DE4; other models are of similar basic design. The main difference between these electronic distributors and their conventional contact-breaker counterparts are in the provision of a timing rotor, a pick-up module and an amplifier module.

Timing rotor

This takes the form of a plastic drum with ferrite coupling rods (four or six according to the number of engine cylinders) embedded vertically and equi-spaced in its periphery. The timing rotor is mounted on the distributor drive shaft beneath the rotor arm (in the position occupied by the contact breaker cam in conventional distributors).

278

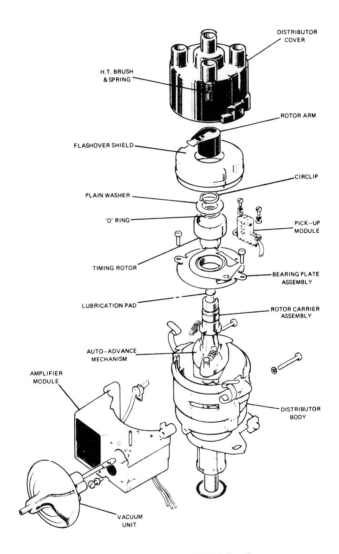

DISTRIBUTOR
COVER

H.T. BRUSH
& SPRING

ROTOR ARM

FLASHOVER SHIELD

CIRCLIP

PLAIN WASHER

'O' RING

PICK–UP
MODULE

TIMING ROTOR

BEARING PLATE
ASSEMBLY

LUBRICATION PAD

ROTOR CARRIER
ASSEMBLY

AUTO–ADVANCE
MECHANISM

AMPLIFIER
MODULE

DISTRIBUTOR
BODY

VACUUM
UNIT

Figure 7.24   Lucas Model 45DE4 distributor

Pick-up module

The pick-up module, mounted on the equivalent of a contact breaker base plate, comprises an E-shaped ferrite transformer core with primary and secondary windings enclosed in a plastic housing (for details see under 'Operation' later in this chapter). The pick-up is connected to the amplifier module by a short flat triple cable.

Amplifier module

This unit comprises an oscillator, operating in conjunction with the windings of the pick-up module; a smoothing circuit; an amplifier and a power transistor connected in the ignition coil primary circuit. The oscillator, smoothing circuit and amplifier assembly consists of discrete components mounted on and soldered to a printed wiring board and is encapsulated to provide protection against vibration. This electronic assembly is contained in a metal housing attached to the side of the distributor body. The housing, in conjunction with the distributor body, acts as a heat sink for the power transistor and other heat-dissipating components.

Three external low voltage connections (supply, drive resistor and ignition coil primary are made by pre-wired cables that pass through a grommeted hole in the base of the amplifier module.

Other features

The body is an aluminium die casting. The shank houses two sintered iron bearing bushes which carry the drive shaft. The standard cover is moulded in phenolic material having satisfactory resistance to surface flash-over under normal service environments. Alternatively, the cover can be made in anti-tracking materials which offer superior resistance to flash-over. The cover size and a flashover shield, provide adequate protection against internal electrical flash-over. The high tension outlets accommodate push-in type connectors.

The cover seats on a flange machined around the rim of the body to provide protection against dust and water ingress. It is secured by two short 'non-flop' spring clips. The rotor arm is common to all distributor models.

The centrifugal advance mechanism has interlocked weights and is resistant to torsional and linear vibrations. The design provides quiet operation with high accuracy of timing characteristic both on acceleration and deceleration. A wide range of automatic timing curves, with slopes up to 5° per 100 rev/min and full range up to 24° (nominal) is obtainable.

The vacuum control unit (models 45DE and 47DE) is integral with the electronic module housing.

Operation

Figure 7.25 shows the circuit diagram of the ignition system. The secondary winding L3 is carried on the centre limb of the pick-up E-core; the primary winding is split into two parts, L1 on the lower limb of the E-core and L2 on the upper limb. Windings L1 and L2 are

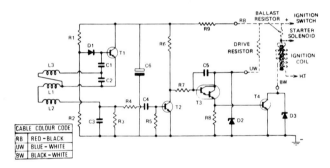

Figure 7.25   Diagram of electronic ignition circuit

oriented so that the magnetic fluxes they produce in the centre limb and winding L3 are in antiphase and such that L1 (having more turns than L2) provides negative feedback to the base of transistor T1, while L2 provides positive feedback. As the timing rotor revolves the ferrite coupling rods bridge the centre and upper limbs of the E-core; the coupling between L2 and L3 and thus positive feedback to the base of T1, is thereby increased.

With the ignition switched on and the engine at standstill, the ferrite rods in the timing rotor will normally not be near enough to the pick-up E-core to cause coupling between L2 and L3. Overall feedback to the base of T1 will therefore be negative and the oscillator will be inoperative. Consequently T3 will have no base drive and so will be in an 'off' state, while T3 and the power transistor T4 will be 'on'. The ignition coil primary winding will therefore be energised via the collector/emitter of T4. The drive resistor provides bias current for the output stage of the electronic circuit.

As the engine is cranked and a ferrite coupling rod passes across the face of the E-core, as described previously, overall feedback to the base of T1 now becomes positive and the oscillator breaks into high frequency oscillation (300–400 kHz). The output from the oscillator, which is sustained for the period the ferrite rod is in the immediate vicinity of the E-core, is smoothed and used to switch on T2, thereby switching off T3 and T4. The ignition coil primary circuit is thus interrupted each time a ferrite coupling rod passes across the face of the E-core, inducing a high voltage in the secondary winding which is distributed to the appropriate spark plug in the usual manner.

High frequency signals induced on the power supply line by the switching of transistors within the electronic unit and which might cause interference with other electronic equipment on the vehicle, are limited by using negative feedback (C5) over the Darlington Pair Stage (T3) to lengthen switch-on and switch-off times.

Diodes D2 and D3 provide protection against voltage transients and reversed battery connections.

Satisfactory operation of the electronic circuit is obtained over a voltage range of 6–16 V.

Installation

Mounting and location

The preferred mounting attitude of the distributor is within 60° of the vertical; horizontal mounting allows oil ingress along the main shaft and also creates a higher operating noise level. The distributor must be positioned where it will be subjected to the minimum of water splash and reasonably clear of any part of the fuel system. The mounting situation must offer immunity from battery acid spray and also enable the distributor to benefit from cooling air while the vehicle is in motion. Additionally it must be mounted clear of heat radiating surfaces, such as exhaust pipes and manifolds, which may result in the maximum temperature (100° C) for satisfactory distributor operation being exceeded.

Electrical connections

The distributor is supplied complete with low-tension cables for connection to the supply, drive resistor and coil primary winding. High tension cable connections are made by push-in connectors attached to the cables. In-line and right-angle connections are available, complete with waterproofing grommets.

Radio interference suppression

Suppression of ignition high tension circuits is achieved by the use of resistive h.t. cable in the usual way. Similarly, the low tension current can be suppressed by a 1.0 μF capacitor connected between the ignition coil +ve terminal and earth. For applications where v.h.f. radio is installed, or glass fibre bodies are used, special suppressors are available.

Timing

The system cannot be timed by the use of a low voltage light due to the switching characteristics of the system. The distributor can, however, be set statically and should then be checked dynamically using a strobe light.

1. *Static timing.* With the engine turned to the static timing position for the appropriate cylinder as determined by the engine timing marks and rotor arm position, rotate the distributor body until the timing rotor to pick-up alignment is as shown in Figure 7.26.

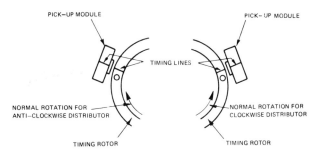

Figure 7.26   Distributor rotation

The timing can be checked by holding No. 1 plug lead 6 mm from an earth point, rotating the engine slowly in its normal direction and observing the relationship of the engine timing marks when a spark is produced.

2. *Dynamic timing.* After setting the timing by the above method, it should be checked by using the h.t. voltage on the specified cylinder to trigger a stroboscope in the normal way.

Electronic tachometer

The dwell angles on this range of contactless distributors differ from the equivalent contact breaker distributors. In general, current sensing tachometers designed for use with contact-breaker distributors will not indicate correctly when used with these distributors.

Voltage-sensing types should be employed, being connected to the coil negative terminal in the normal way. The tachometer manufacturer should be consulted whenever the system employed or the accuracy of indication is in doubt.

The Lucas 'Opus' electronic ignition system

A specific example of a highly successful form of breakerless transistorised ignition systems is the Lucas 'Opus' system suitable for 4-, 6-, 8- and 12-cylinder engines. On Formula 1 racing engines to which this system has been fitted, a sparking rate of 670 sparks/second has been achieved at maximum speed and on a 12 cylinder engine requiring 600 sparks/second the Opus system is capable of producing 800 sparks/ second.

The system is generally similar to that already described except a control unit separate from the distributor and magnetic pick-up unit, contains the amplifier which interprets the electronic timing signals from the distributor and the power transistor in the amplifier unit, then functions as an electronic switch in the primary circuit of the ignition coil. The amplifier unit is connected to the ignition coil via a separate ballast resistance unit and external connecting cables.

The capacitor discharge ignition system (CDI)

This electronic system is widely used on racing and high performance engines. The essential feature of the system is that the ignition energy is stored in the electric field of a capacitor. The magnitude of the stored energy being determined by the capacitance and the charging voltage applied to the capacitor. A transformer is also required with this system whose function is to transfer energy to its secondary circuit and also transform the capacitor voltage into the high voltage required to produce the spark at the plug electrodes, see Figure 7.27.

An advantage of this system is that the secondary voltage rises many times faster than in other systems, which means that the ignition spark is exceptionally strong, though it is only of relatively short duration of the order of 300 ms. The rapidity of the voltage rise does not permit leakage to earth via conductive deposits on the spark plug insulators which can reduce spark energy. The system operates as follows. The storage capacitor is charged to a voltage of about 400 V and is then, at the ignition point, discharged through the primary winding of the ignition transformer by the closing of a solid-state switch. Since the capacitor and the primary winding of the transformer form a resonant circuit, damped oscillations of current and voltage are generated at a frequency about ten times higher than in a conventional or transistorised coil ignition system. As a result, the voltage induced in the secondary winding climbs to the firing voltage about ten times faster than is the case with conventional or transistorised coil ignition systems. The

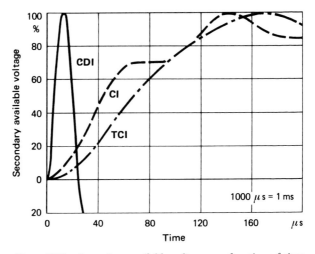

*Figure 7.27   Secondary available voltage as a function of time
    related to the peak voltage (= 100%) of the first half-wave*
CDI   =   Capacitor discharge ignition system
CI   =   Coil ignition system
TCI   =   Transistorized coil ignition system

secondary voltage reaches the spark plugs in the usual way through high tension ignition cables and distributor.

The charging circuit in the capacitor discharge ignition system is an electronic circuit designed to transform the battery voltage into a much higher d.c. voltage with which it charges the storage capacitor. The charging takes place by means of pulses. We distinguish between multiple pulse charging and single pulse charging. The charging time is so short

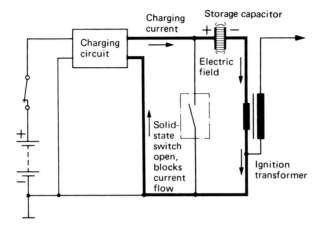

*Figure 7.28   Energy storage through charging the capacitor.
Charging current = primary current*

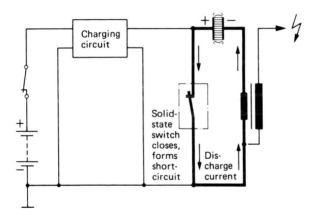

*Figure 7.29   Transferring the stored ignition energy to the
ignition circuit by discharging the capacitor*

that even at the highest sparking rates the energy stored does not fall
off significantly. While the storage capacitor is discharging, the charging
circuitry is short-circuited (see Figures 7.28 and 7.29).

For the solid state switch, a controllable semi-conductor device
known as a thyristor, is used. This device must switch currents of up to
100 A and it blocks, until the ignition point, the discharge current

while the capacitor is being charged; this means that the thyristor must be able to withstand a voltage of several hundred volts. The CD ignition system has therefore also become known as the 'thyristor ignition system'.

At the ignition point the thyristor receives a current pulse, the trigger pulse, through the control electrode. This pulse immediately triggers the thyristor to make it conductive and initiates charging of the capacitor. The thyristor blocks the circuit again only when the storage capacitor is almost completely discharged. Triggering in the CD ignition system may be by means of a mechanical switch or is breakerless.

Since very weak trigger pulses are sufficient to make the thyristor electrically conductive, it is very sensitive to spurious pulses which can cause it to trigger. Such pulses can be generated in practical operation, for example, by the bouncing of the breaker switch contacts and this can result in misfiring. To prevent this from happening is the function of the so-called 'bounce pulse blocking device', which must be installed in every breaker-triggered CD ignition system (see Figure 7.30).

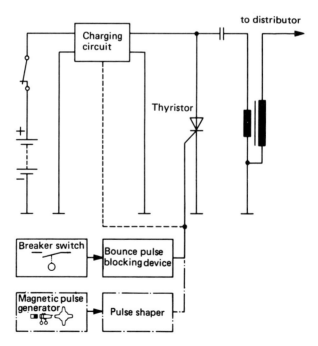

Figure 7.30   Basic circuit diagram of capacitor discharge ignition system breaker-triggered

Breakerless pulse-triggering in this system can maintain a sparking rate of 40 000 sparks/min. which adequately meets present engine requirements.

Capacitor discharge systems may also include alternative charging arrangements, such as a magnetic pulse generator incorporated in a flywheel system as used on smaller engines such as power saws, lawn-mowers, motor boats, snowmobiles, etc. The basic circuit of such a system is also shown in Figure 7.30.

Opto-electronic ignition

In this form of electronic ignition system, the mechanical contact breaker is replaced by a light sensitive pick-up in the form of an opto-electronic bistable trigger. This accurately senses the engine position at any speed and produces a constant crank angle dwell function in the same way as would a perfect set of cam operated contacts.

One example of this system is produced by Lumenition Ltd. In this system a difference in the ignition coil circuit is made from other systems by completely isolating the primary and secondary windings of the ignition coil.

The optical pick-up or triggering device used is a silicon photo-transistor, identical in performance to an ordinary transistor except that its base current is provided by light or more specifically, electro-magnetic radiation in the near infra-red region of the spectrum (9400 Å). A gallium arsenide solid state lamp is used as the source of the infra-red. The output from this trigger is fed into a unique design of amplifier incorporating a chain of four transistors rated such that the complete unit can operate over a wide temperature range and require neither electro-magnetic screening or protection.

The complete system comprises chopper, bistable trigger and power module. The trigger and chopper may be installed in a standard distributor in place of the existing contact breaker.

Another example of an Opto-Electronic ignition system has been developed by Piranha Ignitions Ltd. In this system the existing contact-breaker points, cam and capacitor are replaced by a frictionless, high speed, optically pulsed electronic switch which can be fitted into most standard distributors.

The photo-sensitive transistor used is exposed by apertures in a light scanning disc as it rotates. The output from this photo-transistor is used via an amplifier to switch a drive transistor controlling the ignition coil primary, thus providing the advantages of electronic ignition in a conventional coil ignition system.

CHAPTER 8

MAGNETOS

The magneto has now been completely replaced by the battery-coil or electronic ignition system on almost all types of spark ignited petrol engines. A few racing and stationary and marine engines still use magneto ignition but these are rapidly changing over to electronic ignition systems. 'Flywheel'-type magnetos are still extensively used on engines designed to drive small vehicles and power equipment such as light motorcycles, snowmobiles, lawnmowers, power saws, motor boats, atomisers, agricultural machinery etc. The magneto ignition system is self-contained and is independent of voltage source (external battery) and is relatively insensitive to high and low temperatures and can therefore be used in all climatic conditions.

THE DEVELOPMENT OF THE MAGNETO

The history of the development of the magneto industry in Europe is of considerable interest and goes back to 1898 when Simms and Bosch developed a low tension magneto. The design of this machine was of special interest as the form of armature was subsequently used with the addition of a secondary or high tension winding when the high tension magneto was evolved. It would appear that a Frenchman M. Bonderville was the first man to conceive a high tension magneto, but unfortunately omitted to use a condenser for eliminating sparking at the contacts, although this practice had been in use many years before on a Ruhmkorff coil, by a fellow countryman named Fizean.

The Bosch Company of Stuttgart subsequently established that a high tension magneto could be manufactured on a commercial basis to give reliable and efficient ignition in practice. Robert Bosch was the founder of the European magneto industry and up to the year 1914 the British motor industry depended almost entirely on the Bosch Company for its ignition apparatus. Also many of the principles laid down at that time are still in use.

SOME MAGNETO DESIGNS

An example of a motor-cycle magneto, fitted with automatic timing control is shown in Figure 8.1.

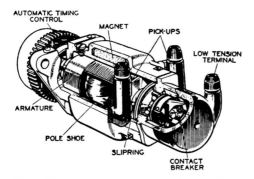

Figure 8.1   Lucas magneto with automatic timing control

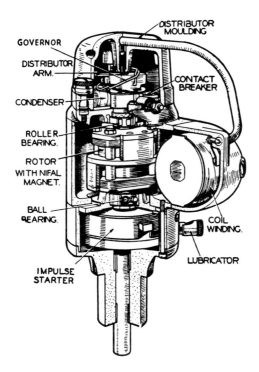

Figure 8.2   Lucas Type 6VRA (vertex) magneto

About 1931 the Scintilla Company introduced the Vertex magneto which was built as an axial unit with a vertical shaft to take the place of a distributor head as used in the conventional coil ignition system. This design gave quite exceptional performance for its weight, was also developed by other companies and found widespread use on engines such as fire pumps and field generators during the 1939–45 war and was later used in America on various forms of sports cars. Figure 8.2 shows the construction of a Lucas Vertex magneto and clearly indicates the design of this camshaft drive machine.

FLYWHEEL MAGNETOS

As mentioned earlier, this type of magneto is in current use on a wide range of engines. The magneto consists of two basic parts:

1. A rotating steel flywheel fitted with permanent magnets attached to the engine crankshaft by a hub member;
2. A stationary aluminium armature base-plate mounted concentrically with the flywheel and which supports the ignition armature, the contact breaker with its capacitor and a lubricating felt.

The contact breaker is usually operated by a cam track ground on to the flywheel hub, see Figure 8.3. Some designs have a separate cam ring or are triggered from a specially shaped cam track ground on the crankshaft. This type of magneto can also incorporate an a.c. generator for lighting purposes or for battery charging.

Operation

The ignition portion of a flywheel magneto generator operates in a similar way to the conventional permanent magnet magneto by producing the voltage necessary for ignition of the fuel-air mixture in the engine cylinder. The magneto works completely independent of any external power supply, the energy being derived from the permanent magnets built into the flywheel. The magnetic system comprising magnets of high coercive force and rectangular cross section which are arranged radially in the flywheel together with corresponding pole shoes.

Rotation of the flywheel produces a large number of reversals of magnetic flux to be induced in the primary winding of the stationary ignition armature. Only one such reversal per revolution is utilised to develop the ignition voltage dependent upon the actual construction of the magnetic system. When the breaker contacts are closed, an induced current flows in the primary winding of the ignition armature; at the instant of ignition this current is interrupted at its maximum value by the opening of the breaker contacts. As a result, the magnetic flux in the armature core immediately reverses direction and induces a high voltage in the secondary winding which produces the spark at the spark plug electrodes.

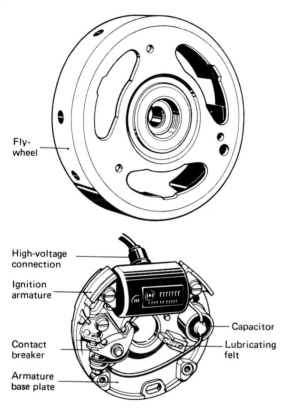

Fly-
wheel

High-voltage
connection

Ignition
armature

Capacitor

Contact
breaker

Lubricating
felt

Armature
base plate

*Figure 8.3   Bosch flywheel magneto showing the two major
components (1) flywheel (2) baseplate assembly*

It is general practice to use a single-lobe cam so that one ignition spark per set of breaker contacts is produced per revolution. Since the magneto rotates with the crankshaft to which it is fitted, this meets the requirements of the two-stroke engine.

So that the spark may be produced at the precise time required by the engine, it is essential for the flywheel to be accurately positioned on the crankshaft and also bear the exact relationship required with the opening of the contact points. A capacitor is connected across the contacts to suppress any arcing at the contact points and so greatly reduce contact erosion.

CHAPTER 9

SPARKING PLUGS

The sparking plug is a vitally important unit in any ignition system, as the initiation factor in petrol engine combustion. The efficacy of whatever form of spark generator is used is intimately dependent on the design, operating characteristics and durability of the sparking plug.

It is interesting to record that a Frenchman named Lenoir devised and constructed a sparking plug illustrated in Figure 9.1 which possessed all the main features of the modern sparking plug and also solved many of the problems encountered in meeting the sparking plug design and construction parameters referred to in Chapter 2.

SPARKING PLUG DESIGN

The many different classes of internal combustion engines now in use, and the divergence in the behaviour of individual engines in each class, make it necessary for sparking plug manufacturers to standardise a range of designs, suitable for giving the best results under the diverse operating conditions which have to be met in practice.

For trouble-free service the plug for any given engine should operate at certain definite temperature conditions, in order to avoid the two extremes of pre-ignition and fouling. Pre-ignition — or ignition by hot-spot effect before sparking occurs — takes place if any part of the plug reaches about 750° C at this stage in the stroke. Fouling — or deposition of carbon, other products of combustion, and ultimately even oil on the insulator, causing the electrical discharge to leak to earth instead of sparking across the electrodes — occurs if the average insulator temperature over the cycle drops to the region of 300° C. The optimum average cycle temperature for the insulator is in the range 500–600° C, which burns off sufficient deposit to ensure freedom from shunting over reasonable operating periods and at the same time avoids the danger zone of pre-ignition. Instantaneous plug temperatures will, of course, fluctuate during the engine cycle between the average.

Because engines vary widely between the extremes of, say, cool-running, low-efficiency, stationary types probably utilising paraffin or vaporising oil, and high-efficiency, hot-running sports and racing units, the heat developed per cycle which the sparking plug has to withstand varies equally widely. In designing ranges of plugs for general industrial use, therefore, manufacturers have to regulate by insulator shape, internal gas space variation, etc., the rate at which the plugs pass heat to the engine cooling system, in order to ensure that in service the correctly matched plug for any given engine will operate within the optimum temperature range mentioned above.

It is usual to refer to 'heat' or 'thermal' value in relation to sparking plugs, which in effect is a measure of each plug's ability to adjust itself to optimum temperature conditions, thereby relating one plug with another through the range. Plugs which maintain their proper operating temperature with relatively small heat input are referred to as *hot* or *soft* plugs. These are suitable for cool-running engines of low thermal

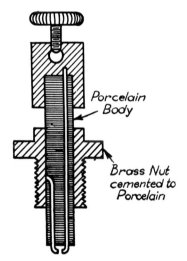

Figure 9.1   *The original sparking plug*

efficiency, and where below-normal conditions exist through deterioration. Plugs which maintain proper operating temperatures with relatively large heat input, and are therefore suitable for high-temperature, high-efficiency engines are known as *cold* or *hard* plugs.

Figure 9.2 clearly shows the relationship between plug operating conditions and temperature.

The form of plug classification, based on the ability of each type to conduct heat from the electrodes to the cylinder head cooling fluid is known as the 'Heat Range'. A rule of thumb summarising this may be

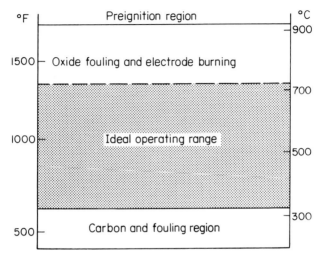

Figure 9.2   Plug operating temperature range

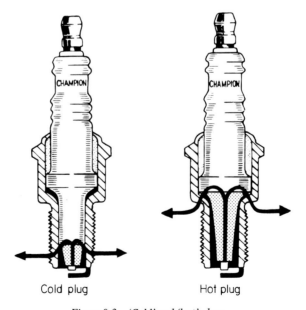

Figure 9.3   'Cold' and 'hot' plugs

expressed as 'cold' plugs for 'hot' engines; 'hot' plugs for 'cold' engines. Alternatively the higher the power output per litre of an engine, the higher the heat range of the appropriate sparking plug. Figure 9.3 illustrates the differences in heat conducting paths of a hot and cold

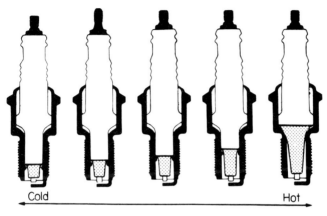

Cold   Hot

Figure 9.4

plug and Figure 9.4 shows constructional differences over the Champion range of plugs from cold to hot.

SPARKING PLUG REQUIREMENTS

From the foregoing it will be evident that the art of selecting the correct plug for a given engine condition lies in matching its heat value to the engine, such that the plug temperature over the complete engine operating range lies within the quoted optimum zone. In addition, however, there are certain broad requirements with which all plugs must comply, and as well as being matched to an engine for heat value, the ideal plug for the application should possess the following characteristics:

(A) Mechanical
   1. Gas-tightness.
   2. Durability.
   3. Adequate electrode projection, thread reach and length of barrel.
   4. Suitable electrode design and freedom from pre-ignition.
   5. Resistance to rust and corrosion.

(B) Electrical
  6. Resistance to current leakage and electrical puncture.
  7. Design of electrodes giving suitable electrical characteristics.
  8. Freedom from short-circuiting or misfiring.

These various requirements are considered in the following paragraphs.

(1) Gas-tightness

This is obviously one of the most important requirements of any plug, because gas leakage not only entails loss of compression and thereby engine power, but the escaping gases produce overheating and damage to the plug.

Unfortunately, all electrical insulators become more or less conductors at very high temperatures. With a leaky plug the region of low insulation resistance corresponding to these temperatures is quickly reached and the inevitable misfiring occurs. Admittedly, this problem of securing gas-tightness under the many differing service conditions is one of the most difficult confronting the plug designer. The general excellence of the product now available reflects great credit on the sparking plug industry.

(2) Durability

Durability in service can only be secured by the most careful selection and testing of the materials used in the construction, and by the adoption of a robust design.

(3) Adequate electrode projection, thread reach, and length of barrel

The rate of propagation of the explosive wave in the combustion chamber is a factor governing engine efficiency. The nearer the plug electrodes can be brought to the centre of the chamber, the better chance there is of the complete ignition process being speeded up. The worst possible location for a plug is in the corner of the chamber or in a pocket. The electrode projection of the plug must be sufficient to prevent the electrodes from being pocketed by the sides of the hole in the cylinder walls, but it must not be excessive, otherwise overheating is liable to result from the poorer electrode heat conductivity.

Thread reach varies widely from about 9 mm to 20 mm for popular types of plug. Prevalent reaches in different countries appear to have evolved through custom and practice rather than for basic technical reasons. In the USA the shortest reaches are the most popular and medium reaches are almost universal on the Continent, and also largely in England, although the latter has an increasing tendency towards long reaches.

There are various factors which can be argued as controlling reach, e.g. the shortest reaches make for quicker fitting and removal and the longest reaches allow cooling water passages to be brought closer to the plug; long reaches are also preferable in light-alloy cylinder heads from strength considerations. The length of the barrel must be such as to permit the easy use of a spanner without danger to the relatively fragile insulator.

(4) Suitable electrode design

The term 'pre-ignition' is used to denote ignition of the explosive mixture *before* the correct moment for ignition, as determined by the occurrence of the spark discharge. A contributory cause of pre-ignition is the presence in the cylinder, during the compression stroke, of glowing particles of carbon deposit ignited by the last explosion.

Pre-ignition may also result from overheated plug electrodes; that is, they may become incandescent and of sufficient temperature to cause ignition without the aid of a spark. The design must be such, therefore, that neither the insulator nor electrodes reach a dangerously high temperature under working conditions.

(5) Resistance to rust and corrosion

The high operating temperature, and the frequent exposure of plugs to the elements, make them susceptible to corrosion.

Plug bodies have been made of brass, but it is general practice to use steel and subject the body to a rust-proofing process.

(6) Resistance to current leakage and electrical puncture

This result is secured by using for the dielectric the best material, and so designing the insulator that adequate sections and ample creeping surfaces are obtained.

(7) Electrical characteristics of electrodes

The sparking voltage is dependent on the shape and distance apart of the electrodes, and on a number of other factors dealt with in Chapter 7. Sharp-pointed electrodes should be avoided to make the impulse ratio as low as possible, and for the very practical reason that gap erosion, at least initially, would be extremely high; in any case, the material used should have a high resistance to spark erosion and electrodes must be sufficiently robust to avoid mechanical damage.

The material of which the electrodes are made also influences the sparking voltage, because there is evidence to show that the electron emission from the hot electrodes — which is a factor controlling the voltage at which the spark discharge occurs is dependent on the nature of the material. However, the scope afforded by this aspect in practice is limited by other factors such as resistance to spark erosion, corrosion by products of combustion, thermal conductivity, etc., which have resulted in high nickel content alloys being almost universally used in modern plugs. Platinum and certain alloys with high platinum content are sometimes used for special purposes, particularly where long electrode life is required, as platinum has a very high resistance to erosion and corrosion.

Silver electrodes have also been used for special plugs with a high heat value.

In the design of the spark plug it is necessary to select the best compromise between the spark discharge voltage and thermal conductivity, burn-off characteristics and machineability of the electrode material. For example, silver has a melting point of 960° C (1760° F), but because of the high thermal conductivity of silver, the service temperatures of a silver centre electrode are far below this temperature. Furthermore, as a precious metal, silver has a low burn-off.

For premium performance for use on motor cycles, a spark plug has been developed by Champion having a small diameter Gold Palladium alloy centre electrode. This design requires a lower ignition voltage than the conventional nickel alloy type of electrode and more clearance is provided within the firing end of the spark plug which permits better scavenging of the fuel deposits and hence less fouling.

(7) Freedom from short-circuiting or misfiring

Freedom from short-circuiting is dependent on the maintenance of an air gap which, where possible, should not be less than 0.4 mm. A good average setting is 0.6 mm, but gaps as high as 0.8 or 0.9 mm are sometimes used, particularly where petrol economy through the use of weaker than normal mixtures is required, as such mixtures are more difficult to ignite and slower burning.

However, the precise gap used in any given engine must be chosen as a compromise between mixture ignition requirements and the voltage output from the ignition generator, with particular reference to that available at starting speeds. It must also allow for a certain amount of gap erosion so that a tolerable interval is given before removal of the plugs and re-setting of the gaps becomes necessary.

Figure 9.5   The effect of burnt electrodes on the spark gap

With all these considerations taken into account, gap settings as low as 0.3 mm or less are sometimes used for high-performance racing engines. The effect of gap erosion is well illustrated by the photomicrograph shown in Figure 9.5. Because of this the plug gap should be frequently adjusted to give a length within the limits specified.

It has already been stated that general plug design and matching to engine should be such that an optimum temperature range of insulator and electrode tips is maintained under operating conditions, in connection with which the design of the electrodes themselves should be such that they do not easily overheat. This is important not only from the aspect of avoiding pre-ignition, but from that of plug life, since chemical corrosion and spark erosion increase rapidly if the electrodes operate at temperatures above the optimum range. In any case, spark erosion occurs mainly from the electrode which bears positive polarity, and since the central electrode, surrounded as it is by an insulating medium, always tends to run hotter than the earth point which is in direct metallic contact with the engine mass and cooling system, ignition systems having an earth electrode of positive polarity tend towards longer plug life.

SPARKING PLUG DESIGN AND SERVICE CONDITIONS

Air-cooled engines − e.g. high-efficiency motor-cycle engines − generally necessitate the use of plugs capable of operating at higher temperatures than plugs for water-cooled engines, and must always be selected carefully as troubles associated with overheating are more liable to occur. Plugs for motor cycles, therefore, are of higher heat value than those for comparable car engines. Twin-cylinder in-line motor-cycle engines frequently give rise to two sets of plug conditions; the front cylinder is often in receipt of less oil than the rear, which can give rise to the front plug overheating while the rear one fouls. The use of different heat values of plug is sometimes necessary to overcome this.

Another phenomenon with twin-cylinder engines where magneto ignition is used is that the central electrode of one plug erodes, whilst the earth point does so of the other, due to the high-tension currents from the magneto being of opposite polarity. The overall life of each plug can often be extended by exchanging them between cylinders periodically, so that both electrodes, in turn, take their share of erosion.

The two-stroke engine is fired at double the usual frequency, and thus the plug has double work to do. As a result plugs in two-strokes cannot be expected to have more than half the life of those in four-stroke engines, and it is not unusual to find a life of considerably less than this, due to the considerably less benefit the electrodes receive from fresh mixture cooling.

A further problem often experienced with plugs in two-stroke engines is that of gap bridging, which invariably stops the engine. The basic cause of the bridging appears to be combustion of the residue from fuel and oil additives laid down by the action of the electrical discharge, the somewhat indifferent charging and turbulence characteristics of the two-stroke cycle providing insufficient scavenging action to clear the deposits away.

There is considerable variation between engines in this respect, and factors such as plug position in relation to cylinder head and porting appear to have considerable effect. So also does spark-gap output level from ignition generator, and to a lesser extent plug electrode geometry. In general terms, a wide spark gap, together with low high-tension current, subject to the obvious limitation of reasonable ease of starting, freedom from misfire, etc., tend to minimise gap bridging trouble. Interference-suppression resistors often help here, as they reduce current peaks.

AUXILIARY GAP TYPES

In some spark plug designs an auxiliary gap is introduced between the terminal post and the centre electrode. The series gap is vented to atmosphere via a hole drilled through the terminal post.

The purpose of this gap is to isolate the secondary winding of the ignition coil from carbon deposits on the insulator nose, which could otherwise form a shunt path and could leak away the voltage. The auxiliary gap gives the coil time to build up sufficient voltage to just break down the series gap and then because of potential available, to fire across the electrode gap instead of shunting to earth. This means that, in certain applications more cold fouling protection is provided.

THE SPARKING VOLTAGE

The voltage required to fire a spark plug depends initially on its design. The material, shape and disposition of the electrodes and their gap spacing are the most important factors, but the voltage requirement is also affected by the nature of the fuel mixture surrounding the electrodes, its pressure, temperature and state of turbulence.

The spark gap between the electrodes is pre-determined to give the most efficient operation under average conditions for individual engines and ignition systems. Under normal operating conditions, this gap increases due to electrical erosion and chemical corrosion, the normal rate being 0.1 mm per 1000 miles of driving.

A new spark plug with 0.7 mm electrode gap working in an engine under ideal conditions might require in the region of 8000—10 000 V to produce a spark. After 5000 miles of service with normal wear, the electrode gap would be 0.8 mm and the voltage required to fire this gap under light load conditions, might be 10 000—12 000 V.

When accelerating or under conditions of severe load, the pressure surrounding the electrodes increases and necessitates still higher sparking voltage — say up to 16 000 V, see Figures 9.6 and 9.7.

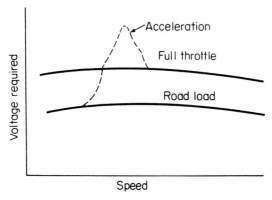

Figure 9.6   Speed/voltage curves

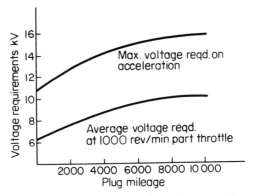

*Figure 9.7   Effects of spark plug mileage on voltage
requirement. It is apparent from the graph shown in
Figure 9.6 that the voltage required to fire the plug
during acceleration increases still further as spark
plug mileage accumulates*

Inter-relating voltage available and voltage required; it can be stated that the voltage available to fire the spark gap decreases with wear and faults in the system, while the voltage required by the spark plug increases over a period of normal operation. Ignition systems are designed to produce a voltage in excess of the highest potential demanded by spark plugs under all normal operating conditions and as long as the ignition voltage exceeds the plug voltage demanded, the system should operate effectively.

SPARKING PLUG CONSTRUCTION

As already mentioned in Chapter 2, the standard form of sparking plug comprises three essential components.

1. *The metal shell or body*, usually made from steel subjected to a rust proofing process.
2. *The insulator* made from ceramic material (manufactured under various tradenames such as Sintox, Pyranit, Corundite) based on aluminium oxide with additions of forms of glass. The material can be moulded or pressed to the required shape and then ground and part glazed and fired at high temperature. Insulators made from such materials have great mechanical strength, high electrical resistance and thus excellent insulating properties and high thermal conductivity essential for satisfactory spark plug operation.

3. *The insulated or central electrode* which usually comprises a steel stem with an electrode made of a special alloy, as already described. This is capable of resisting chemical corrosion arising from the combustion of leaded fuels and electrical erosion caused by the high voltages across the spark gap over the very wide temperature range at which the plug operates. The centre electrode has great influence on the optimum design of the scavenging area and the insulator creepage paths of the spark plug due to its geometrical arrangement in the insulator because heat transfer from the insulator to the colder zones in the rear takes place largely via the centre electrode and consequently can be supported by a high thermal conductivity in the centre electrode material.

DESIGN TRENDS

The general trend of sparking plug design over many years has been in the direction of smaller and more compact plugs resulting in the almost universal adoption of the 14 mm thread size, though 18 mm threads are still used on many large engines and particularly 2-stroke engines. Plugs of 12 mm and 10 mm diameter are used on several high performance engines.

It may be mentioned for historical interest, that the 18 mm diameter and 1.5 mm pitch spark plug was originally used by the De Dion Company of France and was regarded as the standard in this country. In America A.L.A.M. standard $^7/_8$ in diameter thread with 18 threads per inch was widely used, an exception being the Ford Company, which standardised an $^1/_2$ in taper gas thread. Most American engine manufacturers, including Ford, ultimately adopted the 18 mm standard which was largely superseded by the 14 mm diameter.

THE 14 MM TAPERED SEAT PLUG

An interesting and fairly recent spark plug development has been the 14 mm tapered seat plug. The overall diameter of the shell is much smaller than the standard 14 mm design and it has a conical seat instead of the usual gasket seat arrangement. This design of plug is exemplified in the Champion BN-9Y included in this range, compared with the standard plug which uses a gasket.

The ever increasing specific power of today's engines has led to the development of the tapered seat spark plug. By using this type of plug more heat can be dissipated by the engine cooling system, as heat dissipation is more efficient because an increase in the surface of the cylinder head in contact with the liquid coolant can be obtained.

As far as the user is concerned, the tapered seat plug requires more caution when installing it on the engine. Since no gasket is present to absorb any slight over-torquing, wrong installation will invariably result in damage to the threads of the cylinder head. It will be seen from Figure 9.9 that 91% of heat absorbed by the spark plug is dissipated to the engine coolant via the plug threads and plug seat hence the spark plug tightness not only serves the purpose of sealing the combustion chamber but also helps in keeping the plug cool. It is also absolutely essential for the taper sealing to be perfectly clean.

Regarding tightness, the Champion Spark Plug Company recommend the following rule of thumb:

Spark plugs *with gaskets* should be tightened to ¼ turn past finger tight to effect a gas-tight seal and to be sure not to damage threads.

Spark plugs *without gaskets* should be installed as follows:

1. Clean plug hole cavity preferably with compressed air before removing existing sparking plug.
2. Screw down taper seat plug by hand.
3. Use torque wrench to just secure plug. This represents a maximum of $\frac{1}{32}$ of a turn of a plug spanner. If a torque wrench is used a maximum torque of 1.1 mkg (8 lbf.ft is recommended).

BOSCH SPARKING PLUGS

To meet the exacting conditions already described, the current design of sparking plug manufactured by Bosch is shown in Figure 9.8. The spark plug consists of:

1. Leakage current barrier.
2. Terminal stud.
3. Pyranit insulator.
4. Shrink-fitted spark plug shell.
5. Special conductive seal.
6. Internal seal.
7. Permanent gasket.
8. Precision thread with guide.
9. Special non-eroding CR electrode (centre electrode).
10. Wider air gap between CR centre electrode and insulator nose.
11. Thin-walled insulator nose.
12. Scavenging area between plug shell and insulator nose.

SPARK PLUG WITH BOOSTER GAP

Booster gaps produce a sharp voltage increase on the plug electrodes. Therefore, misfiring as a consequence of conducting combustion residues on the insulator cannot occur. Plugs are available with a space of 2 mm between terminal stud and centre electrode; this serves as the booster gap which is vented by a bore in the upper part of the insulator.

The efficiency of a booster gap depends on the stability of its spark discharge voltage. It must be greater than the voltage requirements at the plug electrodes.

Figure 9.10 shows the voltage increase of a spark plug with increasing values booster gap. The ignition system must supply the necessary

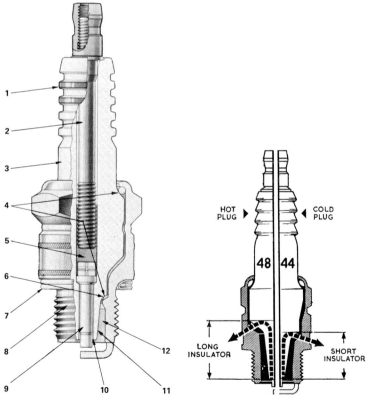

Figure 9.8   Cut-away view of spark plug

Figure 9.9   Illustrating the AC sparking plug heat range system

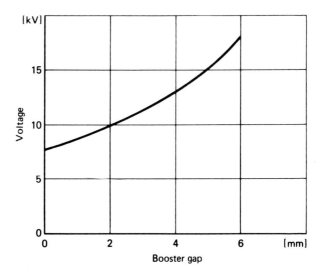

Figure 9.10   Increase in voltage requirements by booster gaps

ignition voltage for any operating state. Sometimes this requires special types of ignition systems, for example, transistorised systems.

AC-DELCO SPARKING PLUGS

A complete range of sparking plugs suitable for all purposes is manufactured by the AC-Delco Division of General Motors Ltd.

The A.C. heat range system facilitates the selection of a suitable plug by the identification marking on the plug. The first digit (see Figure 9.9) signifies the thread size, in this case 14 mm and the second digit indicates the heat range. The higher the second digit the hotter the plug will operate. Therefore, to cure persistent fouling the plug in use should be replaced by one with a *higher* second digit, or with a *lower* digit if there is overheating with tendency to pre-ignition or rapid electrode wear.

A notable feature of the A.C. sparking plug is that the 'hot tip' which is provided by a specially recessed thin section insulator tip designed to follow the engine temperature and so burn off deposits as they form without the risk of pre-ignition.

CHAPTER 10

ANCILLARY EQUIPMENT

The information in this chapter covers the wide range of ancillary equipment which is available for fitting to vehicles. This includes:

Cruise control equipment.
Tachographs.
Radio receivers and tape recorders.
The solenoid-operated actuator.
Reverse battery and high transient voltage protection.
Prevention of starter engagement whilst engine is running.
Testing and diagnostic equipment.
Anti-theft devices.
Drive unit for battery-driven vehicles.

CRUISE CONTROL EQUIPMENT

A 'cruise control' device is a system for attachment to a motor vehicle in order to maintain a pre-set vehicle speed regardless of road conditions and irrespective of wind or gradients. With such a device fitted, the driver is relieved of the effort of holding his foot at the particular angle on the accelerator pedal that is required to conform to a particular speed limit or to maintain a desired cruising speed. Long journeys on motorways or trans-continental trunk roads can therefore be made in greater comfort and with less driver fatigue. This makes for increased road safety and fuel economy largely due to the vehicle running as near as possible to optimum performance realised under steady speed conditions.

The system comprises a pneumatic actuator powered by manifold depression which is connected to the existing throttle linkage. An electronic control unit receives an input signal of engine speed or road speed, dependent upon whether the vehicle has manual or automatic transmission, this then adjusts the throttle actuator to maintain the set road speed. Controls consist of an 'on/off' switch and an engage/resume control. An electrical connection with the brake lamp switch is used to disconnect the unit as soon as the brake pedal is touched. The operation of the 'resume' switch brings the control unit back into operation.

An example of a British design of 'cruise control' device is the AED Econocruise by Econocruise Ltd, (a part of the Associated Engineering Group) which is described below. The author acknowledges the help of the manufacturers in providing the technical data quoted.

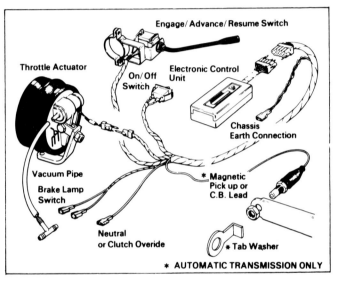

Figure 10.1   Components of cruise control system (Econocruise Ltd.)

In this design, which is shown in Figure 10.1, the pneumatic actuator unit is powered by manifold depression via a vacuum pipe, and the linear axial movement in this unit produced by variations in manifold depression is utilised to effect movement in the engine throttle linkage to which it is directly mechanically connected. On a normal installation it will provide full allowable throttle opening with a depression of about three inches of mercury. The air throughout is about 0.1 cu. ft. per minute at a steady 50 m.p.h. cruise. This increases to about 0.2 cu. ft. per minute at full 'advance'.

The speed pick-up may take various forms according to the type of transmission incorporated in a particular vehicle such as:

(a) Wire from ignition CB for vehicles with manual transmission.
(b) Opto-electronic speedometer cable sensor, for front wheel drive, and other suitable vehicles, usually with automatic transmission.

(c) Remote inductive pick-up module, with magnets strapped to the propeller shaft, for vehicles with automatic transmission.

(d) Inductive transducer, referred to as MAG, with magnetic tabs on propeller shaft, for vehicles with automatic transmission.

The 'OPTO' speedo cable pick-up is mounted either behind the speedo head or in the middle of a split speedo cable. Inside the pick-up, the infra-red emission from an opto emitter transmitting radiation to a photo transistor is interrupted by a vane driven by the speedo cable. This is arranged to give pulses at about the same frequency as those from the inductive pick-up.

The *remote* inductive pick-up can be mounted either close to the metalwork of the vehicle floor pan, under the carpet, on the propeller shaft tunnel, or under the rear seat squab. Two pieces of magnetic material are strapped to the propeller shaft, with a nylon tie. They are mounted in the general area beneath the pick-up and are arranged opposite one another to maintain shaft balance. The total weight added to the shaft is about 12 grams.

The MAG inductive speed pick-up is mounted on a rigid bracket, designed for the specific vehicle, so that it can sense two magnetic tabs as they move past it. The tabs, which are made from mild steel strip with patches of magnetised plastic tape, are bolted under opposite propeller shaft universal joint fixing bolts.

The electronic controller

This unit senses the input signal from the speed pick-up and uses this to accurately adjust the throttle actuator to maintain the set road speed. The frequency of the signal is usually of the order of two pulses for each revolution of the propeller shaft, or about 100 Hz at 50 m.p.h. A memory circuit in the controller stores the value of the input at the instant of 'Engage' and output pulses at about 20 Hz are fed to the actuator to hold the engine throttle at the required opening.

To cover a wide range of vehicles, the electronic controller unit is produced in three basic forms for use with the various types of speed pick-ups or speed sensors mentioned above, these are:

(a) CB
(b) + (c) OPTO
(d) MAG

Wiring to the controller is made via a multi-pin rectangular connector to a cable loom connecting the various components as shown in Figure 10.1. It is important to locate the controller behind or beneath the facia within the passenger compartment as it is not designed to operate in an under bonnet environment.

Control switches

The driver's controls consist of an on/off switch and a simple control switch for the engagement advance and resume function. When the control switch is put into the resume position after a disengagement, the memory circuit of the controller is activated to bring back the speed control to the pre-set speed. Figure 10.2 shows the control switch as fitted to a Ford Granada and the under-bonnet parts of the installation as fitted to a Jaguar XJ6 4.2 litre saloon is shown in Figure 10.3.

Operation

Operation of this system is extremely simple, its operating mode is selected either by a column mounted control switch or by two push-buttons, depending on the installation involved.

With the master switch in the ON position the car is driven to the required cruise speed and a single stroke of the control switch engages the system. The car will continue at this set speed until the foot brake

Figure 10.2   The Econocruise control switch on the Granada saloon

Figure 10.3 Under-bonnet installation of the Econocruise actuator fitted to the Jaguar X564.2 litre saloon

is applied whereupon the system becomes completely inoperative, known as the 'disengaged' mode. From the disengaged mode another new set speed can be selected, or alternatively a 'resume' feature can be activated whereby the car is accelerated smoothly back to its previous cruise speed.

An 'advance' facility enabling one's cruise speed to be progressively raised is obtained merely by retaining the control switch or push-button in the engage position.

THE TACHOGRAPH

This instrument is of very great interest to bus, coach and commercial vehicle users, particularly where specific legal requirements have to be met such as EEC tachograph regulation No. 1463/70.

A tachograph is a speedometer and mileage counter fitted with a clock and recording mechanism. Instead of filling in a log sheet the driver writes his name, the date and other information onto a 'chart' and inserts it into the tachograph which then does the rest. The chart, in the form of a circular disc, is rotated by the clock mechanism and is marked by three sapphire-tipped styli which bear against it. As shown in the diagram (Figure 10.4) the innermost stylus records the distance travelled. Every 10 km the stylus oscillates once so that by counting the peaks the journey length is measured. The middle stylus indicates whether the vehicle is moving and hence records hours at the wheel. By turning a knob the driver can also record how the rest of his time has been apportioned between other work, e.g. loading and rest periods.

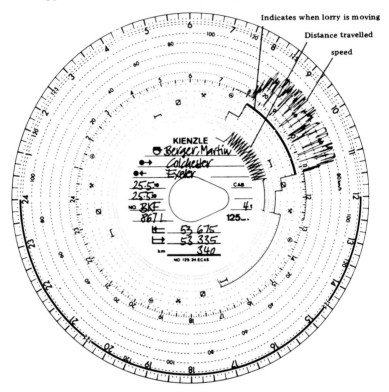

Figure 10.4   The Tachograph record card

The outermost stylus records speed and a jagged, as opposed to a smooth trace, indicates heavy use of the brake and accelerator. A warning light goes on when the pre-set speed is exceeded.

The Lucas Kienzle EEC tachograph

This is an example in extensive use in many countries. This instrument is a development from the earlier 'Journey Recorder' required by law in Germany as far back as 1939 on all buses in the interests of passenger safety and follows the practice of having recording instruments in aircraft and on many railway locomotives.

The instrument is produced in various forms; as a one-driver instrument or for a two-man crew, each arranged for indication and recording of road and engine speed or for road speed in km/h. On the combined (one and two-man) instrument, road and engine speeds are recorded on two identical charts.

Figure 10.5   The Tachograph showing line group selector

Figure 10.6 shows how a chart is inserted after opening the instrument lid and retaining clip and Figure 10.5 indicates how selection of a particular time group is achieved by operating the knob at the top of the instrument. At the start of a journey the driver has to set the operation knob to the position ⊘ (driving time) and at the end of each driving period the knob needs to be reset to the appropriate time group symbol.

⊘   = Driving time

▨   = All other work and presence for work time

⊢⌐ = Breaks and rest periods

The time groups recorded on the chart correspond to those in the log book (see Figure 10.7).

When the vehicle is stationary, all time group recordings are shown as a thin line. When the vehicle is in motion, the time group stylus for driver No. 1 vibrates and this records a broad line. Even though an incorrect time group may have been selected, the broad trace will be recorded indicating 'vehicle in motion'. The opening and closing of the instrument is indicated on a chart by short radial traces as shown in Figure 10.8 made by the time group and speed styli.

The 'function' warning light indicated in Figure 10.9 lights up when the styli are not recording on the chart or when no chart has been inserted. The warning lamp also lights up momentarily when opening and closing the instrument indicating that the warning lamp is in working order.

Figure 10.6   Inserting the card

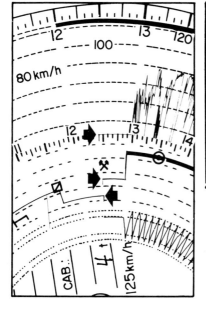

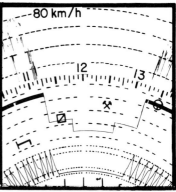

Figure 10.7 (left)
Figure 10.8 (above)

Two basic types of instrument are available, one driven mechanically and the other driven electrically. The mechanical type is driven from the vehicle's gearbox by means of a flexible shaft as shown in Figure 10.10 through adaptor gearing to give the correct instrument calibration relative to the vehicle's rear axle ratio and tyre size. The instrument is

Figure 10.9   Function warning light

operated by a mechanical (spring) clock mechanism. The electrical type is operated by pulse signals from a small multi-pole impulse generator driven by the gearbox speedometer drive pinion. This generator has easily changeable gears for calibration relative to rear axle ratio and tyre size to enable a calibration to be standardised at 1000 rev per mile/km.

The pulse frequency generated by the impulse generator is transmitted to the tachograph through a three-core flexible electrical cable, where it is converted in a frequency/voltage converter to a voltage which is linearly dependent upon frequency. This voltage is used to control a servo system comprising a d.c. servo motor coupled to a potentiometer and a servo amplifier. The servo system is connected so that the servo motor seeks a balance between the variable voltage from the frequency converter and the voltage from the potentiometer. In seeking this balance the system drives a speed stylus and speed indicating pointer.

A separate single impulse per revolution contact in the impulse generator is used to drive a separate servo motor system to give a distance travelled indication.

The chart drive clock mechanism has a combined electronic escapement and drive system which provides greater time accuracy than mechanical clock movements.

Connections are made between the impulse sender for road speed and rev/min and the back of the indicating/recorder instrument by

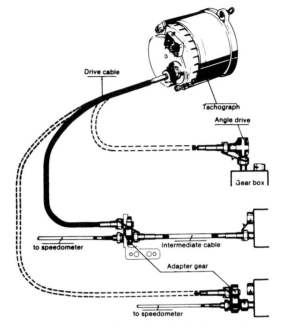

Figure 10.10   Tachograph mechanical connections

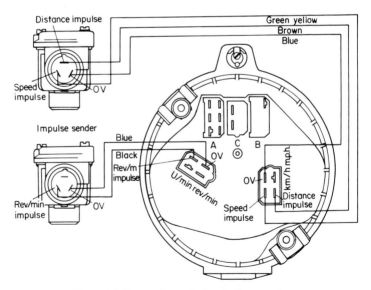

Figure 10.11   Tachograph electrical connections

multi-strand 1 mm² three-core cable with silver plated connectors and plugs and sockets fitted. The circuit diagram for these connections is shown in Figure 10.11.

A battery supply must also be provided for the dial illumination, warning lights, electronic clock and electronic system.

RADIO RECEIVERS AND TAPE RECORDERS

Many cars are now equipped with a radio receiver as original equipment and most manufacturers make provision for a radio on the facia board or on a centre console below the facia.

Stereo cassette or cartridge players are also in general use designed for under-dash mounting. Many of these are provided with slide type controls for volume, balance and tone. Such units may be combined with the radio receiver or arranged to operate through the same amplifier/speaker system as the radio, which may have full manual tuning or push button tuning, usually for one long and three or four medium wave stations. Some radios will receive VHF wavebands, which give a better quality of reproduction and stereo, but only in areas of good reception where the signal is not shielded by hills or buildings.

By using transistorised circuits, current consumption is kept down to about 1 amp or less and high quality powerful sound output is obtained. The physical dimensions for the average radio set is approximately 180 × 45 × 130 mm deep and for a combined quadraphonic and eight track stereo tape player average dimensions are 180 × 90 × 200 mm. It is general practice to fit an 'in-line' separate fuse in the supply lead to the radio/cassette unit. Most sets are arranged for 12 V negative earth.

Special equipment is available for public service and commercial vehicles and on coaches a robust metal boxed unit combined with public address equipment is generally used.

For car use, speakers may be fitted in the front fascia and on or adjacent to the rear shelf. Compact box enclosures containing separate speakers for bass, middle and top frequency may also be mounted on the rear shelf, and are sometimes coupled with fascia mounted speakers. An alternative, though now less popular system, is for flush mounted speakers to be mounted in the driver and passenger door panels.

Suppression of radio interference is now 'built-in' on most vehicles. This subject is covered in detail in Chapter 16.

Radio aerials

Various forms of aerials, both fixed and telescopic, are available. These can be mounted on either the front wing, boot or roof. For visual appeal and corrosion resistance, brass components are nickel and chromium plated. It is important for an aerial to be kept clean and free from corrosion.

To ensure maximum signal transference from the aerial to the radio receiver, low capacity high density copper braided co-axial cable should be used and standardised universal plug fittings for connection from aerial to receiver. It is also necessary to match the aerial to the receiver.

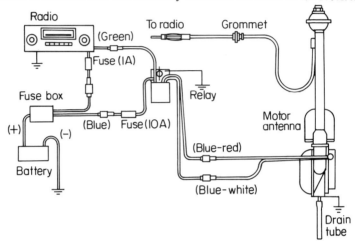

Figure 10.12   Circuit arrangement for motor-driven retractable aerial (Javelin Electronics Ltd.)

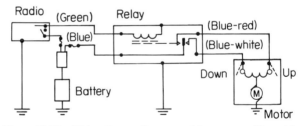

Figure 10.13   Wiring diagram for motor-driven retractable aerial

Motor-driven electrically retractable aerials are being used in increasing numbers, either switched manually by a two-way switch for 'up' or 'down' movement. Alternatively, the switching may be connected to the on/off switch of the receiver so that when the switch on the radio set is turned to the 'on' position the motor is energised in the direction to erect the aerial and when the radio is turned off the motor rotates in the opposite direction and the aerial is retracted. A clutch mechanism isolates the drive from the motor when the end of the aerial travel is reached. Alternatively, timers or limit switches are used.

The current consumption of a typical motor is about 5 A, it is therefore necessary to use a suitable relay for energising and changing the direction of the motor which should be separately fused and wired as shown in Figures 10.12 and 10.13. The time to raise or lower a retractable aerial may be 3—4 sec with some forms of drive and 6—7 sec with others.

THE SOLENOID-OPERATED ACTUATOR

Solenoid operated actuators are being used throughout the world on commercial and special purpose vehicles for a wide range of applications. The functions include: door and ramp actuation; operating hydraulic valves and fuel shut-off valves; applications requiring quick release and cut-off mechanisms designed for fail-safe operation.

The actuator consists of an electrically operated solenoid whose magnetic field draws a plunger inwards when the solenoid is energised. This plunger is provided with a spring loaded link end suitable for connection to the mechanism to be operated.

The CAV 263 solenoid actuator

This unit, which is produced for operation on 12- or 24-V d.c. as standard with alternatives up to 75-V d.c. The unit can be mounted in various ways and angles to suit particular installations and the design of the link arrangement permits end movements up to 6° in any plane to compensate for any small misalignment between the actuator and the apparatus being operated.

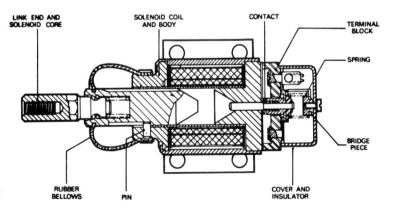

Figure 10.14   The solenoid-operated actuator

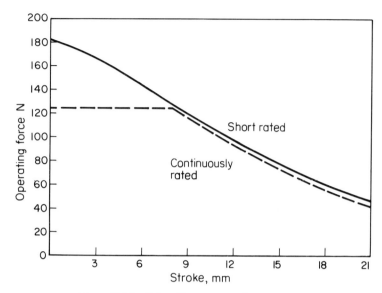

Figure 10.15   Solenoid actuator performance curves

Two basic forms of the unit are available; a short rated version for intermittent operation and a continuously rated version. The short rated version has a single high power pull-in winding and a plunger. It is designed mainly for straight ON-OFF operation since the winding should not be energised for periods exceeding 15 sec duration.

Design of the continuously-rated version differs in that, in addition to the high power pull-in windings, there is a low power economy hold-in coil and an internal set of contacts operated by the plunger as it approaches the end of its stroke. Opening these contacts by the plunger cuts off the current to the pull-in coil but leaves the low power hold-in coil energised thus retaining the plunger in the drawn-in position.

On both types of actuator the maximum plunger travel is limited by an internal stop pin. Normally they are not supplied with a plunger return spring so provision must be made for this in the external linkage. The tension of this spring and such installation factors as position (vertical, angular, inverted or horizontal), details of the linkage and whether or not the solenoid plunger pull is gravity assisted must be taken into account when calculating the solenoid loading.

The plunger travel may be up to 20 mm and the ambient temperature range is from −40° C to +65° C. A sectional view is shown in Figure 10.14 and the performance data for the unit is shown in Figure 10.15 with winding characteristics given in Table 10.1.

Table 10.1. PERFORMANCE DATA FOR SOLENOID-OPERATED ACTUATORS (LIGHT-DUTY)

Type	Rating	Volt	Pull-in winding characteristics at 20° C and nominal voltage		Hold-in winding characteristics at 20° C and nominal voltage	
			Ohm	Amp	Ohm	Amp
263/23	Continuous	24	1.3	18.5	39.0	0.62
263/24	Continuous	6	0.161	37.5	1.725	3.5
263/25	Continuous	12	0.45	26.6	9.2	1.3
263/31	Short	24	2.5	9.6	–	–
263/33	Short	12	0.513	23.4	–	–

The CAV 368 solenoid actuator

A heavy duty version of this type of actuator is the CAV 368 solenoid actuator shown in Figure 10.16. This actuator has a high power pull-in winding and a low power hold-in winding. When current is supplied both windings are energised and the plunger is drawn into the core of the solenoid against the pressure of the return spring. As the plunger nears the end of its stroke it operates a set of contacts which open to cut off the current supply from the pull-in winding. Current supply is continued to the low power hold-in winding to keep the actuator in the operated condition.

The actuator incorporates a delay device facility which can be used in addition to its normal function. If an external relay coil is connected across terminals B+ and AUX it will be short-circuited until after the

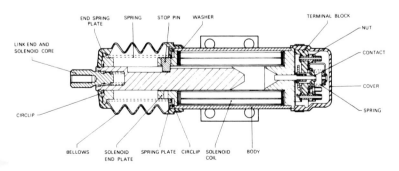

Figure 10.16   CAV 368 heavy duty solenoid actuator

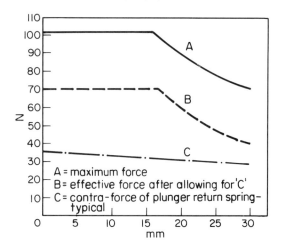

Figure 10.17 Performance characteristics of the CAV 368

solenoid has operated and opened the contacts. The relay coil is then in series with the solenoid pull-in winding and this limits the current.

This unit has a 30 mm plunger movement and the same operating temperature range and provision for misalignment mounting as the 263 model. Performance characteristics are shown in Figure 10.17.

REVERSE BATTERY AND HIGH TRANSIENT VOLTAGE PROTECTION

All vehicle charging systems are vulnerable to mis-use. Two important examples of this are battery connection reversal and generation of high transient voltages. When the battery connections are reversed, the six main rectifier diodes in the alternator are forward biased, applying virtually a short circuit across the supply lines. Depending on the current sharing, one or more of the diodes will be irreversibly damaged, putting the charging circuit out of action.

The second hazard to semi-conductors is the generation of high transient voltages. These can be generated by switched reactive loads and intermittent connections carrying large currents, especially if the battery is disconnected or in a poor state of charge. For example, peak voltages of up to 300 V, well in excess of normal component ratings can be generated.

These problems can be overcome by the use of a polarity conscious relay and a pulse clipper circuit. An example of such a device incorporating these features is the Butec RP4 reverse battery relay and pulse clipper. This unit incorporates a polarity conscious relay which opens the main return lead to the alternator, thus isolating the rectifier diodes from the battery. The charging system returns to a fully operational state once the battery is correctly connected. Reverse polarity protection is also offered to the electronic components of the alternator regulator.

The unit also incorporates a pulse clipper circuit which senses the presence of peak voltages in the charging circuit and momentarily loads the supply lines, using solid state switching to quench incident disturbances. In this way it safeguards the regulator and any other electronic device wired in electrical proximity.

The components are mounted on a printed circuit board and assembled into a metal chassis protected by a flame retardant cover. No adjustment or maintenance is required and all the electrical connections are made via screw type terminals.

Prevention of starter engagement whilst engine is running

On many private cars provision is made to prevent the starter motor being energised whilst the engine is running by incorporating a suitable relay in the starter motor/ignition key switching circuit. The problem is much greater on commercial vehicles and particularly on mid- or rear-engined vehicles where it is not easy to detect a slow running engine whilst the vehicle is stationary. If the starter is energised under this condition, damage to both starter and engine is inevitable.

A unit specially designed to overcome this problem is the Butec TSL1 starter lockout. This unit may be used with most types of 24-V starter and nine diode self-excited alternators. The complete system comprises the TSL1 lockout unit and an associated lockout relay. Complete protection to starter installations is achieved by introducing an inhibited period to the starter control circuit by way of an electronic timer coupled to the alternator and starter motor.

The components are fixed to a printed circuit board and assembled into a metal chassis protected by a flame retardent cover. The unit is factory adjusted and requires no maintenance and may be easily incorporated into existing starter installations.

TESTING AND DIAGNOSTIC EQUIPMENT

The importance of regular servicing for all types of vehicles is now generally recognised and there is a wide range of highly developed equipment available for this purpose. This equipment enables a complete analysis of engine performance and diagnosis of faults to be made and after any necessary adjustments, the engine may be re-checked and tuned to the maker's specification of performance. The equipment may take the form of miniature test meters or ancillary portable testing equipment or electronic test stands complete with a cathode ray oscilloscope.

On some types of performance analysers, diagnosis is made after study and interpretation of oscilloscope patterns and comparison of these with a standard trace. On other types, test results are shown on moving coil meters. Computerised electronic diagnostic equipment is also available on which the test results are shown on a computer print-out. This type of diagnostic equipment has been developed from in-flight aero engine monitoring systems.

Typical examples of such equipment comprise the following.

Test meters (T. I. Crypton Ltd.)

Miniature test meters mounted in moulded polypropylene cases with plastic windows and having shockproof movements to give long life for testing voltage and current regulated charging systems with scales 60—0—60 A and 400—0—400 A for static current testing of light vehicles.

For regulator and starter current testing on heavy commercial and public service vehicles, instruments scaled 500—0—500 A and 1500—0—1500 A.

For checking generator output, regulator adjustment, alternator charging systems, starter operation etc., high grade moving coil voltmeters with dual scales reading 0.3—0—3 V and 3—0—30 V.

Small portable multi-purpose instruments for on-the-vehicle tests incorporating a tachometer for engine speed indication, a dwell angle indicator and a voltmeter for 6- or 12-V systems. Larger pieces of equipment are available such as the Crypton Model 54.

The Crypton Motorscope

Figure 10.18 illustrates the Model 325 which is a highly developed electronic workshop test unit incorporating a cathode ray oscilloscope.

It is designed for programmed testing to cover all types of routine service tests and specific fault finding and trouble-shooting.

The items covered in a typical test sequence include:

Visual:	Fan belt, air cleaner, battery cables, cooling system, fuel system.
Voltmeter (not using remote starter):	Battery volts – ignition ON Coil SW volts – ignition ON (with or without ballast resistor) Coil SW volts – starter cranking Distributor voltage drop Coil SW volts – generator charging
Scope primary and dwell meter:	Coil and condenser condition, distributor contact condition, dwell overlap at 1000 rev/min, dwell angle at 1000 rev/min, dwell variation at 2500 rev/min.
Scope secondary:	Coil polarity, coil windings, coil h.t. lead, plug leads.
Scope kV:	Plug kV, rotor and lead kV, coil kV, acceleration kV.
Timing light:	Basic timing, centrifugal advance, vacuum advance.
Gas analysis:	Air fuel ratio or coemission at idle, acceleration and cruising.
Cylinder balance.	

The 'Autosense' test system (Brown Bros Ltd)

This is an example of a computerised test system which automatically checks such items as starting, charging, ignition, exhaust emission and other items, pinpoints their faults and decides what repairs should be made.

Controlled by a minicomputer, it also prints the test results, vehicle operating specifications and repair instruction codes on a report form which can be easily understood by the motorist as well as the operator/

mechanic. The unit can be programmed to test any petrol engined motor vehicle. It connects to the vehicle either through its own sensors or an adaptor plug to the vehicle's diagnostic socket where available.

The operator connects sensors to the battery, starter, coil and other engine parts. The sensors measure various engine conditions and send information to the computer. Using a hand-held controller with digital keyboard and displays, the vehicle identification and test sequence

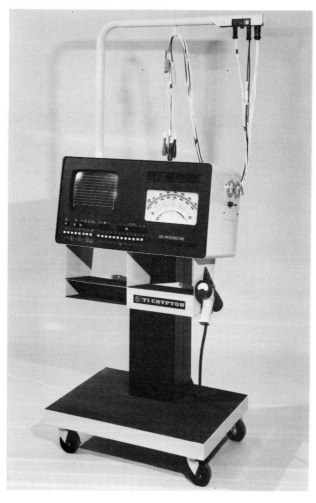

Figure 10.18   The Crypton Motorscope Model 325 (T.I. Crypton Ltd.)

numbers are entered into the system. The computer searches a tape cassette (its permanent memory bank) and transfers the pertinent data, including vehicle performance specifications, high and low operating limits and computer test and diagnostic instructions, to its temporary or 'working' memory.

The computer then takes control of the test. The operator runs the vehicle according to test requirements (ignition on, engine idle or at a certain revolutions per minute) and reads the digital displays to monitor the test. The computer, meanwhile, matches engine conditions against predetermined specifications and types its findings on the printout, marking with an asterisk those conditions that fail to pass the specification limits.

When the test is completed, the computer analyses the results and prints a diagnostic repair code. The operator looks up the number in a manual which contains repair and adjustment instructions. If a hidden problem is uncovered, he calls for additional testing to find and diagnose it. After repairs are made, the car is re-checked for verification purposes. Production engine testing may also be similarly automated with considerable accuracy.

It will be appreciated that the diagnostic equipment referred to may be used without affecting the performance of the various electrical components being tested. There is also available from many suppliers a wide range of instrumented workshop test benches incorporating variable speed motor drives and appropriate test equipment for specific items such as starters, alternators, fuel pumps, etc.

ANTI-THEFT DEVICES

There are various designs of anti-theft devices available which are easily fitted and are energised from the vehicle battery. Most units comprise a pendulum arrangement which operates the horn, should the car be tampered with, together with an immobilising circuit operated from a concealed or key switch.

'Selmar' alarm unit

In this equipment the operating mechanism consists of two balanced pendulums, one inside the other. When the car is rocked or moved, such as in an attempt to open a door, any slight movement causes the inner pendulum to touch the outer. This completes a battery/earth circuit which in turn operates the horn by means of a multi-contact electro-magnetic relay through a thermo cut-out.

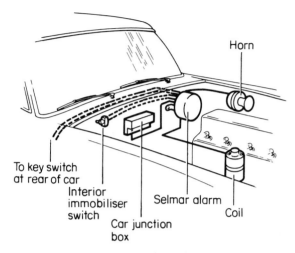

Figure 10.19   A typical 'Selmar' alarm installation

Adjustment is provided on the thermo-switch for duration of the time for which the horn sounds and the sensitivity of the pendulum device may also be adjusted.

The 'Watchdog' Model 5DL

This is an arrangement incorporating its own powerful siren which has a high-speed motorised rotor capable of projecting the alarm signal up to 1000 m (measured in still air). This unit is not connected to the vehicle's horns and should be accommodated in the engine compartment where it is not likely to be tampered with from either outside or underneath the vehicle. A variable finger and thumb control pre-sets the alarm to sound for a period from two seconds to two/four minutes. After this period, the alarm stops and re-sets automatically.

In this system, the control switch is a weatherproof multi-pin radial type lock accessible from outside the vehicle. Self-adjustable door contact switches protect all points of entry. Additional switches may also be linked to the ignition circuit to warn of unauthorised starting and an electrically operated fuel shut-off valve may also be connected to cut off the fuel supply whilst the alarm is 'switched on'.

'Inertial Alarm' System

Another example of an anti-theft device is the 'Inertial Alarm' system suitable for cars and also for use on high security vehicles, goods vehicles, containers and trailers. The device may be arranged to be activated when hitching or unhitching trailers.

The unit is energised from the vehicle's batteries and is set by an external security key operated switch. Using sensors of a design similar to the inertia switches described in the chapter dealing with switchgear (see Chapter 11), the system maintains full sensitivity to tampering with the vehicle or any attempted forced entry.

The inertia switches used ignore low frequency rocking and vibrations of vehicles such as occur in severe weather and traffic conditions, thus

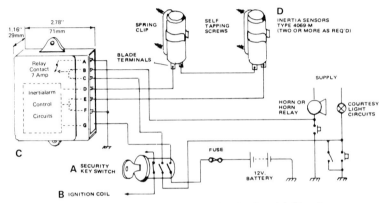

Figure 10.20   Typical installation circuit for the 'Inertial Alarm' system

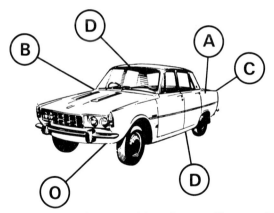

Figure 10.21   Showing position of sensors. The control unit C is connected via security switch A to vehicle fusebox O. The security key switch is also connected into the car ignition circuit B. The sensors D normally fitted behind the trim on door pillars are connected to the control unit

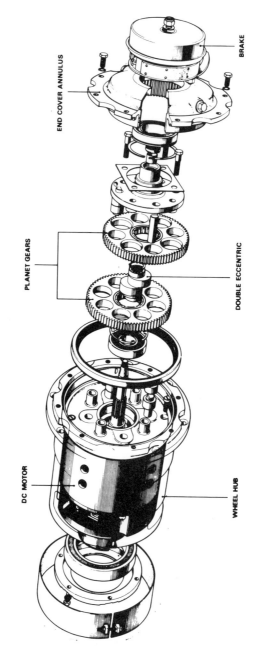

BRAKE

END COVER ANNULUS

PLANET GEARS

DOUBLE ECCENTRIC

DC MOTOR

WHEEL HUB

Figure 10.22   Exploded view of the 'motor-in-wheel' device (CAV)

false alarms are eliminated. Sensors are usually located at the base of the door pillars of cars behind the trim. The control unit may also be linked to courtesy switches fitted to vehicle doors for a further protection. This causes a continuous alarm whilst a door is open. With this system the vehicle's horns or a separate siren may be used and the automatic re-setting takes place after (typically) 10 sec and when the alarm is activated the ignition circuit is inhibited.

Figures 10.20 and 10.21 show a typical arrangement.

DRIVE UNIT FOR BATTERY DRIVEN VEHICLES

An interesting form of drive unit for battery vehicles with great potential for use with present-day batteries and also to run parallel with any future battery developments, is the 'motor-in-wheel' device.

The object of this device is to provide the maximum power at a driving wheel for minimum current consumption. The device comprises a motor and gear reduction mechanism within the hub of a wheel. The gyratory gearing provides a single stage gear reduction that eliminates frictional losses imposed by the conventional gearbox. Braking is pro-vided by a cable operated double-shoe brake acting directly on the motor armature shaft, the braking effect therefore being augmented by the ratio of gear reduction.

The general arrangement of such a device is illustrated in Figure 10.22 which shows the CAV design of motor-in-wheel. In this design the armature shaft of the motor is provided with double eccentrics phased 180° from each other. Each eccentric revolves within a roller race forming the hub of a large planet gear wheel which engages with internal gear teeth formed in the end cover.

The gearwheels are prevented from rotating by the restraining action of four pins fixed to the non-rotating motor housing. Surrounding each pin are two rollers located within circular holes in each gear wheel. The pins are eccentric to the roller centre by the same amount as the throw of the eccentrics on the motor shaft.

In consequence, as the motor shaft rotates, the planet gearwheels are oscillated by the eccentrics in a gyratory motion. This gyratory movement of the planet gear rotates the annular gear in the end cover by an amount equivalent to the difference in the number of teeth of the planet gear and the annular gear. For example, if the annular gear has 92 teeth and the planet gear 85, then the annular gear is rotated a distance equivalent to 7 teeth for each motor revolution – a reduction of 13 to 1. Because the planet and annular gears are so nearly equal in their numbers of teeth, the sliding movement between them is negligible and energy loss due to friction is minimal.

CHAPTER 11

SWITCHES AND SWITCHGEAR

Switchgear requirements

As the automobile electrical equipment increases in complexity, so the demand increases for switches and switchgear which will give all types of driver complete vehicle control under all environmental conditions. Many factors affecting switch operation, grouping and design are considered by the switchgear designer in great detail. These include human factors involving the driver's reaction in an emergency or under stress, location and grouping of switches to give the minimum of driver fatigue and visual distraction. Seat belt restraint has affected the

Table 11.1. NUMBERS OF SWITCH OPERATIONS TAKEN FROM
INSTRUMENTED PROVING LABORATORY CARS

Product	Operations per 1000 miles	No. of operations based on 50 000 miles	Durability test life target
Starter	492	25 000	50 000
Ignition switch	419	21 000	50 000
Brake lights	4317	220 000	250 000
Trafficator switch	1380	69 000	100 000
Headlamp switch	140	7 000	50 000
Dip switch	440	22 000	50 000
Wiper switch	99	5 000	50 000
Screenjet switch	48	2 500	10 000
Horns	456	23 000	50 000

These figures are from a I. Mech. E. Paper on *Automotive Switches and Switchgear* by L. J. Nevett of Joseph Lucas (Electrical) Ltd.

location of switch controls which must also comply with the overall styling of the vehicle facia panel and also meet safety requirements concerning potentially hazardous projections such as projecting knobs or levers which could be a danger to driver or passenger under impact conditions.

Experience has shown that the steering column is the most desirable location for the important signalling functions, including horn-operating push and windscreen cleaning. It is now general practice for these basic primary controls to be operated by two levers operating column-mounted multi-function switches. Details of specific examples of these switches will be given later in this chapter.

DESIGN CONSIDERATIONS

Reliability

Extensive development work is carried out and proving experience gained for long periods of time in order to minimise the risk of switch failures early in service and prevent costly redesign.

Table 11.1 gives statistics obtained from measurements made on specially instrumented cars on which the switched functions were carefully monitored. From such statistics durability targets are set for future practical designs.

The function and operation of the switch

A switch has been defined as 'a device for making and breaking the load current in an electrical circuit'. It may, therefore, be divided into two sections: the contacts which perform the electrical 'make and break' and the mechanical device which moves the contacts together or apart. There are of course a very large number of combinations of contacts and actuating mechanisms to meet the various switching requirements on motor vehicles and in addition to the factors affecting switchgear design already mentioned, the question of cost plays a very large part.

Contacts

The characteristics of contact materials are:

1. Freedom from surface films.
2. High electrical and thermal conductivity.
3. Resistance to electrical and mechanical wear.
4. Low cost.

The materials widely used are copper, brass, phosphor bronze and beryllium copper and where cost permits, silver and certain silver alloys.

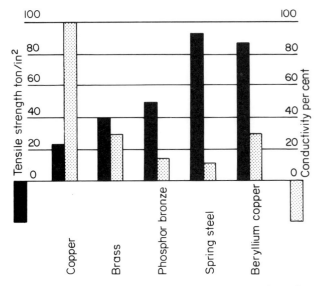

Figure 11.1 Comparative electrical conductivities and tensile strengths

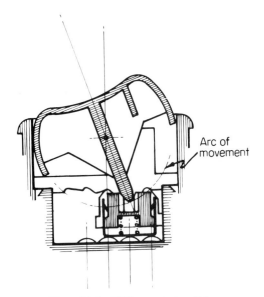

Figure 11.2 Sliding contact switch

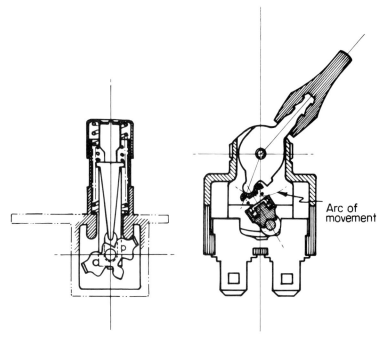

Figure 11.3 Quick-acting design need a fast dip switch

Figure 11.4 Spring-loaded roller-type toggle switch

Contact arc erosion may be considered the largest single problem in switchgear design and to overcome this, the contact material must be carefully chosen to meet the duty of a particular switch such as current-carrying capacity and size of contacts to give adequate thermal dissipation. Adequate contact gap and the correct control of contact movement must also be provided.

In most switch designs the contacts are urged together by some form of spring. When the spring itself is a separate component from the contact and does not carry current it can be made from steel suitably treated against environmental corrosion. When the spring itself carries current, then phosphor bronze is most frequently used, though beryllium copper has advantages of better electrical conductivity and spring properties, also hard rolled brass may be used. These materials are also suitable for the attachment of silver contact tips by resistance welding. Figure 11.1 shows the comparative electrical and mechanical properties of these materials.

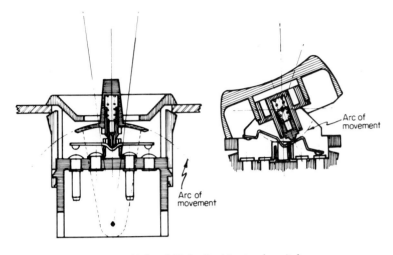

Figures 11.5 and 11.6   Rocking-toggle switches

Actuating mechanisms

There are numerous forms of actuating mechanisms all designed to ultimately close a pair of contacts. To provide efficient and reliable operation over a long period of service, a design of switch must be carefully selected which will meet the requirements of current level, type of load, environmental conditions and frequency of operation.

Figure 11.2 shows an example of a sliding contact type of switch which can be operated either by linear or rotary motion and Figure 11.3 illustrates a quick-acting design used as a foot dip-switch. In this design energy is built up via the operating lever or knob to a critical point past which it is suddenly released, moving the contact at high speed.

There are various ways of creating a quick 'make and break' movement with an over-centre device. This design is exemplified in the conventional spring-loaded roller type of toggle switch shown in Figure 11.4. Two forms of the rocking blade type of toggle switch are shown in Figures 11.5 and 11.6. In the design shown in Figure 11.6 there is no loss of contact pressure immediately prior to break which renders this type of switch particularly suitable for controlling inductive loads. Rocking blade mechanisms are popular with vehicle designers for the light and pleasant 'feel' to their action in addition to their electrical properties.

Multi-function column switches

This design of switch is exemplified in the Lucas Models 163SA and 164SA. These are both multi-function switches and a typical example is shown in Figure 11.7. Model 163SA column-mounted switch is intended for use on cars and light commercial vehicles and combines the functions of direction indicator switch, horn control, headlamp flash and headlamp main/dip beam changeover. It is suitable for steering columns up to 37 mm in diameter. A mounting facility on the switch provides for the fitment of a matching switch on the opposite side of the column to control wiper, washer and/or lighting operations.

The direction indicator switch is operated by movement of the lever in the direction of intended turn, the switch action being cancelled by an integral striker bush when the turn is completed. An intermediate position of the lever, with spring return, can be provided for lane-change signalling of the direction-indicator lamps.

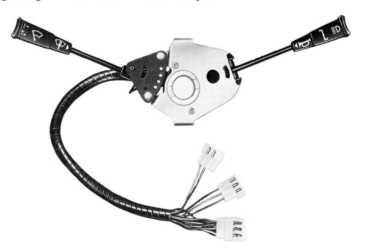

Figures 11.7   Lucas column-mounted switch

Movement of the lever in a plane parallel to the axis of the column selects headlamp main or dip beams. Additionally, an uppermost position with spring return to dip provides headlamp flash (main beams).

A spring-loaded push knob on the outer end of the lever operates the horn(s). Provision can be made for additional contacts to allow the nearside or offside parking lights can be used independently when the vehicle is parked.

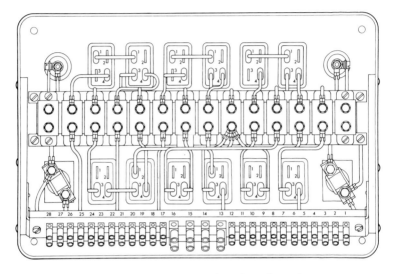

Figure 11.8   Rear view of CAV switch panel

TYPICAL TERMINAL BLOCK CODE

1 Generator Warning Light	10 Start Negative	19 Offside Wiper
2 Oil Warning Light	11 Starter	20 Heater Upper
3 Heater Positive	12 Stop	21 Nearside Wiper
4 Master	13 Battery Positive	22 Heater Lower
5 Start	14 Battery Negative	23 Head
6 Side and Tail	15 Starter	24 Offside Saloon Positive
7 Near & Offside Saloon Positive	16 Stop	25 Side and Tail
8 Nearside Saloon Negative	17 —	26 Nearside Saloon Negative
9 Offside Saloon Negative	18 Demister	27 Fog
		28 —

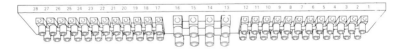

Figure 11.9   Terminal block arrangement of CAV switch panel

Ratings (12 V):

Direction-indicator	63 W
Main beam	250 W
Dip beam	200 W
Horn push	14 A

In Model 164SA the lever-operated switches are intended for use on cars and light commercial vehicles, to control windscreen wiper and electric screenwasher functions. The designs are suitable for use with two-speed self-switching permanent magnet field wiper motors; other types of motor could be accommodated, with the exception of those with depressed parking facility. The switches are designed to match indicator switch Model 163SA on which a mounting facility is provided. Alternatively, these switches can be mounted on the column independently by means of a suitable bracket.

The windscreen wiper switch is operated by angular movement of the lever in a plane at right angles to the steering column. Clockwise movement from the 'off' position gives continuous operation. Anticlockwise movement, with spring return to 'off', provides wiper operation for the period of switch actuation. This facility is known as 'flick wipe' since one flick of the switch is sufficient to give one complete wiping cycle.

On Model 164SA, the knob on the outer end of the lever is pushed towards the steering column. Spring return to 'off' is provided.

Ratings (12 V):

Windscreen wiper	5 A (20 A stalled)
Flick wipe	5 A
Screenwasher	2.0 A.

On commercial and public service vehicles, rocker type switches are used extensively for controlling all electrical services, each service being protected by fast acting self re-setting thermal (bimetal) type circuit breakers rather than wire or cartridge-type fuses. The switches and circuit breakers are arranged in a centralised single switch panel provided with pinch type terminals along its base for quick connection to the vehicles wiring harness and individual switches are clearly marked on a printed indicator below each row of switches. An example of such a switch panel produced by CAV is illustrated in Figures 11.8 and 11.9 which show the rear view and terminal block arrangement respectively.

REMOTE OR AUTOMATIC OPERATING MECHANISMS

Mechanical

Special forms of actuating mechanisms are incorporated within or additional to conventional forms of switches for the automatic operation of reversing lights when reverse gear is engaged. On such switches, which are usually gearbox-mounted and plunger-operated via the gearbox selector rods, it is obviously necessary to provide adequate protection from ingress of water and also prevent gearbox oil from entering the switch via the plunger bore by adequate sealing. Similar types of switch are also used to prevent starting in other than neutral gear.

Combination switches which incorporate both functions on automatic gearboxes are now usually mounted externally and operated via a shaft through the gearbox casing.

Pressure switches

Pressure switches are used on road vehicles mainly for warning lights indicating low oil pressure and for operating stop lights from the hydraulic brake system. Such switches are of the diaphragm type arranged to operate a pair of contacts as a function of pressure on a diaphragm. A typical example of an oil pressure switch is the AC Delco type shown in Figure 11.10.

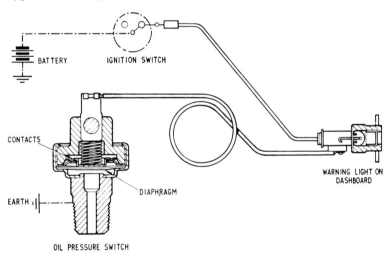

Figure 11.10   The AC-Delco oil pressure switch

It is customary to provide a green or amber light on the dash or instrument panel controlled by a pressure switch. When the ignition is switched on, the word 'oil' on a coloured glass is illuminated by the warning light. With the engine running, the oil pressure, if satisfactory, opens the pressure switch and the light goes out. Should the light remain on when the engine is running above idling speed, then the oil level should be checked, but it should be noted that this warning light is an indication of *pressure* and does not necessarily mean that the oil level in the sump is correct.

Relays

Relays are used in a number of automobile applications where it is required to control equipment taking a relatively heavy current from a lightly rated switching source. A relay can, therefore, be used for remote switching or to switch a function locally and so avoid voltage drop in long cables. A very common use of a relay is for switching twin wind-tone horns where the high current taken by the horns is controlled adjacent to the horns and the horn push button is located on the steering column. This arrangement avoids heavy cables having to pass through the steering column.

The relay is a simple form of electro-magnetic switch in which a relatively light current causes the wound core to become an electro-magnet and exerts a pull on an iron armature and so activates contacts switching a relatively high current. A relay may be wound for intermittent or continuous rating and single or multiple contacts may be arranged to be normally open, or closed, or to provide change-over circuitry. To minimise overall size, various contact materials are used such as silver, or dissimilar materials like nickel silver alloy/copper. The latter combination being used for heavy tungsten filament lamp loads.

ELECTROMAGNETIC DEVICES

Solenoids

The widest use of the solenoid in automobile applications is for the operation of the starter motor. In its simplest form used with an inertia starter, it is an electro-magnetic switch. When operating a pre-engaged starter as described in chapter 4 it performs the mechanical operation of engaging the pinion before energising the starter.

In starter applications, the solenoid must be capable of switching from the normal level of running current to the levels encountered during stall conditions for any type of starter, which could vary from 250 A to approaching 1000 A. The contacts are designed with sufficient bulk of material to allow rapid dissipation of heat. To ensure that contact welding does not take place without impractically large contact areas and high pressures, a feature known as inertia break is incorporated in all heavy duty solenoid designs. After initial contact is made when the solenoid is energised, the plunger is allowed to travel further before the magnetic gap is totally closed. A contact follow-through spring takes up this extra travel and applies pressure to the moving contact in so doing. When the solenoid is de-energised the follow-through spring pulls the shouldered plunger rapidly back causing it to impact against the moving contact, so breaking any welding that may have taken place (Figure 11.11).

Since starter solenoids are usually mounted in very exposed positions they must be proof against water, oil and dust ingress and be adequately protected against salt corrosion.

Follow through or inertia
break spring

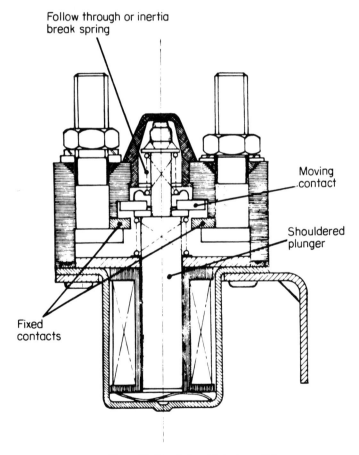

Moving
contact

Shouldered
plunger

Fixed
contacts

Figure 11.11   Solenoid starter switch

Solenoids are also used in differing forms and sizes to actuate over-drive systems, electrical door locking, in emission control systems and for engine manifold venting to prevent run-on, etc.

In some applications, continuous operation rating is required where space and cost is at a premium. In these cases two windings are employed. The first a short-rated one designed with adequate pull/stroke characteristics to perform the specified function. This winding is automatically switched out when the plunger travel is completed and a smaller continuously rated hold-on winding takes over.

Inertia switches

An interesting switch application developed in the interests of road safety and the prevention of fire after crashes, is the inertia type switch. This device is an inertia sensitive switch fitted in the electrical supply to a vehicle's fuel pump. Should a vehicle be subjected to heavy impact forces the switch opens isolating the fuel pump, ensuring that fuel is not pumped into a potentially dangerous situation. The switch can be re-set by pressing a button in the top of the switch accessible through an aperture in the top of the switch cover. The location of the switch should therefore be such that re-setting can easily be done.

On vehicles fitted with central door locking, in the event of a crash an inertia type switch supplies current automatically to the solenoids or motors thus releasing the electric door locks to enable the car doors to be opened or operate a hazard warning. Such a switch is also used as a seat belt sensor when it is used to sense deceleration or acceleration of 0.4 g and will pass a current to operate a solenoid thus locking the belt. A similar application also exists on commercial vehicles where retractor belts are fitted; mechanical sensing is too sensitive with the belt responding to normal driving, cab sway etc. This is overcome by an inertia switch fitted lower to the axle line and provides more freedom for the driver but still retaining maximum safety.

A basic design developed by Inertia Switch Ltd. is shown diagrammatically in Figure 11.12. From this diagram it will be seen that the switch mechanism is operated by a spherical mass normally seated in a conical cavity and restrained by a permanent magnet. When a specified velocity change occurs within a short time, such as under crash conditions, the magnetic 'hold' is broken and the switch mechanism is operated. By varying the cone angle, the magnetic restraint and the mass/geometry relationship of the ball and switch mechanism, a wide range of operating conditions may be satisfied for different types of vehicles. The main body of the switch in which the conical seating is formed and which also houses the magnet, is a plastic injection moulding produced to close tolerances necessary for consistent switch operation and providing a rapid response to crashes, but with a safe threshold against spurious operation. The polar response of the ball and cone system may be varied to produce differing sensitivities to impacts in the various horizontal directions. This is readily achieved by varying the cone angle in different horizontal directions.

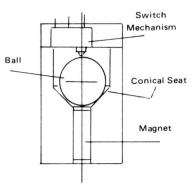

Figure 11.12   Inertia   door   lock switch

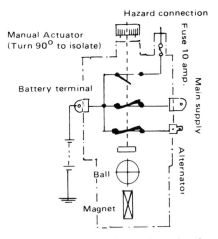

Figure 11.13   Wiring diagram for an inertia switch

Typical impact response characteristics for this type of switch is for the switch to operate as a result of a velocity change in any direction of 15 ft/sec (16.5 km/hr) and impacts of $5\frac{1}{2}$ g $\pm$ 1 g (maximum threshold 9 g or 15 g for specific applications). This type of switch may also be used to automatically isolate the vehicle's electric supply and may also be arranged as shown in Figure 11.13 to simultaneously isolate the electrical supply and automatically connect the hazard warning system. Under such conditions the current ratings would be:

Normally closed battery/ground contacts at 100° C, and	up to 160 A continuous
	up to 200 A four minutes
	up to 300 A two and a half minutes
Normally closed alternator contact	up to 6 A or 10 A continuous
Normally open hazard contact (fitted with 10 A cartridge fuse)	10 A continuous

BI-METAL SWITCHES

These switches are now being used in an increasing number of applications in automobile electrical equipment because of their inherent simplicity and consequent reliability. Also because no actuator or mechanical device, which could involve ergonomic considerations, is required to open or close the switch contacts.

The basic principle of this type of switch is the flexing of a bi-metal strip with a change in temperature. The temperature change may be caused by the passage of current through the bi-metal element or by a change in temperature in the immediate environment in which the switch is located.

Such a design of switch is exemplified by a range of switches, thermostats, thermal cut-outs and miniature circuit breakers produced by The Otter Group of Companies, all of which are produced to the same basic design concept.

PRINCIPLES OF THE OTTER THERMOSTAT

These switches are very sensitive to temperature changes because the relatively large area of thin bi-metal flexes across its width as well as along its length (Figure 11.14).

A built-in snap action is obtained because the crimp draws the outer legs inward, putting an 'over-dead centre' stress in the blade. This is overcome by the flexion when subjected to temperature change. The bi-metal tends to reverse its curvature as its temperature is increased until, on reaching the over-dead centre point, the snap action causes a complete reversal of the original curvature (Figure 11.15).

A high-contact pressure is also obtained because the action of crimping in the snap action forces the bi-metal contact upwards (see Figure 11.16). When assembled as a thermal switch, the fixed contact then forces the bi-metal contact of the 'centre leg' into line with the

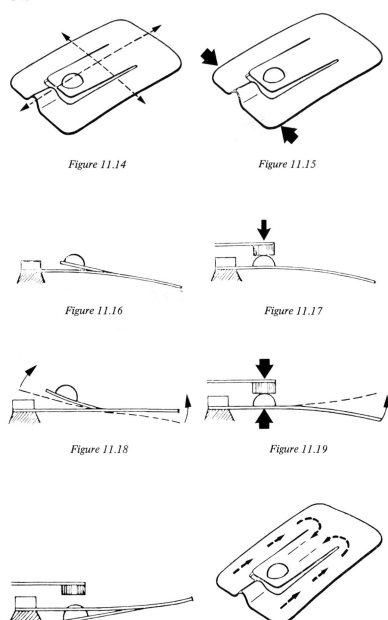

Figure 11.14

Figure 11.15

Figure 11.16

Figure 11.17

Figure 11.18

Figure 11.19

Figure 11.20

Figure 11.21

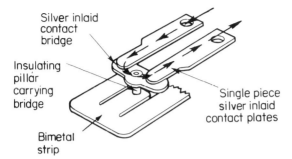

Silver inlaid
contact
bridge

Insulating
pillar
carrying
bridge

Single piece
silver inlaid
contact plates

Bimetal
strip

*Figure 11.22   The Otter thermostat. Current by-passes the
bimetal blade thus eliminating self-heating*

outer legs of the bi-metal so providing extremely high relative contact
pressures, see Figure 11.17.

This arrangement ensures that the highest contact pressure is at the
point of operation. Reference to Figure 11.18 will show that as increasing
temperature causes a change in curvature of the outer legs of the
operating blade, the contact carrying 'centre-leg' is also rising towards
its original unconstrained position, but is of course restrained by the
presence of the fixed contact, such that the contact pressure is steadily
increasing, see Figure 11.19. When the over-dead centre point is reached
by the bi-metal the contacts then snap apart, see Figure 11.20.

Reference to Figure 11.21 shows the path of the electrical current
through the relatively thin operating blade. If required, the bi-metal
can have a relatively high electrical resistance such that the passage of
an electrical current around the path indicated in Figure 11.21 causes
self-heating to occur. By this means the blade is capable of sensing the
current flow to an electric motor or transformer and can break the
electrical supply under current overload conditions due to self-induced
heat and if necessary without any rise in the temperature of electrical
windings etc, or the surrounding air.

The principle of self-heating has the advantage of simplicity in that
consistant and reliable operation is obtained without heat input from
an external source.

For applications where the effect of self-heating is not required the
switch construction is as shown in Figure 11.22.

PRACTICAL APPLICATIONS OF BI-METAL SWITCHES

Current/thermal sensitive safety cut-outs

The Otter V41 current/thermal sensitive safety cut-out was designed
specifically for the protection of permanent magnet automotive electric

motors, in particular, windscreen wiper motors. It is, however, equally suitable for the protection of any other type of fractional horse-power motor where 'on-winding' protection is either difficult or impossible. This design of cut-out is current sensitive, but the cover is designed to allow a substantial pick-up of radiant heat from the motor armature, thus making the cut-out thermally sensitive also. Current rating for this type of service is 20 A max continuous on 12/24 V d.c.

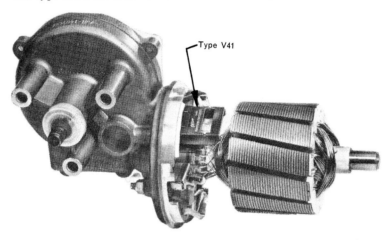

Figure 11.23   Otter Type V41 thermostat fitted to an AC Delco windscreen wiper motor (Courtesy General Motors Ltd.)

Figure 11.23 shows this type of cut-out providing windscreen motor protection. Other similar applications are:

Seat adjustment motor protection.
Electric window lift motor protection.
Electric door locking system protection.

Manual reset thermal cut-outs or miniature circuit breakers

The Otter type R15 is designed for current sensitive applications and is widely used for protection of wiring harnesses for buses and trucks and the protection of a wide variety of circuits. Individual units suitable for manual re-setting are usually grouped together such as on the distribution panel of a bus.

A heavy duty design Type L50 suitable for 30 A max. continuous rating is shown in Figure 11.24. This unit incorporates 'trip-free' ball latch mechanism for manual re-setting.

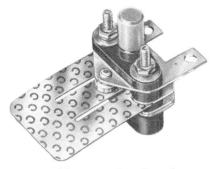

Figure 11.24   Heavy duty thermal cut-out

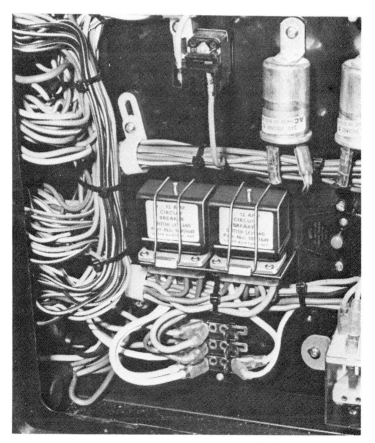

Figure 11.25   Circuit-breakers installed in a car

The bi-metal element and also the strip carrying the fixed contact are sandwiched between self-extinguishing high temperature resistant thermoplastic mouldings. The complete assembly is held together by two setscrews which also serve as fixing screws. Both fixed and moving contacts are composite rivet types having fine silver facings. When operated, the contacts are held apart by a steel ball until re-set by the push button which is designed to protrude through a panel for convenience in re-setting.

The Otter R3 range of circuit breakers offers distinct advantages over normal fusing and virtually eliminates possible fire damage owing to fuses being incorrectly 'repaired' or replaced by wires capable of carrying excessive currents. Figure 11.25 shows this type of circuit

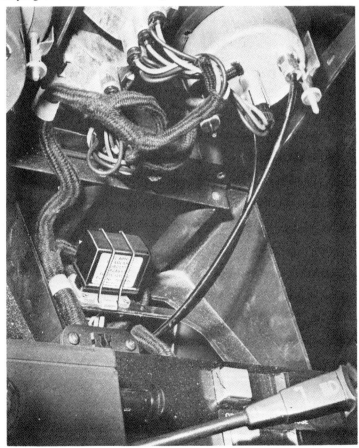

Figure 11.26   Circuit-breaker linked to the wiring harness of a bus

breaker fitted in a car. These plug-in type circuit breakers are produced in boxes of four; each box may contain similar or different current ratings at any rating between 5 and 20 A continuous. Each unit may be manually reset and may be fitted on a vehicle so that neater and more economical cable groupings and runs can be effected.

Headlamp safety change-over circuit breaker

The Otter VAO is a headlamp safety change-over circuit breaker designed to provide alternative lighting in the event of a short circuit occurring in the headlamp wiring.

The alternative source to main beam may be either the dipped beam or, if desired, some other form of auxiliary lighting such as spot lamps. In either case, the automatic changeover facility works to and from both circuits. As the changeover to the alternative circuit is automatic and instantaneous in the case of a short circuit, the driver of a moving vehicle cannot suffer the dangerous and hazardous experience of suddenly being plunged into darkness.

The headlamp change-over unit is designed to fit into the same box as the range of R3 circuit breakers and will also fit the same socket. Construction is of nickel plated 3 mm steel pins, integrally moulded in self-extinguishing high temperature resistant thermoplastic. Silver contacts are used and the units are capable of carrying up to 20 A and of breaking short-circuit currents in excess of 100 A.

Figure 11.26 shows this type of circuit breaker linked to the wiring harness of a bus.

Coolant or air temperature sensing

The basic bi-metal switch already described is incorporated in a sealed metal enclosure to form a very compact switch for sensing coolant temperature and switching on the electric cooling fan.

Other applications on which this type of switch is used in various forms are to operate emission control equipment on petrol engines and also operating automatic chokes, choke warning lights and in some cases switching on carburettor heaters to assist cold start operations. Also for sensing floor temperatures to give warning of overheating in areas around catalysts thus preventing grass fires starting if the car is parked at the time.

A further example of a bimetal type of switch also referred to as a thermostat is shown in Figure 11.27. This switch is used for automatically switching an electric radiator cooling fan and operates through the 'over-centre' flexing of a temperature sensitive bimetal disc. Axial movement at right angles to the plane of the disc operates a pair of contacts via a guided transfer pin. The electrical elements being located in a waterproof switch chamber. The contact operation is adequate to

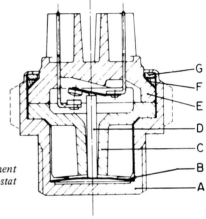

Figure 11.27   Sectional arrangement of bimetal switch or thermostat

handle up to 15 A at 12 V and the switch may, therefore, be used to switch a motor driven fan without any intermediate relay. Normal temperature calibration is between 65° C and 110° C.

SWITCHGEAR APPLICATIONS

An interesting application of solenoid-operated mechanisms is the electro-magnetic door locks fitted to many vehicles whereby all four doors are locked or unlocked simultaneously from one single door. Provision is made for locking or unlocking of all four doors from outside or inside from either front door. Such an arrangement also linked with both inertia switches and thermal cut-outs is now in general use on a range of vehicles.

CHAPTER 12

WIRING HARNESSES

The history of vehicle wiring goes back to around 1910 when some form of short circuiting wire and switch had to be provided for the ignition magneto to stop the engine from inside the vehicle.

The wiring has always been an integral part of the complete electrical system and has evolved together with such equipment as the sterter motor, battery, lighting, generator, windscreen wipers, heater, etc. This wiring forms the means of controlling the various functions essential for starting and running the engine and providing the requisite connecting links between the various components of the electrical system. Failure of the wiring or any of the associated equipment may result in complete immobility.

Before 1914 the wiring was only required to supply power to the ignition, lights and horn. This was usually carried out with twin cables insulated with rubber and covered in cotton braid and where protection against chafing was required the cables were also metal braided or inserted in metal tubing. The cables were individually routed between the components they linked and the bared cable ends were usually connected to terminal posts on the components. Twin cables were replaced by single cables when it became established practice to use the vehicle's chassis as an electrical return.

As more electrical equipment was developed the wiring became much more complex and to meet vehicle production requirements cable looms or wiring harnesses appeared. These were built up on wooden jigs suitably marked to show the positions of the various branches and individual cables where they emerged from where the groups of rubber insulated cables were braided together with black cotton braiding and lacquered. Coloured sleeves or end taping was first used for cable identification which was later superseded by braiding the individual cables in each of the basic colours and also incorporating a second tracer colour in the braiding.

Problems were encountered in tropical countries, when braiding cables together, owing to fungus growths. These were overcome by replacing the cotton braiding with p.v.c. tape, which was spirally

353

wrapped round the groups of cables to provide protection from abrasion. Later p.v.c. insulation was generally adopted for the insulation of individual cables in basic colours and also with tracer colours included. The use of p.v.c. insulation enables bare copper conductors to be used. Prior to this the conductors had to be tinned to prevent any chemical reaction between the sulphur content in the rubber insulation and the copper of the conductors. Further advantages of p.v.c. over rubber are that this material is impervious to petrol, oil and other corrosive fluids to which it might be subjected, also it is non-combustible thereby lessening appreciably the risk of fire on unfused circuits.

On early looms, ring-type connectors or terminal posts were used on the various components and in-line connections were made by 'male' bullet-type terminals inserted into insulated 'female' metal sleeves. By the early 1950's the ¼-in blade terminal became widely accepted. This arrangement is now used for in-line connections or plugged directly into various types of equipment and the range of blade sizes was extended to cover from $\frac{3}{16}$ in to $\frac{1}{2}$ in with matching receptacles. This enables a wide range of electrical loadings to be catered for from small indicator lamps to alternators. Multiway terminations can also be made with blade connectors and where these are used in rigid mouldings, risk of incorrect connection is minimised and a pressure or latching device may be incorporated to prevent accidental disconnection.

Looms tended to become both very large and complex and difficult to handle on the assembly line. Multiway connections either blade or round pin enabled the complete cable harness to be broken down into modules or separate loom sections. This greatly simplified assembly and enabled sections of a vehicle wiring to be easily replaced after damage in service at a much reduced cost.

PRACTICAL APPLICATIONS

There is a natural breakdown of the complete wiring harness into main circuit groups or blocks which are interdependent and are also arranged to serve a specific circuit or vehicle area. All cables must be arranged so that they reach their destinations with the shortest run and thus avoid excessive voltage drop and are also of the required current carrying capacity. Such a wiring breakdown is shown in Figure 12.1. These groups may be listed as:

1. Battery/starter motor circuit.
2. Alternator (or dynamo) battery charging circuit.
3. Circuits essential for running the engine such as ignition system, electric fuel pump, charge-warning light, oil pressure warning light and instrumentation.

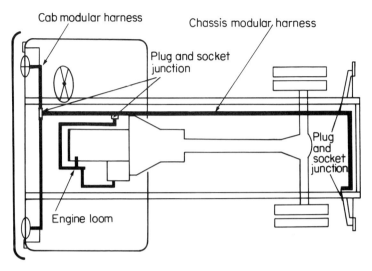

Figure 12.1   Diagram of cable harness grouping

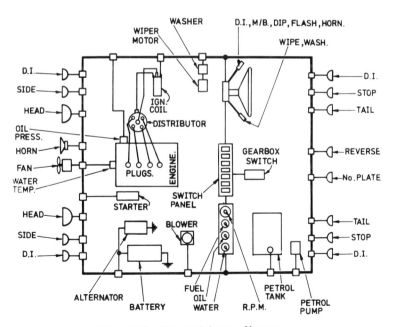

Figure 12.2   'Ring main' type of harness

4. Components required when engine is running such as stop lamps, direction indicators, windscreen wipers, screen washers, reversing lamps, heater blower motor, flashers and horn circuits.

5. Head, side, rear and panel lighting circuits fed from the battery through individual switches.

6. Other circuits required when the ignition is switched off. Interior lighting, cigar lighter, boot light.

An alternative arrangement is to incorporate a 'ring-main' type of harness as shown in Figure 12.2 in which a single heavy-gauge busbar-type of power cable is installed around the vehicle and the various functions are switched or operated by remote relays picking up power from this ring-main. Future trends may also provide for the switching of the individual components by a pulse or frequency operated electronic signalling system. Such systems are referred to as time-shared multi-plexing systems which will provide for various operations by means of sophisticated electronic circuitry as:

(a) Vehicle condition monitoring (VCM). This takes care of component failures such as lamps, etc.
(b) Diagnostic facilities. This will enable diagnostic equipment to be easily plugged in to the vehicle for pre-service or fault checks.
(c) Engine management.
(d) Anti-skid control systems.

LOOM CONSTRUCTION

Circular grouping of cables and flat strip wiring

Looms constructed from p.v.c. insulated cables and grouped in further p.v.c. insulation are in world-wide use and also in a flat form configuration composed of equal and multi-diameter cables which are laid parallel to one another and welded to a flat plastic backing strip. This type of wiring harness is manufactured by Rists Wires and Cables Ltd. under the registered trade name 'Fabrostrip'.

The arrangement has the advantage that it can be laid in areas such as behind trim panels hitherto inaccessible when using a circular wiring harness. Better heat dissipation can also be obtained than with the more conventional bunched cable looms. Figure 12.3 shows details of the various ways in which flat strip wiring may be arranged to follow vehicle contours.

PRINTED CIRCUITS

Printed circuits are in wide use for instrument panel wiring which in turn is linked to the main wiring harness. These circuits comprise flat copper conductors bonded to a rigid or flexible insulating sheet or panel. The circuit pattern is printed on to a copper sheet pre-bonded to a sheet of insulating material and the unwanted copper stripped away by etching, stamping or hand stripping, leaving only the required conducting paths.

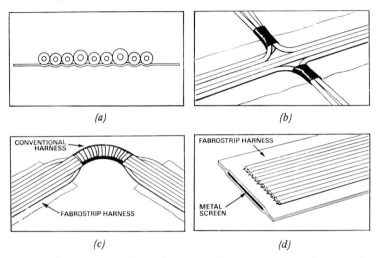

Figure 12.3 Various methods of arranging flat-strip wiring. (a) The strip takes various sized cables. (b) The cables can be broken out at appropriate points. (c) Turning corners. (d) Strip with earth plate

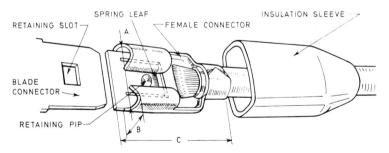

Figure 12.4 Lucar connector

Printed circuitry has advantages in vehicle applications where space is restricted but such circuits are unsuitable for carrying heavy current loads and also where there is risk of mechanical damage. Various types of connectors have been developed for linking the printed circuit panels to the conventional wiring harness.

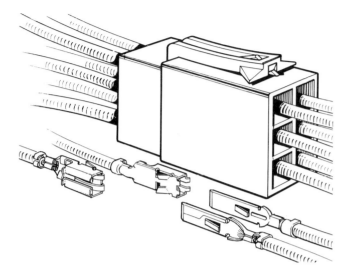

Figure 12.5   Multiway connectors with ¼ in blade and Lucar mouldings (Courtesy Rists Wires & Cables Ltd.)

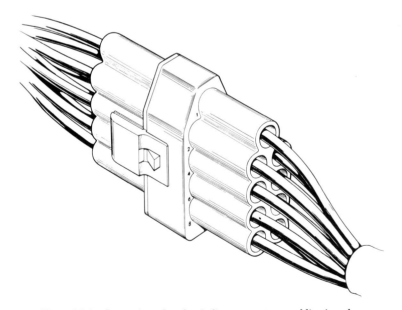

Figure 12.6   3 mm pin and socket in-line connector moulding in nylon

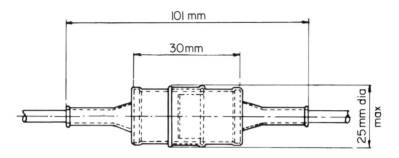

Figure 12.7   5-way 33 mm pin and socket connector (round) (Courtesy Rists Wires & Cables Ltd.)

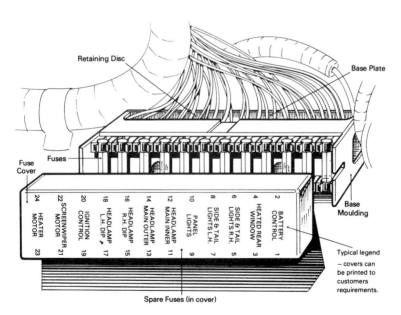

Figure 12.8   Standard fuse box with twelve fuses (Courtesy Rists Wires and Cables Ltd.)

TERMINATIONS

Blade connectors are available such as the Lucar connector shown in Figure 12.4. They are in very general use on individual components and in multiway mouldings, see Figure 12.5. Circular pin multiway connectors (3 mm) are also widely used for in-line and bulkhead connections. These connectors are moulded in circular and rectangular form for individual components, relays, switches and alternators, to meet a wide range of current loadings. Figure 12.6 shows a typical 3 mm pin and socket moulding for in-line harness connections and Figure 12.7 shows a similar 5-way round version.

Protection against corrosion of the actual connector elements is provided for in the design of multiway connectors and special weatherproof versions have been developed for use in locations where this is necessary. Similarly cable loom branches and connections are protected against vibration and abrasion by heavy duty coverings and various forms of anchorage and the use of grommets and retaining clips and fixings at points on metal edges where chafing could occur. Figure 12.9 shows the special features of a typical bulkhead connector described above.

CIRCUIT PROTECTION

To provide protection of the wiring system against damage from short circuits thermal interruptors or fuses and fuseable links are widely used. A simple form of in-line fuse for use in individual cables is shown in Figure 12.10. Figure 12.8 shows a typical fuse box for twelve fuses and five spares.

On public service and commercial vehicles, thermal/current cut-outs and miniature bi-metal type circuit breakers, usually with provision for manual re-setting, are fitted for wiring harness protection, as referred to in Chapter 11 dealing with switchgear.

AUTOMOBILE CABLES TO METRIC STANDARDS

As mentioned earlier in this chapter, the two important factors which must be borne in mind when determining the size or gauge of automobile cables are: current-carrying capacity and voltage drop. The former will be determined by the power taken by the component to be wired and a general guide regarding permissible voltage drop in any circuit (which includes feed cable and earth return path to battery) may be taken as

'The maximum permissible voltage drop must not exceed 10% of normal battery voltage, e.g. 1.2 V for a 12-V system'. However, the maximum resistance value for certain circuits (alternators, for example) is much more critical.

Cables used in automobile wiring systems (excluding the starter and ignition h.t. circuits) are made to BS 6862 *Cables for vehicles (metric units)*. This British Standard includes tables and data relating to all types of cable suitable for use in motor vehicles.

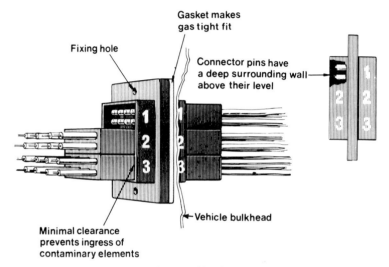

Figure 12.9   Bulkhead connector

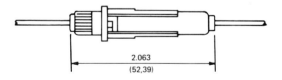

Typical method of representation on harness drawing

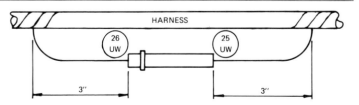

Figure 12.10   In-line fuse assembly

Metric cable sizes

As in the old imperial sizes, metric cable conductors consist of a number of identical wires. For example, the size 28/0.30 indicates a conductor made up of 28 strands each 0.3 mm in diameter. The conductor cross-sectional area is given in square millimetres, i.e. $28/0.3 = 2$ mm.

WIRING COLOUR CODE FOR VEHICLES. BS AU7: 1968. Chart and colour code for vehicle wiring (Automobile Series)

As mentioned earlier when dealing with cable identification in the first types of cable harnesses, colour coding was essential for tracing individual cable runs. British Standard AU7 was first published in 1963 under the authority of the Automobile Industry Standards Committee of the BSI. The specification reproduced here was revised in 1968 to line up with subsequent developments in vehicle electrical equipment, including the adoption of the 'negative earth' system.

Figures 12.11a and b show the layout of the standard circuits available for use in motor vehicles and provide a code for the identification of the circuit by the colour of the cables used in them. Two wiring diagrams are shown, one for the basic circuits including those for the obligatory lighting and signalling equipment (Figure 12.11a), the other for supplementary circuits (Figure 12.11b).

It should be noted that because of variations in the design of certain components such as windscreen wiper motors, it is not possible to cover all switches and internal wiring arrangements. For the purposes of colour coding, basic circuits are shown and where necessary the component manufacturer should be consulted for further details.

Cable colours

The colour key used on the diagrams is as follows:

Key letter	Colour	Key letter	Colour	Key letter	Colour
N	Brown	U	Blue	R	Red
P	Purple	G	Green	LG	Light green
W	White	Y	Yellow	B	Black
O	Orange	K	Pink	S	Slate

Where two colours are shown, the first colour is the main colour and the second colour is the tracer.

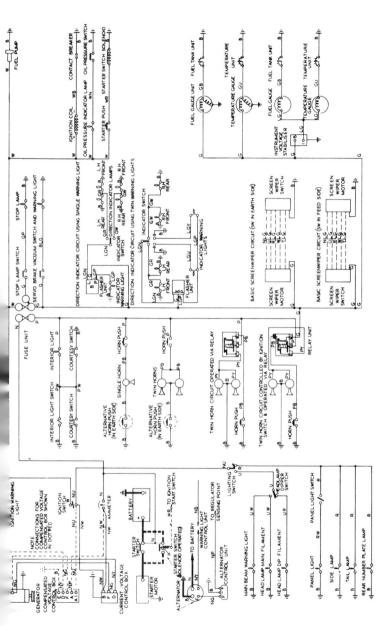

Figure 12.11(a)   Wiring diagram for lighting and signalling equipment

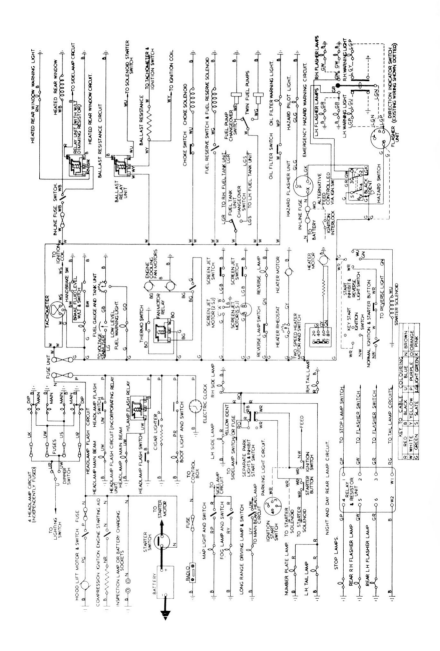

CHAPTER 13

HEATING, DE-MISTING, VENTILATING AND AIR CONDITIONING

HEATING AND VENTILATING

To meet world wide climatic conditions, interior heating, ventilating and windscreen de-frosting and de-misting systems are in general use and complete automatic air conditioning equipment is also available as a built-in facility, or in some vehicles as an optional extra. It may also be fitted as a piece of ancillary equipment.

Most systems provide for heating and the car interior in winter and ventilating it in summer. Supplemented by arrangements for blowing warm dry air on to the windscreen for the purpose of de-misting or de-frosting.

In general, these systems consist of a small heater radiator or matrix heated by engine coolant, through which air is forced by means of a blower fan (the main electrical item in the system) or by the ram effect caused by the forward movement of the car. The fan, driven by a small permanent magnet electric motor, usually wired via the ignition circuit, may have a variable speed control in one, two or three steps, or be infinitely variable by means of a switch, potentiometer, or electronic chopper control. The warmed air is distributed through ducting and manually controlled adjustable outlets to the car interior as required.

When ventilation only is required, the same ducting is used to distribute fresh air at normal ambient temperature picked up from the front of the vehicle. For de-frosting and de-misting, the warmed air is directed on to the base of the windscreen. On some vehicles, the facia end outlets may be aimed to de-frost the side windows or closed to concentrate the airflow at the windscreen.

Heated rear windows

Electrical means of de-misting or de-frosting of rear windows are now in general use and are a great aid to road safety by maintaining rear vision by rapidly heating the rear window to a point at which mist and ice evaporate. This is achieved by a flat strip heater element fitted directly on to the inside surface of the rear window or a wire element 'cast in' to the window glass.

Figure 13.1   Smiths' rear window heater

Current consumption of the heater element is relatively high, it is therefore desirable for this unit to be wired via the ignition switch so that it can only be used when the engine is running and also that some form of light signal is provided so that it is not 'left on' longer than is necessary. Figure 13.1 shows such a heater.

On vehicles fitted with automatic air conditioning, the switching of the heated rear window is linked to the main control switch so that it forms part of the complete system.

Where complete air conditioning is fitted, a refrigeration unit is used to provide the requisite cooling of the incoming air when it is desired to have the in-car temperature below the outside ambient temperature. The car windows and footwell ventilation must be closed before the air conditioning system is used. Prior to the introduction of combined heating and ventilating systems, the de-frosting device consisted of a spirally wound michrome wire heater element mounted horizontally along the bottom of the windscreen and a short distance from it.

Considerable development in both mechanical and electrical engineering work has gone into the design and production of automobile heating and ventilating systems. This includes the mechanical and electrical forms of control, the lightweight flexible tubing from which the air ducting is made and the ingenious metal and plastic flaps, grilles and adjustable air vents fitted at various points inside the vehicle. It is, of course, essential that these components and control units are simple to operate with the minimum physical effort, be clearly marked, readily accessible and have a pleasing non-corrosive finish which harmonises with the interior trim of the vehicle. The complete system must also function efficiently and operate with the minimum of noise.

COMPLETE AIR CONDITIONING

The basis of air conditioning is cooling the air inside the vehicle, but the complete system does much more than simply cooling the air, it controls the temperature to a pre-set value by blending warm and cool air and at the same time reduces the humidity, which has the effect of minimising heater fug and condensation. Thus, by maintaining a supply of fresh air which is an ideal blend of warm air below and cool above within the vehicle under all climatic conditions, the driver is less susceptible to fatigue on a long journey which makes for greater road safety.

An air conditioning system achieves these results by combining the conventional heating and ventilating system with a refrigeration system based on the same principles as the domestic refrigerator. Two fundamental principles upon which an air conditioning system operates are:

1. When a gas is compressed its temperature rises.
2. The latent heat of evaporation of a fluid.

The first of these natural principles is illustrated in the operation of a bicycle pump which gets hot as a tyre is pumped up. The second is the name for the large amount of heat that a liquid absorbs when it becomes a gas or the large amount of heat that it gives up when it changes from a gas to a liquid. It is on this principle that the operation of most domestic refrigerators is based.

For most automobile air conditioning systems, the working fluid is R12 or Freon and a typical amount required is about 1 kg (2 lb) weight. This fluid is circulated around a closed circuit through pipes and various mechanical devices. It reaches the heater-core-like evaporator in a high pressure liquid state (at approximately 57° C and 150 p.s.i.) and enters the evaporator via an expansion valve (a simple needle valve which controls the flow of refrigerant). As the fluid flows into the evaporator it reduces in pressure (to about 25 p.s.i.), becomes gaseous and also expands, each process absorbing a good deal of heat directly from the evaporator core and hence the car's interior.

The low-temperature low-pressure vapour then passes to an engine-driven compressor where it is compressed to about 170 p.s.i. and in this heat laden high pressure state it enters the condenser at the front of the vehicle. As it is at a very much higher temperature than the surrounding air it cools rapidly. The temperature at which it becomes

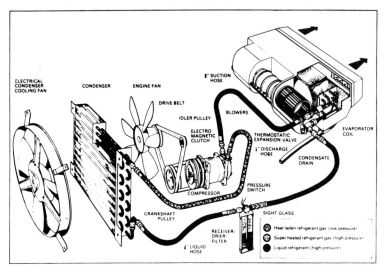

Figure 13.2   Typical air conditioning installation (Alpinair)

liquid is above ambient, since it is under a great deal of pressure. Not only does the fluid give up the heat associated with its fall in temperature, it also, therefore, releases the latent heat of evaporation (or in this case condensation) associated with its change of state. The condensed liquid under fairly high pressure (about 150 p.s.i.) is then dried and temporarily stored until required by the evaporator.

The interior of the car is blown across the very cold evaporator core and its temperature is monitored by a thermostat, pre-set by a dial on the dashboard, which controls the expansion valve and hence the flow of refrigerant around the air conditioning circuit. Since the evaporator core is so cold, water vapour within the air freezes or condenses out and is then drained off via pipes to the outside of the car.

Because the water has been condensed out, the humidity of air supplied is less than that taken in to the evaporator and because all the air flows over a wet core, smells, fumes and dust are filtered out and flow out with the waste liquid. It has been calculated by Alpinair Ltd. that an evaporator will drain off some 32 litres of water a day in conditions of 56 to 70% humidity and that 80% of all impurities are absorbed in a single pass of air. As almost all air conditioners work on a large proportion of re-circulated air with only a 15% fresh air breed, any impurities get absorbed quite quickly.

A practical example of this system which may be adapted for use on a very wide range of vehicles is illustrated in Figure 13.2. This diagram shows the refrigerant flow through the basic air conditioner system.

The simplest form of air conditioner may be fitted under the dash or may be fitted into the centre console of some cars or built into the dash. In the latter arrangement the evaporator unit is hidden within the facia and either feeds a front air chamber from which flexible pipes run to air outlets in or behind the facia, or feeds conditioned air to the existing ventilating system of the car.

AUTOMATIC AIR CONDITIONING

An automatic air conditioning system integrated with the heating and ventilating system is one in which the basic function is to maintain the car interior temperature of both upper and lower systems at levels pre-set by the driver, regardless of the climatic conditions. A typical system is the Rolls-Royce automatic air conditioning system which achieves this basic function with the minimum attention or adjustment by the driver.

This is a highly developed electronic control system electrically connected to sensitive temperature sensors and electro-mechanical servo mechanisms controlling the desired air flows (Figure 13.3). The system has two 'temperature selectors' situated on the facia, one controls the air temperature in the upper part of the car and the other, the air temperature in the lower part of the car.

Any in-car temperature between 17° C (63° F) and 33° C (91° F) can be independently selected with the two air temperature controls situated on the facia, both the air temperatures will then be automatically maintained within the limitations of the system.

A 'function switch' situated directly above the temperature selectors on the facia has five positions and enables the driver to select one of three automatic positions or a defrost position. The system can of course be turned off by selecting the fifth position.

The two 'temperature selectors' and the 'function switch' are the only manual control switches for the automatic air conditioning system. All other operations and functions are performed automatically to give maximum comfort at all times.

The automatic air conditioning system is designed so that the 'upper' system delivers hot and warm air from the windscreen demister outlets and cool or cold air from the circular facia outlets and the rectangular facia outlet. The 'lower' system delivers hot, warm or cool air from the lower outlets in both the front and rear compartments of the car. The lower outlets are situated on either side of the transmission tunnel; in

Figure 13.3   Air circulation diagram (Rolls Royce Motors Ltd.). 1. Windscreen 2. Solar sensor 3. Circular facia outlets 4. System lower outlets 5, 6. System upper outlets

the front compartment they are below the facia and for the rear compartment they are below the front seats. The 'lower system' outlets are closed before the air temperature becomes unpleasantly cold to the occupants' feet, the entire cold air supply is delivered to the 'upper' system outlets in this instance.

An electronically-heated rear window without a manual control is incorporated into the system. The automatic air conditioning system switches on the element when the car is being heated as this is the only time that misting or ice formation is likely to occur.

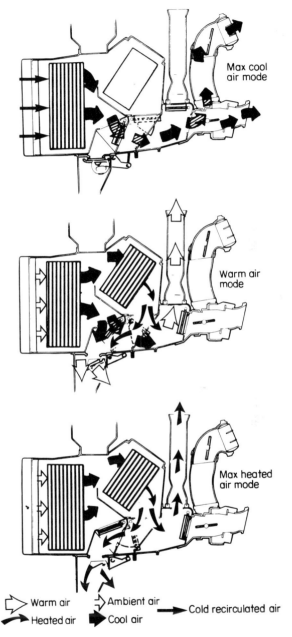

Max cool
air mode

Warm air
mode

Max heated
air mode

Warm air   Ambient air   Cold recirculated air
Heated air   Cool air

Figure 13.4   Air flow through heater and control box

Basic operation of the system

All air entering the system is dehumidified and cooled by passing it through the refrigeration evaporator matrix, before it enters the heater and control box assembly (see Figure 13.4).

The heater and control box supplies hot and cold air to two independent ducting systems. One system is the 'upper' system which comprises the windscreen demister outlets, circular facia outlets and rectangular facia outlet. The other system which delivers air to the lower outlets in both front and rear compartments of the car is the 'lower' system.

Both 'upper' and 'lower' systems have an air temperature (blend) flap which adjusts the ratio of hot to cold air so that the air temperature of the upper system outlets and the air temperature of the lower system outlets can be independently varied from maximum hot to maximum cold.

Air temperature (blend) flaps form the basis of the automatic air conditioning system as they are each operated by an electro-mechanical servo mechanism controlled by an electronic circuit. These circuits are fed with air temperature information from certain parts of the car and acting on this information, drive the servos so that the air temperature (blend) flaps are moved into the correct positions to achieve the required in-car temperatures. The 'upper' and 'lower' systems operate independently, each system having its own set of air temperature sensors air temperature selector, servo and servo electronics.

The system also includes a mechanically operated water tap. The two recirculation flaps, the windscreen/facia mode change flap and the lower quantity flap are operated by electrical actuators. These actuators are operated by signals which are related to air temperature (blend) flap servo positions.

Two fans controlled by an electronic fan speed control, are used to boost the air input to the automatic system.

Driver's controls

The driver's controls for the automatic air conditioning unit comprise three switches positioned one beneath the other on the facia (just below the outside air temperature gauge) and three flap controls situated in the centre of the facia.

Manual switch controls

Function switch

The function switch is located at the top of the console which carries the manually operated switches. The switch is marked 'AIR CONDITIONER' and has five positions, OFF, LOW, AUTO (automatic), HIGH and DEF (defrost); the functions of the system when this switch is turned to the various positions are as follows:

OFF. When the switch is turned to the OFF position ambient air is prevented from entering the car as the recirculation flaps will be in the recirculation position (i.e. open). Turning the function switch to the OFF position from any other position immobilises all the flaps in whatever position they happen to be, except the recirculation flaps which if not already in position, move to the open position (i.e. full recirculation). The fan motors are also switched off.

LOW. With the switch in the LOW position the air conditioning system gives automatic air temperature control with the fan motors running at a fixed low speed.

AUTO. If this position is selected the air conditioning system gives automatic air temperature control and automatic fan speed variation in relation to the position of the air temperature controlling servo mechanism.

HIGH. With the switch in this position the air conditioning system gives automatic air temperature control with the fans running at a fixed high speed.

DEF. When DEF (defrost) is selected both the upper and the lower air temperature (blend) flaps move to their respective full hot position. Air to the lower system outlets is cut off and all available air is therefore directed to the windscreen demister outlets. The fan motors will operate at maximum speed even if the engine cooland is cold and heated air is not available.

Temperature selectors

These two controls are situated on the facia below the outside air temperature gauge and the function switch. The air temperature selectors are marked UPPER TEMP and LOWER TEMP respectively. The upper and lower air conditioning systems are controlled independently of each other by the respective air temperature selector.

Turning both the air temperature selectors fully clockwise results in a stabilised in-car air temperature of approximately 33° C (91° F), whilst turning fully anti-clockwise will give an in-car air temperature of approximately 17° C (63° F). Midway position of the air temperature selectors will produce an in-car air temperature of approximately 25° C

(77° F). It should be noted that the air temperature selector positions result in stabilised in-car air temperatures being achieved and not that the air temperature (blend) flaps will move to the full hot or full cold positions.

The air temperature selectors operate potentiometers which are connected as variable resistances so that current flow through them changes in relation to their angular position. These two independent current signals are used to help determine the required air temperature (blend) flap positions.

Manual flap controls

In addition to the three switches which control the system, three manually operated control flaps are provided to regulate the flow of cold air in the upper system.

Facia outlets

Two circular outlets are located on the facia and supply cool and cold air whenever the upper system is cooling and the car interior and the control knob situated adjacent to each outlet, is withdrawn. The outlets may also be swivelled to direct the air flow as required.

The rectangular outlet situated below the radio on the facia admits cool and cold air into the car whenever the upper system is cooling the car interior and the control handle is pressed down.

Engine compartment components

The engine compartment components are illustrated in Figure 13.5 to the left of the scuttle wall (item 6).

Refrigeration system

The Arcton 12 refrigerant is circulated by the compressor which is driven by twin V-belts from the engine crankshaft. The electro-magnetic clutch incorporated into the compressor is energised in the drive position at virtually all times.

High pressure vapour is pumped from the compressor to the condenser matrix situated in front of the radiator matrix. Ambient air is induced to flow across the condenser matrix by the engine coolant viscous fan

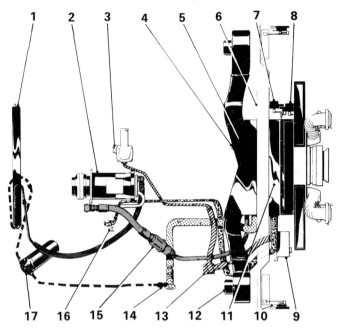

1 Condenser matrix
2 Refrigeration compressor
3 Cut-in switch - coolant
   temperature
4 Refrigeration compressor
   ambient switch
5 Evaporator box
6 Scuttle wall
7 Lower quantity flap
   actuator
8 Mode flap actuator

9 Voltage stabiliser and fan
   speed module
10 Recirculated air flap actuator
11 Heater and control box
12 Fan motor
13 Coolant tap
14 Expansion valve
15 Suction throttling valve
16 Cut-in switch - engine oil
   pressure
17 Receiver/Drier

Low pressure vapour   High pressure liquid
Low pressure liquid and vapour   Coolant - engine to heater
High pressure vapour   Coolant - heater to engine

Figure 13.5   Components of automatic air conditioning system

assembly and the forward motion of the car. This action condenses the
refrigerant from a high pressure vapour at the top of the condenser matrix
to a high pressure liquid at the bottom.

High pressure liquid refrigerant passes from the condenser matrix to
the receiver/drier unit located on the right of the condenser matrix
inside the engine compartment (viewed from the front of the car). The
receiver/drier is a metal cylinder incorporating a sight glass in the top;
the unit contains a moisture absorbing chemical. The functions of the

unit are to absorb any traces of moisture which may be present in the system and to ensure that the refrigerant is passed to the expansion valve in liquid form.

The sight glass is provided for visual inspection of the refrigerant when the system is operating and a steady stream of liquid should be observed. The presence of bubbles or foam usually indicates incorrect operation of the system or insufficient refrigerant, however, it is normal for some foaming to show in the sight glass when the ambient air temperature is 21° C (70° F) or below.

The refrigerant then passes to the expansion valve located on the right-hand side of the engine compartment (viewed from the front of the car).

The expansion valve controls the flow of refrigerant to the evaporator. To enable the expansion valve to function correctly two signals are required; firstly the temperature in the evaporator outlet pipe is sensed by a small phial attached to the outside of the evaporator outlet pipe and secondly, the pressure in the evaporator outlet pipe is sensed by an equaliser line to the suction throttling valve.

The expansion valve meters the flow of refrigerant, in the form of a 'wet vapour' to the evaporator matrix in response to the pressure and temperature signals and keeps the evaporator matrix full of refrigerant. In the evaporator matrix assembly heat is transferred from the air which is passing through the unit to the refrigerant, causing the refrigerant to boil and vaporise.

From the evaporator matrix the refrigerant in the form of a low pressure vapour, passes to the suction throttling valve, the function of which is to throttle the flow of refrigerant so that a constant pressure is maintained in the evaporator matrix regardless of compressor speed or evaporator matrix loading (e.g. maximum cooling).

The cycle of the refrigerant is completed when it passes from the suction throttling valve to the refrigeration compressor as a low pressure vapour.

Heating system

Heated engine coolant is circulated by the engine coolant pump. The take-off for the heating system is located in the engine coolant thermostat housing. From this point the heated coolant passes along a pipe to the coolant tap which is operated by a 'Bowden' cable from the lower servo assembly. From the coolant tap the heated coolant passes through the heater matrix located in the heater and control box.

At this point, heat is transferred from the heated coolant to the air passing through the matrix; the coolant is then returned to the engine via a pipe connected into the coolant pump intake.

Car interior components

The components situated inside the car, are for introductory purposes, divided into two categories namely, the main components and the air ducting with its respective outlets.

The evaporator, heater and control box assembly is fitted into the scuttle wall. This assembly controls the humidity, temperature and flow of the air passing into the car; operation of the assembly is described under the heading 'Air flow'.

Air can flow into the car from the following outlets depending upon the mode of operation of the system, circular facia outlets, rectangular facia outlet, front compartment lower outlet, rear compartment outlets and windscreen demister outlets.

Air leaves the car from either the recirculation flap outlets (only when the system is operating at maximum cold) or the stale air outlet situated behind the rear seats.

The air temperature at the system outlets is controlled by an electro-mechanical servo assembly mounted over the transmission tunnel below the facia; the assembly operates the two temperature flaps. The availability of air at the system outlets is determined by conventional electric actuators mounted adjacent to their respective flaps.

A fan speed module and voltage stabiliser assembly located on the left-hand side of the car (viewed from the driver's seat) supplies a stabilised voltage of 7.5 V to the servo motors and 9 V to the servo modules. The fan speed module and voltage stabiliser assembly also controls the speed of the two fan motors. The speed of the fan motors and mode of operation for the system is dictated by the various positions of the three manually operated control switches on the facia. However, when the AUTO (Automatic) position is selected with the function switch, the fan speed is controlled by the servo positions, as dictated by the air temperature selectors in relation to the in-car and ambient air temperatures detected by the air temperature sensors.

There are five disc type thermistor temperature sensors, three located inside the car, these are the knee roll sensor, top roll (solar) sensor and the cantrail sensor. The fourth and fifth sensors are the ambient sensors which are mounted on a common board, located behind the right-hand side portion of the rear bumper (viewed from the rear of the car).

Air flow

Intake air

Ambient air is drawn from outside the car through two scuttle intake grilles into the scuttle wall aperture, the air travels along the scuttle aperture, past the fan motors and into the evaporator box through moulded ducting.

All air passes through the refrigeration matrix, reducing both the humidity and temperature; the air then passes into the heater and control box.

Heater and control box

From this point, the air in the system is divided into three sections. The two outer sections of the heater and control box are similar and basically contain cold air in the lower passages and warm air in the upper passages. These outer passages contain air for the upper part of the system (i.e. windscreen, centre facia outlets and rectangular facia outlet). The centre section contains air for the lower part of the system (i.e. front compartment lower outlets and rear compartment outlets), the upper passage contains warm air and the lower passage cold air.

Except for certain extreme circumstances all modes of operation of the air conditioning system require a blend of hot and cold air to provide varying degrees of warm or cool air. Therefore, in both outer passages there is an 'upper system air temperature (blend) flap' and in the centre passages there is a 'lower system air temperature (blend) flap'. Basically, these flaps control the temperature of the air which is passed into the car by closing the cold air passage while opening the warm air passage and vice versa.

A mode flap is used to ensure that the requirements of the upper system are satisfied, i.e. extremely cold air must not be directed onto the windscreen and only cool or cold air is available from the two circular facia outlets and the rectangular facia outlet. The mode flap is situated in the distribution box and remains closed to all facia outlets unless cool or cold air is required, it then moves position, closing the passages to the windscreen and opening the passage to the facia outlets.

Stale air

An outlet for stale air is provided in the centre of the parcel shelf behind the rear seats, this allows air to pass from the interior of the car through the luggage compartment to atmosphere.

CHAPTER 14

FUEL INJECTION EQUIPMENT

The advantages of fuel injection over the conventional carburettor have been appreciated for many years. The quest for improved engine performance linked with fuel economy together with legislation regarding the control of exhaust emission has enormously increased the use of fuel injection systems. The incorporation of electronic control systems has also considerably helped in the development of very efficient and commercially viable systems.

The most important advantages of fuel injection in a normal Otto cycle engine over other fuel supply systems are higher horsepower per unit of displacement, lower specific fuel consumption, fewer unburned components in the exhaust gas, higher torque at low engine speeds, greater flexibility and more uniform combustion in the individual cylinders. These advantages do not arise as a result of injection of the fuel alone, instead, they result from the fact that a fuel injection system gives the engine designer greater freedom. For example, optimum structural design of the intake channels in the engine between the air filter and the cylinder head become possible. These channels can be designed and matched for each cylinder so that the pulsations in the flow of air produce a supercharging effect which results in improved volumetric efficiency. In addition, when fuel injection is employed, various possibilities exist for matching the quantity of fuel injected to the many different operating conditions of the engine.

The Bosch company have received world-wide recognition for their pioneer work in the development of fuel injection systems. The author is indebted to this Company for provision of technical data on the various fuel-injection systems on which the following information has been based.

THE BOSCH CONTINUOUS FUEL-INJECTION SYSTEM

As mentioned in chapter 2, earlier mechanical forms of engine-driven injection pumps injected a metered quantity of fuel at high pressure directly into the engine cylinders. Present-day practice is to spray the fuel into the intake manifold or into the intake port. Such a mechanical system is exemplified in the Bosch continuous injection system which does not require a separate or external drive. Instead, in this system the fuel in continuously metered as a function of the volume of air drawn into the engine cylinders and is injected into the intake manifold at about 3 bar overpressure.

In this system the fuel is pumped by an electrically-driven roller cell fuel pump During operation, the quantity of intake air is metered by an air-flow sensor installed in front of the throttle plate. Depending on the position of the throttle plate, i.e. depending on the position of the accelerator pedal, more or less air is drawn into the engine. Then, depending on the volume of air metered, a fuel distributor apportions a quantity of fuel to the individual cylinders through the associated injection valves which produces an optimum air-fuel mixture with regard to engine power, fuel consumption and exhaust gas composition.

The air-flow sensor and the fuel distributor are combined into one assembly, the mixture control unit. The precisely metered quantity of fuel is fed to the injection valves which continuously spray the fuel in finely atomised form into the intake manifold in front of the cylinder intake valves. From there the fuel is drawn into the engine cylinders together with air when the intake valves open. This principle of operation is shown in the block diagram Figure 14.1 and the wiring diagram Figure 14.2.

The air-flow sensor operates on the suspended body principle. It is installed in front of the throttle plate and consists essentially of a round air-flow sensor plate mounted on a lever and suspended in an air funnel in such a way that it is deflected upward from its base position a

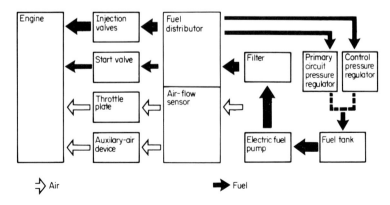

Figure 14.1   Block diagram of Bosch fuel injection system

distance proportional to the volume of air drawn into the engine through the funnel. The function of the fuel distributor is to feed a quantity of fuel to the engine that is proportional to the distance the air-flow sensor plate in the air flow sensor, rises from its base position. By means of a control plunger, narrow rectangular slits, one for each of the cylinders in the engine, are opened or closed varying amounts and thus the discharge openings for the fuel (these metering slits) are made larger or smaller in terms of open cross-section. The control plunger is supported on the lever in the air-flow sensor and therefore follows the movements of the air-flow sensor plate (with a different lever ratio). As a result of this arrangement, the amount the control plunger rises is proportional to the amount of fuel fed to the engine.

The components primarily responsible for ensuring that this is the case are the differential-pressure valves, one of which follows each metering slit. As a result of these differential-pressure valves, a constant pressure differential of 0.1 bar prevails at each metering slit (about 4.7 bar overpressure in front of the slit, about 4.6 bar behind it). As long as this pressure differential exists, the quantity of fuel fed to the engine is proportional to the open cross-sectional area of the metering slit.

The position of the control plunger is influenced by the control pressure applied to it. After the engine has reached the normal operating temperature, this pressure is about 3.7 bar overpressure. At such a time the force applied to the control plunger is thus much less than when the engine is warm. As a result the air-flow sensor plate in the air-flow sensor is raised a greater distance for the same volumetric rate of air flow and a larger amount of fuel is fed to the engine.

The function of the warm-up regulator is to increase the control pressure after the engine has warmed up; this is accomplished by a bimetal strip in the cold condition acting against the delivery valve spring which determines the control pressure. When the engine is turned on, however, this bimetal strip is electrically heated and as it becomes warmer the delivery valve spring becomes more and more effective, increasing the control pressure.

Compensation for the greater frictional load (frictional resistance) during the warm-up period can be made by increasing the volumetric efficiency of cylinder filling by means of an auxiliary-air device installed in a line bypassing the throttle plate. The amount of air which passes through the auxiliary-air device is controlled by a sliding plate in which a specially-shaped hole is drilled; the position of this plate and hence the degree to which the hole opens the auxiliary-air channel, is controlled according to the temperature of an electrically-heated bimetal strip.

Other possibilities for modifying the air-fuel ratio to compensate for varying operating conditions are provided by suitable design of the air funnel in the air-flow sensor (basic adjustment) or by additional accessories. The start valve. controlled by the thermo-time switch, sprays additional fuel into the common intake manifold during the engine starting operation. The engine idle speed is adjusted with the idle speed adjusting screw on the throttle plate fitting.

The idle mixture adjusting screw accessible from outside on the mixture control unit is provided for adjustment of the idle air-fuel mixture.

THE ELECTRICAL CIRCUIT

The wiring diagram for this circuit is shown in Figure 14.2 and constructional features and operation of the electrically-operated components of the system are described below.

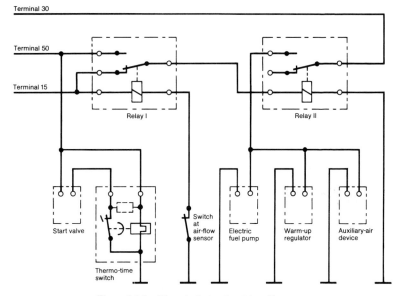

Figure 14.2   Electrical circuit wiring diagram

The air-flow sensor plate in the air-flow sensor activates a switch which is closed when the engine is turned off. When the air-flow sensor plate rises from its seat, however, the ground line from relay 1 is opened.

When the ignition is switched on (terminal 15), relay 1 is energised. Relay 11 remains at rest, however, and the electric fuel pump remains switched off.

When the engine is started (terminal 50), the control current for relay 11 flows through the working contact of relay 1 and the electric fuel pump is switched on through the working contact of relay 11. At the same time, current starts to flow to the warm-up regulator and to the auxiliary-air device. The start valve is also switched on at this time through a thermo-time switch.

As soon as the engine draws in air, the air-flow sensor plate in the air-flow sensor rises from its seat and opens the ground line from relay 1. Relay 1 is then de-energised and returns to its off-position. Relay 11 remains energised however, and the electrical fuel pump continues to operate.

If the engine comes to a stop as a result of exceptional conditions, the electric fuel pump is also automatically stopped even though the ignition is still turned on. This results from the switch at the air-flow sensor closing. This switches relay 1 to the working position and interrupts the control line leading to relay 11. At this point there is no longer a connection from terminal 30 to the electric fuel pump.

THE ELECTRIC FUEL PUMP

The electric fuel pump shown in Figure 14.3 is a roller cell pump driven by a permanent magnet electric motor. The rotor disc, mounted on the motor shaft is fitted with metal rollers in notches around its

Figure 14.3   Roller cell electric fuel pump

periphery which are pressed against the eccentric bore pump housing by centrifugal force and acts as seals. The fuel is carried in the gaps between the rollers and is then forced into the fuel injection tubing.

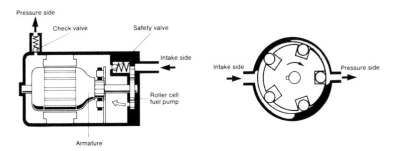

Figures 14.4 and 14.5   Fuel flow direction

In this system, as indicated in Figures 14.4 and 14.5, the fuel flows directly around the electric motor. There is no danger of explosion, however, because there is never a combustible mixture inside the pump housing. This pump delivers several times the quantity of fuel actually required, so the excess fuel is diverted off at the primary circuit pressure regulator and it flows under no pressure back to the fuel tank.

THE WARM-UP REGULATOR

The warm-up regulator construction is shown diagrammatically in Figure 14.6. The electrical connection is made by a plug connector and the operation of this unit is as follows:

During the warm-up period of an engine, two compensations are basically required compared with conditions at normal operating temperature:

1. Compensation for condensation losses on the cold walls of the combustion chamber and intake manifold.
2. Compensation for power lost as a result of greater friction.

Regarding (1), compensation is made for condensation losses by a richer mixture. This function is performed by the warm-up regulator. It lowers the pressure applied to the control plunger during the warm-up period, as a result of which the air-flow sensor plate is lifted a greater distance from its rest position by the same rate of air flow and the open cross sectional area of the metering slits in the barrel assembly is therefore enlarged.

The change in the control pressure during the warm-up period is illustrated in Figure 14.6 and takes place as follows. As long as the engine is cold, a bimetal strip presses against the delivery valve spring. As a result, the pressure on the diaphragm is reduced, the discharge cross section is enlarged and the control pressure is thus lowered. When the engine is started, the electrical system designed to heat the bimetal strip is switched on. This strip warms up, relaxing the pressure on the

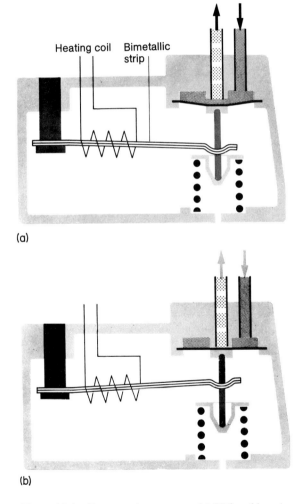

(a)

(b)

Figure 14.6 Compensation systems. (a) With cold engine.
(b) With warm engine

delivery valve spring and at a certain temperature it rises up completely from the spring plate. This means that after the warm-up period the delivery valve spring is fully effective. A fairly high control pressure must develop before the discharge cross section opens, i.e. the control pressure assumes its normal value of about 3.7 bar overpressure.

Regarding item (2) above, compensation is made for the power lost as a result of greater friction by feeding a larger volume of the air-fuel mixture to the engine than corresponds to the position of the throttle plate. This is done by bypassing the throttle plate with an auxiliary air device in which the cross sectional area of the channel open to flow of air is controlled by a pivoted blocking plate with a specially shaped hole; the movement of this plate is dependent on an electrically heated bimetal strip. At normal operating temperature this extra air channel is closed.

THE START VALVE

In the electrically-operated start valve, a typical sectional view of which is shown in Figure 14.7, a helical spring presses the movable armature in the magnetic circuit together with the seal against the valve seat and closes the fuel inlet. When the armature is drawn back, however, the fuel inlet is opened. The fuel then flows along the sides of the armature to the swirl nozzle. There a swirling motion is imparted to the fuel and it leaves the nozzle in finely atomised form.

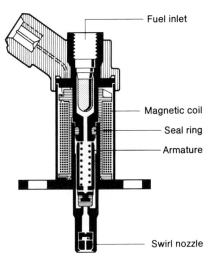

Figure 14.7   Sectional view of electrically-operated start valve

THE THERMO-TIME SWITCH

The thermo-time switch shown in Figure 14.8 limits the length of time that the start valve remains open and at higher temperatures it prevents the start valve from opening at all. This switch is therefore

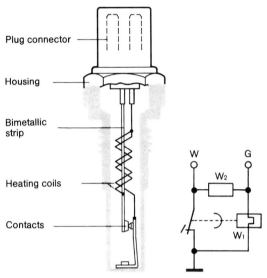

Figure 14.8   Sectional view and circuit of thermo-time switch

mounted on the engine in such a way that it can absorb heat from the engine.

Figure 14.9 indicates typical installations on a water-cooled (a) and on an air cooled engine (b).

Auxiliary starting assembly

The auxiliary starting assembly consists essentially of an electro-magnetically operated start valve which is switched on at the beginning of the starting process and a thermo-time switch which limits the duration of time that the valve is open or at higher temperatures prevents this valve from opening at all. The start valve, located as shown in Figure 14.10 sprays additional fuel into the common intake manifold. Because of the relatively high primary circuit pressure of 4.7 bar overpressure, the fuel is well atomised by the swirl nozzle.

The thermo-time switch shown in the schematic diagram, Figure 14.8, closes or opens the power circuit leading to the start valve depending

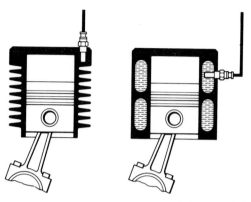

*Figure 14.9   Thermo-time switch on (left) air-cooled
engine and (right) water-cooled engine*

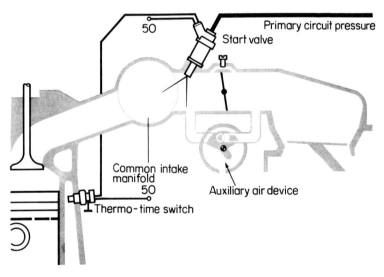

Figure 14.10   Operation of auxiliary starting assembly

on the engine temperature. During cold starting, the power circuit is
interrupted depending on the temperature of an electrically heated
bimetal strip. The thermo-time switch has either a single heating coil
(W_1) or if rapid heating is required, two heating coils W_1 and W_2.
When the switching temperature is reached, the contacts are opened,
cutting off power to the start valve. This switches heating coil W_2 off.
Heating coil W_1 holds the contacts open until the end of the starting
operation.

Auxiliary air device

The auxiliary-air device already referred to regarding compensation for frictional resistance in a cold engine is mounted by means of a two-hole flange at some point on the engine where the temperature is characteristic of the engine's operating condition. A plug is provided for the electrical connection.

Hot or warm starting

As a result of suitable design of the various valves and of the fuel accumulator in the system, sufficiently high pressurisation is assured even after the engine has been turned off for an extended length of time. This prevents formation of vapour bubbles in the lines.

ELECTRONICALLY-CONTROLLED FUEL INJECTION SYSTEM

The advantages of an electronically-controlled fuel injection system compared with a normal carburetted engine are similar to those already described for the mechanical continuous injection system. Experience has also shown that in comparison with the fuel system employing a carburettor, consumption of fuel by an electronically-controlled injection system is less during general driving and particularly during heavy traffic conditions and high performance driving.

An electronic fuel injection system functions by the rapid and accurate assessment of data received from the various sensors fitted to the engine and by responding automatically to the slightest throttle pedal movement. Such a system is exemplified by the Lucas EFI system.

A schematic layout of this system is shown in Figure 14.11. Information concerning manifold air pressure, engine speed and crankshaft angle, plus a number of secondary factors, is collected by sensors and relayed to the electronic control unit. The control unit uses this data to calculate the required opening time of the fuel injectors and sends the equivalent electrical pulse to each injector. As the injector opens, pressurised fuel is sprayed around the inlet valve, where it mixes with air before entering the cylinder to be compressed and spark ignited. The system consists of two main component groups; the fuel circuit and the electronic control circuit.

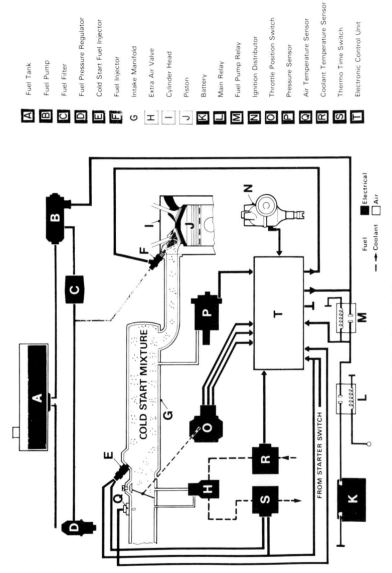

A Fuel Tank
B Fuel Pump
C Fuel Filter
D Fuel Pressure Regulator
E Cold Start Fuel Injector
F Fuel Injector
G Intake Manifold
H Extra Air Valve
I Cylinder Head
J Piston
K Battery
L Main Relay
M Fuel Pump Relay
N Ignition Distributor
O Throttle Position Switch
P Pressure Sensor
Q Air Temperature Sensor
R Coolant Temperature Sensor
S Thermo Time Switch
T Electronic Control Unit

Figure 14.11 The electronically-controlled fuel injection system

THE FUEL CIRCUIT

Pressurised fuel is supplied by an electric pump, the pressure being maintained at 28 lbf/in^2 (2 kgf/cm^2) by a pressure regulator. A filter removes any particles of foreign matter from the fuel. Branches of the

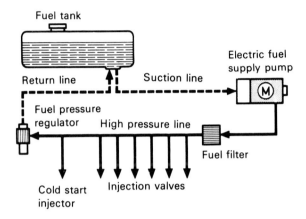

Figure 14.12   Fuel circuit

fuel line are connected directly to the injector for each cylinder and also to a cold-start injector in the engine inlet manifold (Figure 14.12).

THE CONTROL CIRCUIT

At the heart of the control circuit is the electronic control unit which assesses data received from the various sensing devices. The principal data comes from a pressure sensor connected to the air intake manifold and from a set of auxiliary switches in the ignition distributor. These sensing devices provide data relating to pressure of the air in the intake manifold, engine speed and crankshaft angle respectively.

Other data assessed by the control unit provides for mixture enrichment in cold-start conditions, engine warm-up, full load running, acceleration and fuel cut-off during engine over-run.

The system also takes account of fluctuations in air density created by temperature changes and the admission of additional air to the manifold under cold idling conditions.

COMBUSTION REQUIREMENTS

Oxygen from the air is a requirement of every combustion process. Complete combustion in a spark-ignition engine requires air and fuel in the ratio of 14:1; this is known as the Stoichiometric air/fuel ratio. (See further detailed discussion under heading 'Air-fuel ratio' later in this chapter.)

In practice, whether the engine is set to operate on the lean or rich side of this ratio depends on the combustion characteristics of a particular engine. This can only be determined by extensive engine and vehicle testing, covering performance, economy and exhaust emission. Once the best setting has been established, it is important to maintain accurate control on the air/fuel ratio within close limits under all operating conditions, to avoid over-richness (causing incomplete combustion and excess of harmful emissions) or over-leanness (causing loss of power and high operating temperatures). An electronic fuel injection system ensures that the necessary close control of air/fuel ratio is always achieved.

THE PRINCIPLE OF ELECTRONIC FUEL INJECTION

The amount of fuel supplied to the engine is controlled by three factors; injector orifice cross section, fuel pressure and duration of opening of the injectors.

Injector orifice size is a fixed function for a particular injector design. Constant fuel pressure is provided by the electric pump and pressure regulator. Thus the variable factor in electronic fuel injection lies in controlling the opening period of the injectors. This is achieved by translating the data supplied to the electronic control unit from the various sensors into electrical pulses which are in turn relayed to the solenoids which operate the injectors, thus determining the moment and duration of fuel injection.

Fuel pump and filter

The fuel pump is of the roller vane type as described earlier in the section dealing with the CIS fuel injection system and illustrated in Figures 14.4 and 14.5.

A paper element fuel filter, housed in an aluminium body in the pressurised fuel line, ensures removal of any foreign particles.

Pressure regulator

An adjustable pressure regulator (see Figure 14.13) in the fuel line maintains fuel pressure at or below 28 lbf/in^2 (2 kgf/cm^2). Should the pressure rise above this value, a spring-loaded diaphragm is operated and the fuel return port is opened, allowing excess fuel to return to the tank.

Injectors

The solenoid-operated fuel injector (Figure 14.14) consists of a valve body and needle valve to which the solenoid plunger is rigidly attached. Fuel is piped to the injector under pressure from the electric fuel pump passing first through a filter and then into the injector valve body. The needle valve is pressed against a seat in the valve body by a helical spring to keep the valve closed until the solenoid winding is energised.

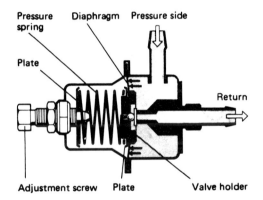

Figure 14.13   Pressure regulator

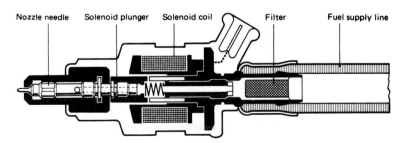

Figure 14.14   The solenoid-operated injection valve

When a current pulse is received from the electronic control unit a magnetic field builds up in the solenoid coil which attracts the plunger and lifts the needle valve from its seat. This opens the path for pressurised fuel to emerge as a finely atomised cone of spray.

The injectors are mounted on rubber seals to provide efficient insulation against high engine temperatures and to limit noise.

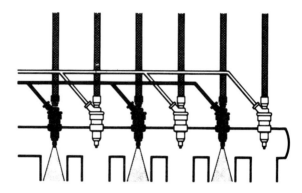

Figure 14.15   Injection valve groups

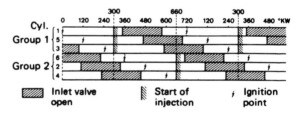

Figure 14.16   Injection timing for 6-cyl. engine

To minimise electronic complexity, the electronic control unit energises the injectors in two groups (e.g. in the case of 6-cylinder engines in two groups of three (Figures 14.15 and 14.16) selecting the appropriate group according to a signal received from auxiliary switches in the ignition distributor. The injectors of each group are connected in parallel, thus injection takes place simultaneously into each inlet valve port for the selected group of cylinders.

Some inlet valves are closed and some open at the moment of injection. In the latter case the fuel mixes with the air in the valve port and remains there until it is drawn into the cylinder on the next induction stroke.

Ignition distributor

The distributor is fitted with either two sets of auxiliary contact breaker points or two reed switches spaced 180° apart. These triggers are operated by either the normal mechanical cam action or by a magnet set into the rotor arm to provide information relating to camshaft angle and engine speed.

The electronic control unit

This unit which is the key item in the electronic fuel injection system is responsible for metering the quantity of fuel supplied to each cylinder. The unit contains a number of printed circuit boards on which are mounted a series of transistors, diodes and other electronic components which make up the vital data analysing circuits responding to the various input signals. After processing of the input data, power output circuits within the electronic control unit generate current pulses which are transmitted to the solenoid injectors to effect opening of the injector valve. The unit operates at battery voltage and there is a built-in correction to ensure that normal voltage variations which occur during the running of the vehicle do not cause incorrect output signals to be generated.

All the electrical units of the electronic fuel injection system are connected to a cable harness terminating in a 25-pole plug which connects to the mating socket on the sealed metal case of the electronic control unit. Hence, as soon as the ignition switch is actuated the fuel pump is switched on and the fuel injection system becomes operational.

Pressure sensor

Normal atmospheric pressure prevails in the intake manifold in front of the throttle valve, whilst a depression exists on the engine side. This vacuum level changes with changes in throttle opening and is measured by the pressure sensor, which references vacuum against atmospheric pressure.

The resultant 'absolute manifold pressure', expresses engine load. The pressure sensor is basically an inductive transducer, housed in one of two chambers separated by a diaphragm and operated by manifold pressure acting on a pair of evacuated aneroids in the same chamber. The aneroids expand or contract in response to pressure variations, moving an armature in or out of the magnetic core of the transducer to change its inductance.

As the throttle is closed, the absolute intake manifold pressure decreases. The aneroids expand, moving the armature out of the magnetic core of the transducer and lowering its inductance. The result is that an electronic time switch in the control unit shortens its pulse and the injectors release a reduced amount of fuel.

With the throttle open, the absolute intake manifold pressure is high and the aneroids are compressed. The armature is deep in the magnetic field and inductance is high; a longer pulse from the time switch and a correspondingly increased charge of fuel results.

On some installations, to avoid vibration created by fluctuating manifold pressure on the aneroid/armature assembly, a damping measure is provided by means of a throttling orifice at the sensor's inlet port. By-passing the orifice is a relief valve which opens when manifold pressure exceeds a pre-determined level, as in rapid acceleration. This avoids fuel/air mismatch which would otherwise occur through delay caused by the throttling orifice.

Correction factors

In addition to the primary factors of intake manifold pressure and engine speed, a number of secondary factors must be considered to obtain optimum engine performance. These are: cold-start enrichment, warm-up enrichment, full-load enrichment, temporary acceleration and idling. Changes in air density due to ambient temperature variation must also be taken into account.

The following ancillary items of equipment are included in the system for the monitoring of these parameters.

Temperature sensors

Cooling temperature sensor

Engine coolant temperatures provide an approximation of the temperatures produced by combustion and are measured by means of a sensor which incorporates a NTC thermistor. NTC (Negative Temperature Coefficient), means that the thermistor's electrical resistance decreases in response to increased temperatures. The installation of this sensor is as shown in Figure 14.9 included in the description of the CIS system.

Air temperature sensor

A similar NTC thermistor is situated in the air intake between the air filter and the throttle valve (Figure 14.17). This sensor monitors air temperatures so that the fuel mixture can be corrected for changes in air density.

Cold start injector

Mixture enrichment for easier cold starting is provided by a cold-start injector mounted in the intake manifold. The solenoid-operated cold-start injector (Figure 14.18) is energised through the starter motor feed and the thermo-time switch so that it is operated only when the engine is cold and is being cranked.

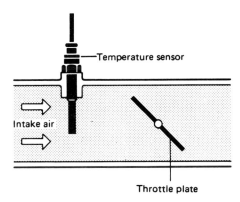

Figure 14.17   Temperature sensor in intake manifold (intake air)

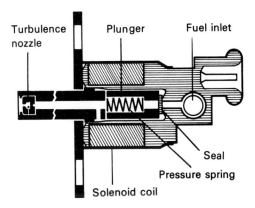

Figure 14.18   Cold-start injector

Thermo-time switch

The thermo-time switch controlling the cold-start injector is an electrically heated bi-metal switch, with heater windings and contacts (in series with cold start injector windings) connected to the starter motor feed. The switch is mounted in the engine coolant.

In cold starting conditions the switch contacts are closed when the engine is first cranked, so that the cold-start injector is energised. After a short interval the heated bi-metal strip opens the contacts to de-energise the cold start injector. The switch is maintained in the open position by the rising engine water temperature.

Extra air valve

Smooth idling in cold conditions is obtained by the provision of an extra air valve (Figure 14.19) which allows pre-determined quantities of air, under the control of cylinder head coolant temperature, to by-pass the throttle.

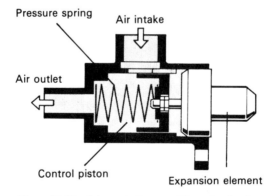

Figure 14.19   Schematic diagram of the extra air valve

The air valve comprises a profiled air inlet, an outlet port, a spring-loaded piston with a metering edge and an active element incorporating a wax capsule. The valve is connected to the air cleaner and manifold and mounted with the active element immersed in the engine coolant.

The wax capsule expands and contracts with variations in engine temperature, causing the piston's metering edge to move past the profiled inlet port, thus varying the amount of air allowed to by-pass the throttle valve. With a cold engine an increased amount of air is supplied to the manifold; the pressure sensor responds to the consequent pressure change and more fuel is supplied to the engine. At the same time the engine temperature sensor prompts the control unit to allow more fuel to be injected and an enriched mixture results.

As engine temperature rises, the valve permits correspondingly less air to enter and the pressure and temperature sensors respond to the changed conditions to reduce the amount of fuel supplied. The profiled inlet port is closed at about 70° C (158° F). When the engine is hot, the idling speed can be controlled by means of an adjustable air by-pass system.

Throttle switch

The throttle switch (Figure 14.20) provides the control unit with data on idling, over-run, acceleration and full-load running.

Indication of idling or over-run (both closed throttle positions) is supplied by wiping contacts operated by the throttle spindle; when the

Figure 14.20   Throttle position switch

throttle is within 1–2° of the closed position the switch contacts are also closed. The auxiliary distributor contacts indicate whether the engine is idling or in over-run. In the idling condition the control trims the fuel/air mixture and cuts of the exhaust gas recirculation system (when installed). If the engine is over-run the fuel supply can be temporarily cut off, re-commencing when engine speed drops below a certain level or when the accelerator is depressed.

Acceleration is indicated by the movement of a further wiping contact over a set of comb-like strip contacts. Movement of the wiping contact over the strip contacts generate additional injector pulses in the control unit. These additional pulses temporarily lengthen injector opening period to provide acceleration enrichment.

A further wiping contact, which closes after a given throttle opening, indicates to the control unit that full load conditions prevail. The control unit then effects fuel enrichment and cuts off exhaust gas recirculation (when installed).

EXHAUST EMISSION CONTROL

Additional features matched to the electronic fuel injection system to comply with US emission legislation include an exhaust gas recirculation system and a depression limiting valve.

Exhaust gas recirculation system

By recirculating a percentage of the hot exhaust gases emitted from the engine through a control valve into the inlet manifold to mix with the ingoing petrol/air mixture, combustion temperatures are reduced and the production of oxides of nitrogen is minimised.

A solenoid operated valve mounted with its outlet port on the inlet manifold has its inlet port connected to the exhaust manifold. With the solenoid de-energised the valve is open, allowing exhaust gases to recirculate.

To prevent recirculation at idling and full load, signals from the appropriate contacts in the throttle switch are relayed through an amplifier unit to trigger the solenoid and close the valve.

When the engine is cold, a bi-metal thermo-switch inserted in the engine coolant has the same effect.

Depression limiting valve

The function of this valve is to reduce the depression in the inlet manifold under rapid deceleration thereby correcting the fuel quantity to give the best possible balance between emission results and vehicle response.

The valve is connected by flexible pipes between the air filter and the inlet manifold on the engine side of the throttle. It is opened by atmospheric pressure operating against a spring when manifold pressure falls to a predetermined level.

THE BOSCH EFI-D AND EFI-L ELECTRONICALLY-CONTROLLED FUEL INJECTION SYSTEMS

These are both intermittently operating low-pressure injection systems. The principle of both systems is shown in the block diagram Figure 14.21.

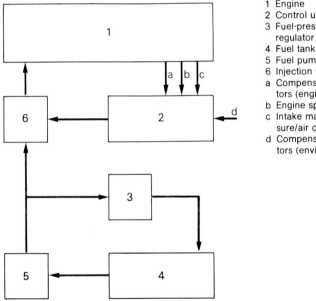

1 Engine
2 Control unit
3 Fuel-pressure regulator
4 Fuel tank
5 Fuel pump
6 Injection valves
a Compensation factors (engine)
b Engine speed
c Intake manifold pressure/air quantity
d Compensation factors (environment)

Figure 14.21   Principle of the EFI-D and EFI-L electronically controlled systems

In both systems the respresentative quantities of the fuel requirements of the engine are sensed by detecting elements and converted to electric signals which are sent to the electronic control unit. This control unit processes these signals and derives the fuel requirement of the engine from them. The control unit generates electric pulses which are sent to the injection valves; these pulses correspond to the quantity of fuel calculated. The injection valves are electro-magnetically opened and spray the amount of fuel calculated in front of the intake valves in the engine.

The systems consist of the following basic components:

Fuel system	Delivery, pressure generation, pressure regulation, cleaning, injection valves.
Detecting elements	Collect all engine operating data necessary for calculation of exact fuel metering.
Electronic control unit	Processes the data provided by the detecting elements, determines the duration of injection, controls the injection valves.

This design of the injection system offers the advantages of not only being able to operate at low injection pressure, but also of making injection of fuel possible to groups of valves at the same time, thus considerably reducing equipment costs.

When the EFI-D system is used in 4-cylinder engines, two groups of two injection valves each are formed. The valves in a given group are connected electrically in parallel and open simultaneously. The same principle applies for 6- and 8-cylinder engines (two groups of three valves and four groups of two valves respectively).

When the EFI-L system is used, all injection valves are combined into a single group. This means an additional important simplification in the injection system. In order to attain sufficient uniformity in the distribution of the fuel mixture despite this grouping, half of the amount of fuel required for one operating cycle is injected twice during each rotation of the camshaft.

Solenoid-operated injection valves

The injection valves in the EFI-L system differ from the designs used in the EFI-D system only by a smaller opening cross-section. This reduction is necessary because the fuel is injected twice per rotation of the camshaft in the EFI-L system whereas it is injected only once per camshaft rotation in the EFI-D system.

The air intake system and also the enrichment of the mixture during cold starting and the warm-up period are as already described and common to both EFI-D and EFI-L systems.

On the other hand, compensation is carried out in different ways in these two systems for the following operating conditions:

Enrichment of the mixture during acceleration.
Full-load enrichment.
Matching the mixture during overrun.
Compensation for the intake air temperature.
Compensation for the effect of altitude.

Operating principles of the two systems

In the EFI-D system, fuel injection is controlled primarily by the intake manifold pressure and engine speed.

In the intake manifold, atmospheric pressure prevails in front of the throttle valve but behind the throttle valve a lower pressure prevails which, depending on the position of this valve, is variable. In order to determine the most important item of information on engine operation, namely the engine load, this lower absolute pressure in the common intake manifold is used as a measurement quantity. The pressure in the common intake manifold is a measure for the volume of the intake air and is thus a measure of the engine load. Information on the pressure in the common intake manifold is provided by the pressure sensor.

In the EFI-L system, fuel injection is controlled by the amount of air drawn into the engine. Reduction of the noxious constituents in the exhaust gas and simplified construction of the injection system were the primary objectives pursued in the development of this system. These

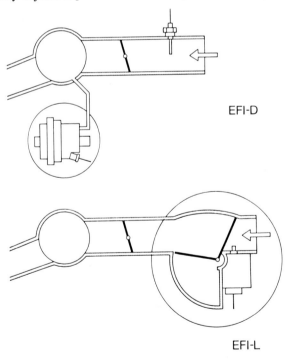

EFI-D

EFI-L

Figure 14.22   Comparison of the two systems; (top) pressure sensor; (below) air-flow sensor

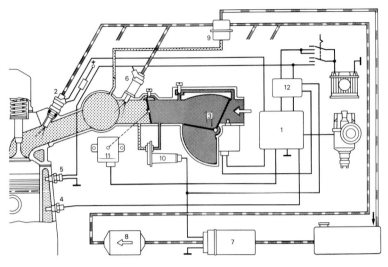

1 Electronic control unit
2 Injection valve
3 Air-flow sensor
4 Temperature sensor
5 Thermo-time switch
6 Start valve

7 Electric fuel pump
8 Fuel filter
9 Fuel-pressure regulator
10 Auxiliary-air device
11 Throttle valve switch
12 Relay set

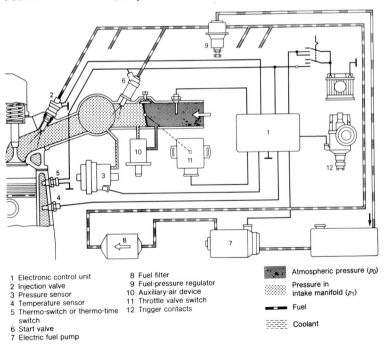

1 Electronic control unit
2 Injection valve
3 Pressure sensor
4 Temperature sensor
5 Thermo-switch or thermo-time switch
6 Start valve
7 Electric fuel pump

8 Fuel filter
9 Fuel-pressure regulator
10 Auxiliary-air device
11 Throttle valve switch
12 Trigger contacts

Atmospheric pressure (p_0)

Pressure in intake manifold (p_1)

Fuel

Coolant

Figure 14.23 Components of Bosch electronically-controlled fuel injection systems; (top) Type EF1-D; (bottom) Type EF1-L

different objectives are fulfilled by the new principle of air-flow sensing. Also, advantages result from the use of integrated circuits in the electronic control unit and by the simplified design of individual dual system components.

The amount of air drawn into the engine is measured with an air-flow sensor developed specially to meet the requirements of the vehicle. This sensor transmits the most important information for fuel metering to the control unit. For this reason, this system is designated 'air-flow sensitive' or 'EFI-L' for short because the German word for 'air' begins with the letter L. Figure 14.22 shows diagrammatically a comparison of the two measurement systems and Figure 14.23 illustrates schematically the components in the two systems.

AIR-FUEL RATIO

Proper ignition and combustion of the air-fuel mixture can only take place within a certain range of air-fuel ratios. When petrol is used, the average ratio for full combustion of the fuel (the so-called stoichiometric air-fuel ratio) is 14:1. This means that about 14 kg of air are required for complete combustion of 1 kg of fuel. In this stoichiometric air-fuel ratio, the air factor, $\lambda = 1$.

This value, λ (Lambda), is determined as follows:

$$\lambda = \frac{\text{actual volume of air drawn into engine}}{\text{theoretical requirement of air}}$$

$\lambda = 0.9$ — means a rich mixture, i.e. a deficiency of air; the intake of air is less than the theoretical requirement.

$\lambda = 1.1$ — means a lean mixture, i.e., a surplus of air; the intake of air is greater than the theoretical requirement.

Spark ignition engines develop the greatest power at a 0–10% air deficiency ($\lambda = 0.95–0.9$) and consumes the least fuel at about a 10% air surplus ($\lambda = 1.1$).

When there is an air deficiency, the fuel is not utilised adequately and the concentration of unburned noxious constituents in the exhaust gas is higher. With a surplus of air, the power developed by the engine is lower and the temperatures of the engine and exhaust gas are higher because of the slower combustion.

The air-fuel ratio of the mixture drawn into the spark ignition engine must lie between $\lambda = 0.7$ and 1.3 regardless of whether the engine is fitted with a carburettor or a fuel injection system.

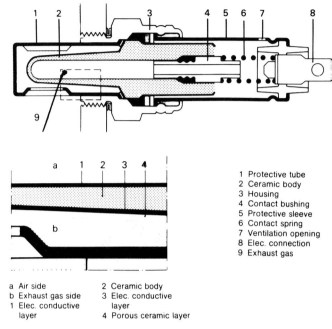

1 Protective tube
2 Ceramic body
3 Housing
4 Contact bushing
5 Protective sleeve
6 Contact spring
7 Ventilation opening
8 Elec. connection
9 Exhaust gas

a Air side
b Exhaust gas side
1 Elec. conductive
   layer

2 Ceramic body
3 Elec. conductive
   layer
4 Porous ceramic layer

Figure 14.24   Cross-section of the Lambda probe

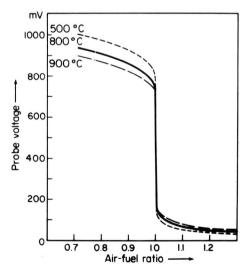

Figure 14.25   Probe output characteristics

CONTROL OF AIR-FUEL MIXTURE IN RELATION TO EXHAUST GAS CONSTITUENTS

Regulations limiting the concentration of noxious constituents of the exhaust gases emitted from internal combustion engines are constantly becoming more stringent and this dictates continually more accurate metering of the air-fuel mixture.

Bosch have developed a probe which measures the oxygen concentration in exhaust gases. The output signal from this probe is used to regulate the air-fuel mixture and as a result makes it possible, used together with special exhaust gas catalyzers, to lower the concentration of noxious constituents of the exhaust gases.

Construction of the Lambda probe

Figure 14.24 shows the construction of the Lambda probe. The ceramic body of the probe is fitted in a housing which protects the ceramic body against mechanical influences and serves for installation of the probe. The outer part of the ceramic body is positioned in the stream of exhaust gases, while the inner part is in contact with the ambient air.

The ceramic body consists basically of zirconium dioxide. Each of its surfaces (inside and outside) is coated with an electrode made of a thin layer of platinum permeable to gas. In addition, a porous ceramic layer is applied to the surface exposed to the exhaust gases. This layer protects the surface of the electrode against contamination caused by combustion residues in the exhaust gas stream and assures that the characteristics of the probe do not change. Probe output characteristics at various temperatures are shown in Figure 14.25.

Principle of operation

Operation of the Lambda probe is based on the fact that the ceramic material used becomes conductive for oxygen ions at temperatures of about $300°$ C and higher. If the concentration of oxygen inside the probe differs from that outside the probe an electrical voltage is developed between the two surfaces because of the special characteristics of the material used. This voltage is a measure for the difference in the oxygen concentration on the two sides of the probe.

The exhaust gases emitted from an internal combustion engine contain residual components of oxygen even when combustion takes place with an excess of fuel. For example, with $\lambda = 0.95$, these oxygen components amount to 0.2–0.3% by volume. The residual oxygen concentration depends greatly on the composition of the air-fuel mixture which is fed into the engine for combustion purposes.

This dependence makes it possible to use the oxygen concentration in the exhaust gas as a measure for the air-fuel ratio and thus for the air factor, λ. By suitable selection of probe materials, the highest sensitivity of the probe is in the range of $\lambda = 1$.

Operation in a vehicle

The special sensitivity of the λ probe in the range of $\lambda = 1$ makes it possible to feed the output signal from the probe as an actual value to the control unit in the EFI-D or EFI-L. As a result, it is possible to construct a closed loop (see Figure 14.26). This means that the system itself can monitor whether the specified air-fuel mixture actually results in combustion that emits exhaust gas low in noxious constituents. If the mixture deviates from the specified value, this is sensed by the λ probe on the basis of the residual oxygen concentration in the exhaust gas and this condition is communicated to the control unit in the form of an electrical signal. The control unit processes this signal to change the duration of injection and thus corrects the air-fuel mixture practically inertia free.

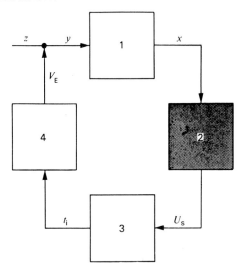

1 Controlled system (engine)
2 Sensing element (λ probe)
3 Regulator (control unit)
4 Regulating element (injection valves)
x Regulated condition (oxygen concentration in exhaust gas)
U_s Probe voltage (function of x)
t_i Regulating pulses for injection valves
V_E Amount of fuel injected
Z Disturbance factor (for example changed operating conditions)
y Air-fuel mixture

Figure 14.26 Simplified schematic block diagram of the Lambda closed loop for regulation of the air fuel mixture

Using the λ closed loop and the EFI-D or EFI-L, it is possible to achieve the degree of accuracy in regulating the composition of the air-fuel mixture which is required for effective operation of exhaust gas Catalysers for detoxication of exhaust gases.

CHAPTER 15

ELECTRICAL CONTROL IN TRANSMISSION SYSTEMS

The demand for fully automatic transmission systems for many types of vehicles is continually increasing, particularly for public service vehicles. The increase and predicted worsening of traffic density conditions impose increasing strains on drivers and any means whereby these tensions can be reduced can only result in greater efficiency and safety.

Whilst electrical control of semi-automatic and automatic gear changing systems has never been by any means universal, there are several highly successful applications in use giving fully automatic gear changing without the use of a clutch pedal, using the accelerator and brake pedals only. Careful use of the accelerator pedal and when necessary the selector lever, allows the driver to choose his own method and style of driving.

The essentials of such a transmission system consist of:

1. A torque converter.
2. A gearbox with epicyclic gear train.
3. Electrical/hydraulic or pneumatic control.

The use of a torque converter coupled to an epicyclic gear train enables a continuous and varying torque to be applied to the road wheels during starting and accelerating and enables quick gear changes to be made under load which, absorbed by the converter, give this form of transmission an immense smoothness of operation. In the case of public service vehicles ease of handling is a feature appreciated by all types of driver.

Early systems such as the Laycock de Normanville overdrive, and the CAV electrical gear change system are briefly described for their historical interest.

LAYCOCK DE NORMANVILLE OVERDRIVE

An early example of the use of the electrical system for transmission control was provided for the automatic operation of the overdrive mechanism on the Laycock de Normanville overdrive. The main components of the system were a centrifugal switch and a solenoid unit. The former was driven at a speed proportionate to car road speed, and when this reaches a predetermined value the switch contacts close. Provided that the driver elected to utilise the overdrive mechanism – by operation of a manual switch – the solenoid unit would be energised, and movement of its iron plunger would actuate the overdrive operating shaft, so effecting transition into overdrive.

The control gear also incorporates an indicator lamp which remains alight so long as the electrical circuit is operative, and an interlocking device which prevents a return to the normal forward gear when coasting with the throttle closed. The circuit diagram of a typical installation (Figure 15.1) shows how the various items of the control gear are connected.

Before transition to overdrive can be made, three conditions must first be satisfied. These are:

1. The manual switch to be closed by the driver.
2. The car to be in top gear, thereby closing the contacts of a switch associated mechanically with the gear lever.
3. The road speed to have reached a predetermined value.

When these conditions exist, current flows from the supply through top-gear switch, centrifugal switch and manual switch and via the relay coil to earth. The electromagnetic effect of the current in the relay coil caused relay contacts A and B to close.

The effect of contacts A closing is to complete the electrical circuit to the overdrive operating solenoid and warning light. The former consists of a closing coil of low resistance, a holding coil of high resistance, a switch with normally closed contacts in series with the closing coil, and a soft iron plunger linked externally to the overdrive operating shaft. When the solenoid unit is energised, both coils exert an electromagnetic effect on the soft-iron plunger. Movement of this plunger causes the change into overdrive to be made and also opens the switch contacts in series with the closing coil; the magnetic effect due to the holding coil is sufficient to keep the plunger in the overdrive position with only a small current consumption.

It is important that a return from overdrive to top gear is not made with a closed throttle and a fast road speed and the purpose of relay contacts B and the throttle switch is to prevent this from happening. The throttle switch, linked mechanically to the accelerator pedal, is arranged to be closed when the throttle opening is less than approximately one-fifth. If a period of fast driving in overdrive is followed by a period of coasting with the throttle closed, the car will deccelerate and the centrifugal switch contacts may open, but transmission will continue

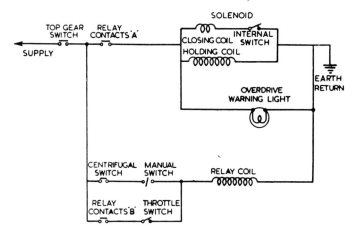

Figure 15.1   Circuit diagram of control gear for Laycock de Normanville overdrive

to be through overdrive because the throttle switch contacts will now be closed, providing an alternative path for the relay coil energising current.

If, now, the throttle is slowly opened, power will steadily and increasingly be applied from the engine to the road wheels. The centrifugal switch contacts may therefore close before the throttle switch contacts open, so that overdrive will be maintained. If, on the other hand, the throttle switch contacts open before the centrifugal switch contacts close, a return to top gear will be made until the car has attained sufficient speed to operate the centrifugal switch. At throttle openings above approximately one-fifth, overdrive can be selected or switched off at will, providing the road speed is high enough to cause the centrifugal switch to operate.

CAV FULLY AUTOMATIC ELECTRICAL GEAR CHANGE SYSTEM

This system was designed primarily for public service vehicles on busy city routes requiring some 4000 gear changes daily with the normal gearbox. It relieved the driver completely of all gear changing operation. It comprised, for pneumatic control, the following parts:

1. A gear selector unit connected by a cable to the control unit and normally mounted on the steering column.
2. An electro-pneumatic valve unit mounted on the gearbox.

3. A speed-sensitive alternator mounted on the gearbox and driven from the transmission shaft.

4. A torque-sensitive switch actuated by the accelerator pedal and taking current from the battery through the control unit.

5. The control unit which translates electrical signals from units 3 and 4 into the required gear change actions.

The gear selector could be set for manual or automatic control, and in the former case gear changing was effected as in a normal gear change operation, but through the medium of the selector switch and pneumatic valve circuit. Under automatic control, when a start is made from rest with the engine running, depressing the accelerator pedal brings into operation the torque-sensitive switch to engage the starting gear, which can be either the first or second gear ratio, depending on the starting conditions. From then onwards, all gear changes, up or down, and including the return to neutral, are completely automatic, being selected and controlled in accordance with the road speed and engine torque as indicated by electrical signals transmitted to the control unit by the speed-sensitive alternator and the torque-sensitive switch. The control unit integrates these signals on relays which actuate the gear change.

A special timing arrangement in the control unit, providing about three seconds delay in signalling throttle changes in the lower range of load conditions, eliminates unnecessary gear changes, which otherwise would occur in heavy traffic each time the load condition changes momentarily from 'normal' to 'light'. Other features are incorporated to ensure maximum safety, efficiency, and reliability.

THE RENAULT SYSTEM

An automatic transmission developed by Renault incorporating electrical controls fitted to several of their highly successful car models and in widespread use, has many noteworthy features. The complete transmission unit, on the front of the engine incorporates a hydraulic torque converter in place of the conventional clutch, a gearbox provided with epicyclic gear trains and the necessary electronic/hydraulic controls, a differential housing enclosing two reduction gears, crown wheel and pinion and the differential.

Fully automatic gear changing is provided without the use of a clutch pedal, but over-riding is obtained by the driver by the use of a selector lever mounted on the steering column. Speeds at which gear changes are made are selected using an electronic and hydraulic system in which a computer receives signals from various points including an electro-magnetic governor. A kick-down switch is linked to the accelerator pedal which automatically selects a lower gear for rapid acceleration. This switch is closed only when the accelerator pedal is floored.

LEYLAND G2 AUTOMATIC TRANSMISSION SYSTEM

An example of an automatic transmission control system for public service vehicles developed by British leyland over many years of operating and manufacturing experience from a family of multi speed epicyclic transmission systems is the Leyland G2 system. This system embodies constructional and functional features referred to earlier which have been found desirable to meet the most varied and exacting conditions.

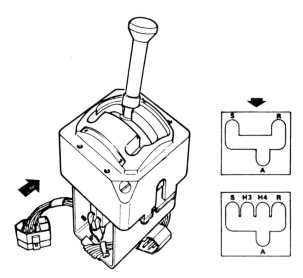

Figure 15.2   Hand-operated gear selector switch. Inset shows alternative gate positions

The Leyland G2 automatic transmission control system affords the driver fully automatic gear changing on five speed epicyclic gearboxes. He is then free to drive on the accelerator and the brake and the vehicle will always be in the correct gear for prevailing conditions.

A performance switch operated by the accelerator pedal together with the gear selector and a perception head fitted in the gearbox, combine to transmit signals to the electronic control unit which translates these to activate an electro-pneumatic valve which in turn activates the gearbox.

The following notes on the system have been completed with the assistance of British Leyland, Heavy Vehicle Division.

The principal units of the systems comprise the following:

1. Gear selector switch

This carries the driver's requirements to the system and provides a signal to a warning lamp module which warns the driver of a speed signal failure. It is a hand-operated switch as shown in Figure 15.2 with either three or five positions.

(a) *A neutral/start circuit interlock position* – S is provided, which must be selected before attempting to start the engine.
(b) *'Automatic'* – by placing lever in slot marked A.
(c) *Reverse gear* – by positioning lever in slot R.
(d) H3 – by placing lever in this slot the automatic range will be limited up to 3rd gear.
(e) H4 – this position limits the automatic range up to 4th gear.

2. Perception head or speed signal generator

This comprises a multi-toothed disc coupled to the gearbox output shaft and a flange mounted radially disposed magnetic transducer which transmits a periodic signal or frequency proportional to shaft speed (Figure 15.3).

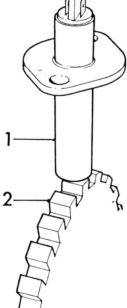

Figure 15.3   Reception head. 1. Transducer.
2. Toothed wheel

3. *Performance level or kick-down switch*

This switch ensures that the best advantage is taken from the transmission for both economy and performance and is fitted into the throttle system to modulate the gearshift pattern in accordance with the power output of the engine. A switch is located beneath the accelerator pedal which if actuated:

(a) From stationary, engages first gear, or
(b) Whilst second gear is engaged, below a pre-determined road speed, results in an immediate down-change to first gear. Furthermore, upward changes occur at higher speeds than normal — nearer maximum engine power.

4. *Electro-pneumatic valve unit*

This contains a set of six solenoid operated air valves connected by pipes to the brake-band operating cylinders in the gearbox (Figure 15.4).

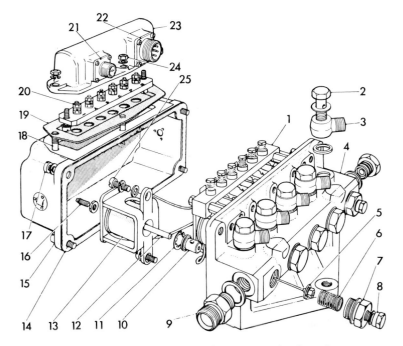

Figure 15.4   Five-way electro-pneumatic valve unit

5. *Throttle dipping valve*

When up-shifting, it is desirable to cut the engine fuel supply momentarily, thus minimising wear on the gearbox brake bands and assuring smooth gearshifts. This is achieved by closing the throttle by the fuel pump governor. An EP valve, Figure 15.5, which receives its drive signal from the translator is used for this purpose.

6. *Low air pressure protection unit*

Prevents the engagement of a starting gear until adequate air system pressure is reached (Figure 15.6).

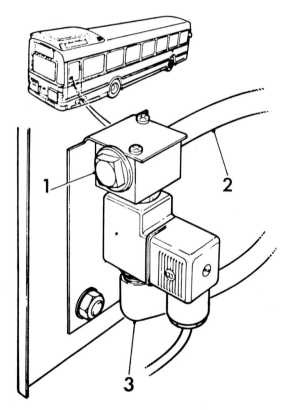

Figure 15.5   Throttle dip valve

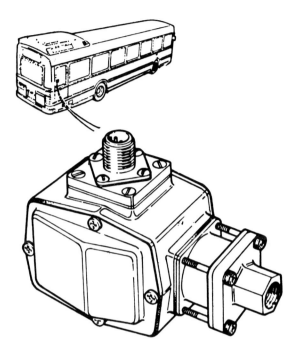

Figure 15.6   Low air-press protection switch

7. Translator

This solid state electronic and electro-magnetic unit houses the sensing and switching units circuits as shown in interior view.

The complete G2 translator analyses incoming signals from the gear selector, performance switch and perception head, subsequently energising the appropriate solenoids in the electro-pneumatic valve unit. For ease of servicing, plug-in colour coded modules are employed. These comprise potted miniaturised electronic components with printed circuitry which can instantly be removed and replaced. Thus an area of failure may be quickly located and the malfunctioning module is replaced as a whole and no attempt need be made to determine what part of the module is at fault. Reverse polarity protection is inbuilt, in addition to line fuses for protection against external circuit overloading.

The translator can be divided into two main sections, primary and secondary.

Primary

The primary consists of seven basic solid state components:

1. *Filter module* (Grey). This is a two-line network, ensuring a comparatively ripple-free supply to modules.

2. *Voltage regulator module* (Brown). Provides a smooth stabilised 18 V supply to all sensing parts of the circuit and has built-in overload protection.

3. *Frequency-voltage converter module* (Black). Converts frequency produced by the perception head into a variable d.c. voltage which is proportional to output shaft speed.

4. *Level switch module* (Yellow). These switches are voltage sensitive and turn 'on' and 'off' at pre-set but variable levels to give the required gearshift pattern and provide signals to the secondary relays.

5. *Warning switch module* (Red). The warning switch is a comparator of similar design to the level switch. Its purpose being to energise its associated relay W, should the speed signal be lost or, a rapid drop in signal frequency occur, indicating an emergency condition. The translator is inhibited from making sudden down-changes under input signal conditions and as the vehicle speed increases will make up-changes, ultimately 'locking' in the highest gear achieved.

6. *Timer module* (Blue). A two-stage timing device used for the control of throttle dip and neutral gap operations. Both periods are adjustable by external potentiometers during manufacture of panel.

7. *Inhibitor module – kickdown* (Green). A device used in conjunction with the performance level switch. When the signal from the performance switch is removed, this unit continues to provide a feed to the T relay for a number of seconds. This interval is adjusted by the value of resistor R1 selected on assembly.

8. *Diode module* (Brown). Comprises seven encapsulated diodes.

Secondary

The output stages comprise fourteen plug-in double-wound coil, four-pole twin contact change-over relays. Their clear polycarbonate covers permit relay action to be observed, obviating the need for cover removal.

OPERATION OF THE G2 AUTOMATIC CONTROL SYSTEM

The operation of this unit can be clearly seen from Figure 15.7.

Forward gears

Upon selection of 'automatic', normal depression of the accelerator pedal engages second gear (or first gear on certain vehicles) from stationary. Current is fed via terminal 5 energising relay coil D; contacts D8 and 9 isolating reverse, whilst D15 and 16 feed the primary circuit, activating the throttle dip timer TD1. Relay C is then energised, contacts 6, 7, 9 and 10 feeding the throttle dip EP valve. Simultaneously, current is fed to the contacts 8 and 9 of relay N via the contacts of the series connected relays 5S, 4S, 3S and 2S to the solenoid on the gearbox EP valve. This arrangement, i.e. the output of 5S being connected to the input of 4S and so on (cascaded), prevents the engagement of two gears simultaneously. The vehicle will then move off when the brakes are released and accelerate as the throttle is progressively opened. The output from the speed perception head is carried to the frequency/voltage converter which produces a d.c. voltage proportional to vehicle road speed. During acceleration the output voltage increases and at a pre-set level, switch LS2 triggers, operating relay 3S of which contacts 14 and 15 open whilst 15 and 16 close, thus releasing 2nd gear and engaging 3rd. At the same time, contacts 6 and 7 of relay 3S close, energising throttle dip timer TD3 which energises relay C, which again momentarily reduces the engine power output to provide optimum change characteristics.

This sequence is then repeated; at the appropriate speed, level switch relay LS3 and relay 4S operate. Contacts of 4S release third gear and engage fourth gear, a further pair energise timer TD4, dipping the throttle. In succession as the speed rises, LS4 relay disengages fourth gear and 5S switches the fifth solenoid, during which time engine power is similarly reduced.

A reduction in road speed results in a corresponding drop in output from the frequency/voltage converter, releasing in turn LS5, LS4 and LS3 switches, the secondary relays 5S, 4S and 3S following, until the starting gear is reached. At every downchange contacts 8 and 9 of each relay close, in sequence, actuating neutral gap timers NG4, NG3 and NG2 which in turn energise relay N, whose contacts open removing the supply to the solenoids. Thus a neutral period is created to permit synchronisation and ensure a smooth engagement of the lower ratio.

The throttle dip EP valve closing duration is adjustable via potentiometers PH, PK, PM and PP. Neutral gap timer NG2, NG3 and NG4 periods are adjustable via potentiometers PB, PC and PD respectively.

420

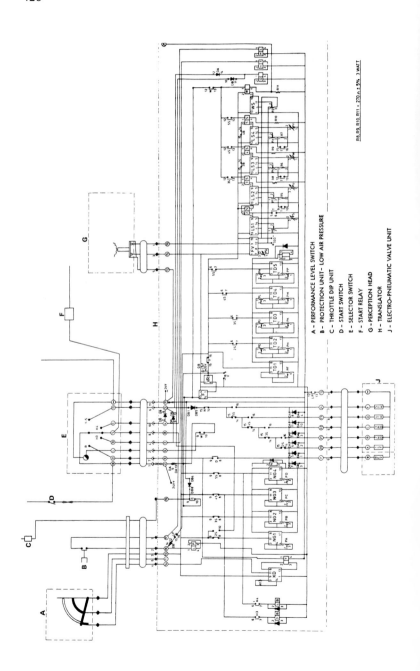

A - PERFORMANCE LEVEL SWITCH
B - PROTECTION UNIT - LOW AIR PRESSURE
C - THROTTLE DIP UNIT
D - START SWITCH
E - SELECTOR SWITCH
F - START RELAY
G - PERCEPTION HEAD
H - TRANSLATOR
J - ELECTRO-PNEUMATIC VALVE UNIT

R8,R9,R10,R11 = 270 Ω ±5% 3 WATT

Kick-down

When greater acceleration is required, as in overtaking or hill climbing the gear change can be delayed by depressing the accelerator pedal to an over-travel position to actuate a plunger operated micro-switch located beneath the pedal. Thus, up-shifts occur at faster speeds than normal, this is explained below under the heading 'Duel level speed pattern'.

Furthermore, actuation of the kick-down switch when second gear is engaged and the road speed is below a pre-determined value will result in an immediate down-change to first gear. Current is applied via terminal 14 to the green kick-down inhibitor and relay T is operated. Contacts 5 and 6 on the T relay open, releasing relay 2S and its contacts − removing the feed to second gear and applying it to first.

Dual level speed pattern

The up and down shifts between any two ratios are adjustable during panel manufacture by potentiometers PR, PT, PV, PX, PS, PU, PW and PY to suit the particular vehicle performance characteristics.

Study of the circuit, Figure 15.7, will also show that resistors R5, R6 and R7 are effective only when the associated T relay contacts 9−10, 12−13 and 15−16 are open. When the kick-down facility is used, the T relay is energised and R5, R6 and R7 are 'shorted out'.

Variations

A hold-on facility may be provided on third and fourth gears in order that use of the transmission may be made for braking when descending steep gradients, or, for elimination of gear hunting under adverse climbing conditions. Selection of H3 will limit the automatic range to third gear and H4 will limit the range to fourth gear.

This limitation is achieved by relays 3B and H3 for third gear and by relays 4B and H4 for fourth.

Selection of H3 applies current via terminal 3 to relay coil H3 of which contacts 6 and 7 close energising relay 3B, whose contacts 14 and 15 open removing the feed to relay 4S. This prevents the operation of relay 4S and therefore ratio 4 is inhibited. H4 operates in a similar manner.

The system is interlocked − if a 'hold' position was selected at a speed at which an immediate down-change would result in the engine over-speeding − a down shift will not occur. Application of the vehicle brakes would be necessary to reduce speed to which a down-change could be safely accomplished. The appropriate gears are then eliminated from the range until automatic is reselected. All gearshifts remaining in the range occur automatically.

Reverse gear

When reverse is selected, it remains engaged until manually shifted, despite any speed or throttle variation. Current is fed directly to terminal 9 and contacts 8 and 9 of relay D to reverse solenoid of the EP unit.

Safety features

In addition to protection against polarity reversals and external overloads, the following characteristics are incorporated in the system:

Signal failure

With automatic selected, the continuous engagement of a particular gear is dependent upon a steady uninterrupted output from the perception head. Should this fail, the instrument panel mounted warning lamp 'Gearbox' is illuminated and the transmission will be 'locked' into the gear prevailing — via the warning switch and relay W to pins 1 and 4 of the particular relay (this condition will occur should the rear driving wheels become locked during braking) and may only be released by isolating the system from the supply. The vehicle should be halted, neutral selected and the start noved to Off. The reset, pause and then move start switch to 'Aux' position and selector to S.

A safety interlock is provided to prevent reverse gear being engaged whilst the vehicle is moving forward at too high a speed. When reverse is selected current is fed via terminal 9 and the normally closed contacts 8 and 9 of the D relay to the reverse gear solenoid. The D relay is operated when automatic is selected; reverse cannot be engaged when the D relay is energised. Contacts 9 and 10 of relay 2S, when operated, provide a feed to the hold coil on relay D. Relay 2S energises when speeds of second gear and above are attained, thus preventing the D relay releasing until low speeds are reached.

Note that all potentiometers are bench set during translator manufacture with great precision and should not be changed or adjusted unless full consequences of such adjustments are appreciated.

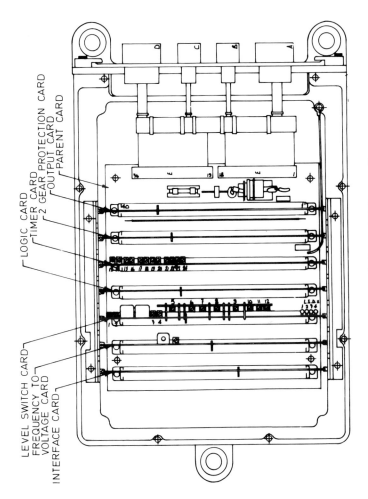

LOGIC CARD
TIMER CARD
2 GEAR PROTECTION CARD
OUTPUT CARD
PARENT CARD

LEVEL SWITCH CARD
FREQUENCY TO
VOLTAGE CARD
INTERFACE CARD

Figure 15.8   The Butec AGC5 controller

The Butec AGC5 electronic controller

A further development of the Leyland G2 automatic transmission system includes the Butec AGC5 electronic controller or translator in place of the G2 translator already described. This unit may replace the original G2 translator. The difference between the two forms of controller is that the standard G2 translator employs solid state logic in its primary circuits and electro-mechanical relays in its secondary circuits to drive the translator outputs, whereas the AGC5 translator is a wholly solid state unit.

The AGC5 unit meets the full requirements of the Leyland automatic control system for pneumocyclic gearboxes. The unit is mechanically and electrically compatible with the existing relay translator and is housed in a cast aluminium case with anti-vibration mounting, see Figure 15.8. It is of modular construction with the complete electronic sub-assembly removable from the case and comprises a parent printed circuit card with seven plug-in printed circuit cards. The complete sub-assembly forms a mechanically stiff structure which is clamped by the cover of the case.

Diagnostic facilities are provided to facilitate fault identification and location which comprises:

A speed signal simulator, selected by a push button and controlled manually. Indefinite dwell at any speed enables particular faults to be located.

Visible light emitting diode indicators on the speed level switch to aid fault location within the panel. Voltage monitoring points on all input and output signals which aid fault location within a panel or in the peripheral control system.

This modular design allows detailed changes in the control system functions to be made without system re-design.

THE CAV 488 AUTOMATIC TRANSMISSION CONTROL SYSTEM

Another example of an automatic control system for four- or five-speed air-operated direct-acting epicyclic gearboxes is the CAV 488 system which requires no effort or attention from the driver in connection with gear changing, leaving him free to drive with two pedal control – accelerator and brake.

The complete system is shown in the schematic block diagram in Figure 15.9 and the main components of the control equipment comprise the following.

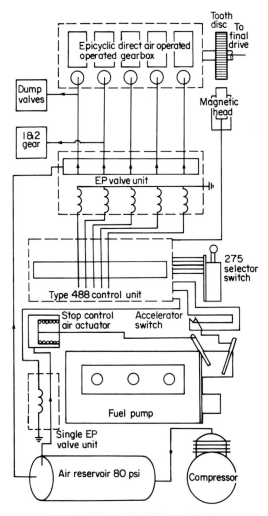

Figure 15.9   Schematic diagram of CAV Type 488 automatic gear-change control system

Control unit

This is suitable for use in a wide range of ambient temperatures, the 488 control unit is housed in a compact splash-proof container weighing approximately 3.4 kg (7.5 lb).

Separated into six functional modules for easy servicing, each module has a printed circuit base, employs silicon transistors and diodes and plugs into a printed circuit parent board. Servicing is effected by substitution of a new sub-unit. A screw-secured multi-pin plug connects the unit to the electrical system.

The fully transistorised circuit has 'built-in' safety circuits and is designed to function on a fail-safe basis, is programmed to give flexible control, smooth gear change and manual override when required. Optional features include two level gear change speed selection; inhibitor circuits to prevent automatic doors on public service vehicles opening at vehicle speeds in excess of three miles per hour; driver controlled 'lock-on' in first or second gear, which provides a hold facility under critical traffic, load or speed conditions.

Gear selector switch

The selector switch is mounted in a convenient position on or near the steering column and is operated by movement of the lever through the gate. Three gate positions are provided: one automatic, manual third and reverse. A warning light indicates neutral position of the lever and a retractable stop prevents accidental selection of reverse gear.

Toothed wheel and pick-up

The toothed wheel and magnetic pick-up are mounted in the gearbox or on the transmission final drive shaft.

As each tooth passes over the pick-up head it disturbs the magnetic field emanating from the pick-up. This disturbance is detected and the signal fed back to the control unit, whose transistorised circuitry uses this to actuate the gear shift solenoids at the predetermined speed and in the correct sequence.

Electro-pneumatic valve unit

This unit, which is mounted on the gearbox, admits compressed air to the appropriate gear actuating cylinder and at the same time releases air from the cylinder previously used. The valve block consists of a bank of five double valves, each pair of which, when actuated by the plunger of a solenoid, diverts the compressed air from a common gallery to one of five outlets corresponding to a specific gear actuator. Provision is made for filtering and removal of any moisture in the incoming compressed air.

Operation of the system

With the gear selector switch in the neutral position, no signals are generated and the system holds the gearbox in neutral. The warning light on the switch is illuminated and serves as a warning to the driver that neutral has been selected. The driver now starts up the engine and selects 'Automatic'. Whilst the vehicle is at rest, the engine will continue to idle in first gear until the accelerator is depressed, when the vehicle will move off smoothly, being already in gear.

When changing gear automatically, the throttle dip and gear delay circuits are initiated at the same instant, the delay being between the dis-engagement of one gear and the engagement of the next.

The throttle dip is only involved on 'Up' changes other than top, remaining operative for a short period after the gear delay has creased, ensuring that the incoming brake band is fully home before power is restored, thus avoiding brake band slip.

The gear delay circuits operate on upward and downward power changes, providing a momentary 'no gear' condition. This ensures a degree of synchronisation between the increasing engine speed (no load) and the incoming gear, thus promoting a smooth gear change.

All gear changes, upward or downward and the return to rest are now completely automatic, being selected and controlled by the control unit. This is programmed to operate strictly in accordance with the required road speed, as indicated by the electrical signals supplied by the toothed wheel and magnetic transducer system.

CHAPTER 16

RADIO INTERFERENCE SUPPRESSION

Interference with television and radio reception caused by various items of equipment fitted to automobiles can be a serious problem. The television screen may be covered with bright white streaks, caused by interference, completely obliterating the image on the screen. At the same time sound reception can be marred by continuous clicks corresponding to the spark frequency of the equipment on a vehicle. This interference will affect radio equipment carried in the vehicle and also television and radio equipment located at some distance from the vehicle, certainly up to a distance of 10 m from the source of interference.

Radio frequency energy is generated in the normal operation of many items of motor vehicle electrical equipment. This energy is distributed over a wide range of frequencies. It differs from a radio signal emanating from a transmitting station in that while the latter contains information, the random radio frequency interference does not. If the energy values of the transmitted signal and the random radio frequency interference are comparable, the information content of the former may be lost. To avoid this, the 'unwanted' radio frequency interference must be suppressed to such a level at which it will not adversely affect:

1. The reception of domestic radio and television and the functioning of other short wave communication apparatus not associated with the vehicle.

2. When applicable, receiving apparatus within the vehicle itself.

PROPAGATION OF INTERFERENCE

When a changing current flows in a conductor it produces around that conductor varying magnetic and electric fields. Interaction between these fields results in energy – taken from that available in the conductor – being radiated outwards with the velocity of light and at the frequency of the exciting current. If the latter is of radio frequency, then the radiation will take the form of radio frequency waves.

428

The strength of the radiated field depends upon the current flowing in the radiating conductor. The amplitude of a signal induced into a conductor or aerial in the field will depend on its distance from the radiating conductor, as well as its orientation and effective impedance. Radio frequency interference can be introduced into receivers by conduction and radiation. The conducted component either flows directly from the source through power supply leads, or is induced into the power supply leads by capacitive and mutual coupling of cables and equipment in close proximity. A cable carrying radio frequency interference currents, passing near to an aerial, might result in interference being conducted to the receiver even though the cable and aerial are not electrically connected.

Radiated noise, apart from being picked up by the receiver aerial, may also be picked up by cables and metal fittings which act in turn as radiating aerials. In some instances the value of the distributed stray capacity and inductance of a cable may result in a harmonic of the impressed radio frequency interference being generated. This may then be radiated, adding to the radio frequency interference spectrum, although it need not necessarily increase the amplitude of the noise in the frequency band under consideration.

SOURCES OF INTERFERENCE

Sudden variations or interruptions of current which generate radio frequency interference occur in the normal operation of many items of electrical equipment on the vehicle. These include the ignition system; machines having commutators or slip-ring brushgear such as d.c. generators, alternators, windscreen wiper motors and fan motors; current and voltage regulators, instrument voltage stabilisers and petrol pumps. In addition, semiconductor devices such as silicon diodes and transistors must not be overlooked, since they perform functions similar to switches or vibrating contacts and are thus liable to create radio interference. These devices are used in automobile electrical equipment in alternators, electronic output control units and electronic ignition systems.

The major source of interference on a petrol-engined motor vehicle is the ignition system. Interference is generated when the spark gaps break down and set up oscillatory currents, which result in resonance between the secondary circuit capacitance of the ignition system (including spark plug, cable, distributor) and lead inductances. This occurs as each plug fires. Energy is stored in the ignition secondary circuit capacitance of 50–100 picofarads, which is charged to a voltage in the region of 10–25 kV and this discharges through the plug and distributor gaps in an oscillatory manner for a few milliseconds, during which time the current may attain peak values of 200 A or more.

The magnitude of these oscillatory currents means that fields can be set up to cause serious interference to radio communication systems over a wide frequency range. The radio frequency energy is radiated from the high tension circuits and also from the low tension wiring due to mutual and direct coupling with the high tension system. The amplitude of the interference increases with frequency to a maximum in the region of 40–100 MHz, and then falls off, although remaining appreciable up to at least 600 MHz.

Interference from other items of vehicle electrical equipment is often troublesome to radio equipment carried on the vehicle itself or on other vehicles in close proximity. For example, some electric motors and generator control units can produce appreciable interference fields at a distance from the vehicle and can be troublesome to radio reception if left unsuppressed.

The magnitude of the radio interference field radiated from a motor vehicle depends on several other factors in addition to the item of equipment forming the source. The layout of the installation – for example, the length and positioning of the wiring – can have considerable influence on the 'noise' field. Also, the effectiveness of the screening which results from the method of construction of the vehicle body has a pronounced effect and one which is extremely variable and unpredictable, since it depends on the manner in which the various metal panels are held in contact. The electrical screening depends on how well joints are welded, or if bolted, the number and positions of bolts, whether serrated washers are used and how much insulation in the form of paint is included in the joint. The effectiveness of the body work as an electrical shield will also vary with frequency due to radio frequency resonances within the panelling arrangement.

CONTROL BY LEGISLATION OF RADIATED INTERFERENCE FROM VEHICLES

The problems associated with radiated interference from high voltage ignition systems have been studied in many countries and national and international legal standards now exist which place maximum permitted limits on the radiated field. Legislative control over the limits of radiated interference from motor vehicles is instrumental in ensuring that broadcasting, television and communications systems are not affected to any great degree by the very large number of vehicles in use today.

Early legislation varied from country to country. In the UK, the first legislation regarding the suppression requirements to be met by motor manufacturers was based on British Standard BS 833: 1953 and was passed as Statutory Instrument No. 2023 (1953) of the Wireless Telegraphy Act of 1949. It was aimed at limiting radiated interference having frequencies which affected medium and long wave domestic radio reception, short wave communications and television reception in Band 1 (40–70 MHz).

In 1964 a common international standard was recommended by a group of experts comprising a Working Party of the Comite International Special des Perturbations Radioelectriques (CISPR). The technical recommendations of the standard (i.e. CISPR Recommendation No. 18) were later adopted by Working Party 29 set up under the auspices of the United Nations Economic Commission for Europe to study constructional aspects of motor vehicles. As a result Regulation No. 10 of the E.C.E., namely 'Uniform Provisions Concerning the Approval of Vehicles with Regard to Radio Interference Suppression' was issued December 1968. The equivalent European Economic Community regulation is EEC 72/245.

EEC Regulation 10 has now been implemented by many countries, for example all the EEC countries, plus Spain, Sweden and Czechoslovakia, etc. In the UK, the British Standards Institution has issued an updated standard (BS 833: 1970) which, unlike the previous standard, includes recommendations for achieving the specified limits. New legislation (Statutory Instrument No. 1271: 1973) based on the new British Standard became law in April 1974.

The new radiated interference limits are based on quasi-peak measurements at 10 metres distance from the source and are as follows:

1. 50 μV/m in the range 40–75 MHz.
2. 50 μV/m at 75 MHz increasing linearly with frequency to 120 μV/m at 250 MHz.

This represents an extension of the previous legislation in force in the UK under the Wireless Telegraphy Act 1949 which specified only item (1).

Item (2) covers the limitation of radio interference frequencies likely to cause annoyance to reception of domestic v.h.f. radio and Band II and III television.

In addition to a vehicle manufacturer having to comply with the above limits, owners of vehicles purchased since the new legislation came into force must by law maintain the same ex-works standards of suppression.

A prototype sampling techniques and production conformity tests ensure that the performance of ignition suppression equipment fitted to 80% of a vehicle manufacturer's production is consistent and complies with the legal requirements. These procedures, which have been employed in the UK since 1970, together with the knowledge gained from the batch sampling techniques of measurement used for the previous seventeen years, have provided greater understanding of the pattern of interference from vehicle ignition systems and the subjective effects on radio communications.

METHODS OF MEETING RADIO INTERFERENCE SUPPRESSION LEGISLATION

Resistive suppression of the ignition high tension circuit is usually all that is necessary to meet the statutory requirements (see Figure 16.1). The following technical data has been kindly provided by Joseph Lucas Ltd.

1. Preferred method

The preferred method is by use of resistive high tension cable for all high voltage leads. Such cable has a core of graphite-impregnated stranded and woven rayon or silk, enclosed in extruded p.v.c. or synthetic rubber insulation. Resistive high tension cable is available in two ranges, one having a resistance value of 4000–7000 ohms/ft (13000–23000 ohms/m) and the other of 7000–16000 ohms/ft (23000–35000 ohms/m). The lower value is intended for longer high tension cable runs and the higher value for shorter runs. As a guide, it is recommended that for high tension leads up to 1 foot (30 cm) in length, the higher resistance value should be used, but above this length, the lower resistance value should be used).

A variety of end connectors enable fitment to be made to all types of ignition equipment terminations.

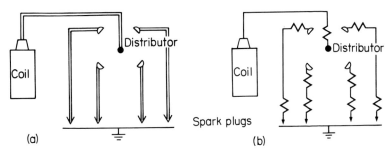

Figure 16.1   Comparison of methods for complying with legislation regarding suppression of radio interference

In the great majority of instances, no further suppression equipment is required to meet statutory requirements, so that this preferred method represents the most economic means of compliance. It is also the method used on most vehicles in current production.

In a small number of instances (for example, where fibreglass is employed in the vehicle bodywork) further suppression may be essential to meet severe ignition interference conditions and additional screened resistors at the spark plugs may be required. In extreme situations, it may also be necessary to provide resistors at each distributor high tension terminal. These resistors must be of the wire-round type which have been developed and are now available to meet the increasing need for more effective suppression at very high frequencies. The resistors exhibit low d.c. resistance but high impedance to very high frequencies.

2. Alternative method

If wire-cored high tension cable is employed, it will be necessary to fit the wire-wound resistors referred to above at all spark plugs *and* distributor high tension terminals in every instance to meet the statutory requirements. Consequently, this method is more costly in the great majority of applications.

The solid carbon resistors which have been widely employed hitherto for ignition interference suppression exhibit a reducing impedance characteristic in the higher frequency region and thus may be inadequate in meeting the new international suppression standards. Moreover, they are not suitable for use with modern push-in distributor high tension terminals, not for use with resistive cable. Consequently, while a range of such resistors may be available for service and maintenance purposes they cannot be recommended for use in new installations.

The practice of replacing resistive high tension cable by wire-cored cable in service will not result in better ignition performance and, unless appropriate precautions are also taken, will contravene the regulations regarding maintenance of the original degree of suppression.

SUPPRESSION OF INTERFERENCE WITH RADIO EQUIPMENT CARRIED IN THE VEHICLE

This consideration can most conveniently be divided into the following categories, which represent progressively greater suppression requirements:

(a) For car radio reception of long wave and medium wave signals.
(b) For reception of v.h.f. signals in the vehicle (f.m. car radio).
(c) For the use of v.h.f. radio communication apparatus in the vehicle.

As the requirements become more severe, the degree of suppression necessary to ensure satisfactory performance from the radio equipment increases. In determining acceptable suppression standards, economic considerations are naturally of considerable importance. For instance, if all vehicles could be suppressed to a standard suitable for category (c), then the other categories would be automatically catered for; this arrangement, however, would be quite out of the question economically. Consequently it is necessary to compromise by determining levels of suppression which will result in adequate performance and yet be economically acceptable.

(a) For reception of long wave and medium wave signals by a receiver carried in a vehicle

To determine whether interference originates externally within the vehicle, the following simple test may be applied. If interference persists with the vehicle stationary, engine stopped and all electrical equipment except the radio switched off, then the source is external.

If interference ceases, the source is internal and is being fed to the radio either through the power supply cables or through the surrounding case, due to inadequate screening of the set, or may be picked up directly on the receiver aerial.

1. Ignition interference with LW and MW car radio reception will to a great extent be limited by the suppression measures adopted to meet the statutory requirements. On vehicles equipped to the standard described in the preceding paragraph, any further interference with radio reception at these frequencies can normally be reduced to an acceptable value by fitting a 1 μF capacitor between the ignition coil primary supply terminal and earth (Figure 16.2).

If a car radio is being installed on a vehicle already in service having wire-cored high tension cable and carbon suppressor resistors at the centre distributor terminal (or resistive high tension brush in the distributor) and spark plugs, again fit a 1 μF capacitor as above. If ignition interference still persists, fit appropriate carbon resistors at the distributor cap outlets to each plug.

2. Interference from a d.c. generator or alternator can be recognised as a whine or growl which varies in pitch with engine speed. The remedy is to connect a 1 μF capacitor between the output terminal and the earthed frame of the generator (Figure 16.3). In the case of insulated return machines, a capacitor between each output terminal and frame should be connected. Under no circumstances must capacitors be connected to field terminals or damage to the generator output control unit may result.

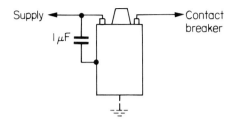

Figure 16.2 Fitting capacitor to ignition coil primary circuit

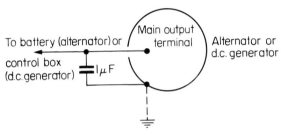

Figure 16.3 Fitting capacitor to alternator or d.c. generator

3. Interference with a car radio emanating from the generator output control unit can be recognised by periods of continuous crackle and is caused by normal regulator action. This interference can only be eliminated or reduced by electrical filtering – no attempt should ever be made to cure the interference by the indiscriminate fitting of capacitors or by tampering with the regulator contacts.

Filter units are available for the suppression of compensated voltage control units used with d.c. generators and can be connected immediately adjacent to a unit causing interference. Some manufacturers produce replacement control units with built-in suppression.

Electronic regulators used with alternators have radio frequency suppression as a built-in feature of their design.

4. The low-to-medium pitched whine from motor-driven components such as windscreen wipers, heater fans etc., can usually be reduced or eliminated by connecting a 1 μF capacitor between the insulated terminal of the unit and frame.

(b) For reception of v.h.f. signals within the vehicle

The increased popularity of the v.h.f. wavebands for entertainment broadcasting has resulted in the appearance of car radio designs which can receive frequency modulated transmissions on very high frequencies (88–108 MHz). Most transmissions in the v.h.f. wavebands are polarised horizontally, that is the electro-magnetic component of the radio wave travels in a horizontal plane. The car radio aerial on the other hand, is generally mounted vertically because of mechanical considerations, mainly ease of fixing to the vehicle body. The cross-polarisation of the receiving and transmitting aerials results in a wide variation in the received signal strength even within the service area of the broadcasting transmitter. While quite satisfactory reception is possible under these conditions, it is obvious that under conditions of weaker signals, interference from electrical apparatus on the vehicle can cause annoyance unless the correct suppression measures are adopted.

Suppression for v.h.f. receivers at acceptable cost is a more difficult proposition than for short, medium and long wave receivers. The addition of external suppression to existing equipment at v.h.f. frequencies is much less effective than in the long and medium wavebands and the conventional types of suppression capacitors already mentioned are not efficient at these higher frequencies. Furthermore, practically all items of equipment capable of causing interference on a vehicle need suppression, whereas for long and medium waveband receivers such items as screenwiper and fan motors, voltage regulators and so on need suppression treatment only occasionally. As a result, the cost of suppression for v.h.f. sets is likely to be considerably greater than for former types of receivers.

The following notes indicate the lines of approach to these problems and the suppression components which are available.

1. Mention has already been made of wire-wound resistors for inclusion in the ignition high tension circuit. These resistors exhibit low d.c. resistance but high impedance to very high frequencies. The spark plug resistor has partial screening, the screen taking the form of a metal sleeve which fits over the plug and is earthed to the plug hexagon. The capacity between resistor and sleeve forms a resistance-capacity filter, providing additional suppression over the resistor alone. The distributor suppressor is designed for use only with push-in high tension terminals.

In the case of vehicles using resistive ignition high tension cables, for v.h.f. reception it may be necessary to fit these resistors either at the spark plugs alone or also at the distributor high tension terminals. For vehicles with wire-cored high tension cables and suppressed to the new standard, these resistors will already be fitted to meet the statutory requirements.

In the case of earlier vehicles with wire-core ignition cables, for v.h.f. reception fit wire-wound resistors at the spark plugs and distributor, if necessary changing the distributor cap to one of the push-in terminal type to accommodate the resistors.

2. The usual $1.0\,\mu\text{F}$ capacitors are limited in their usefulness to relatively low frequencies on account of inherent characteristics. The ideal capacitor has a steadily decreasing impedance with increasing frequency, but in practice the conventional capacitor has appreciable series inductance by virtue of its connections and construction and its impedance decreases with rising frequency only up to the point at which the capacitive and inductive reactances are equal. This usually occurs at a frequency of a few MHz. After this the effective impedance increases due to inductive reactance and the capacitor becomes less effective.

Capacitors are available in which these undesirable characteristics have been virtually eliminated for the frequencies under consideration. These are variously described as feed-through and bushing capacitors. The main circuit current flows through a central conductor through the capacitor so that, in effect, the capacitor is wound on the main current conductor, thus eliminating external and internal leads and their unwanted inductance. Such capacitors can be made with no pronounced resonance peaks and have a steadily decreasing impedance with increasing frequency up to at least 300 MHz.

3. Filter units comprising inductors and capacitors are frequently used, but at very high frequencies where a moderate degree of suppression is required, suppression by inductors alone is often recommended.

This technique has proved particularly useful in suppressing small machines such as screenwiper motors, fan motors, etc. A most convenient form of inductor for this purpose consists of a dust-iron core with a single layer winding. The inductance value is chosen so that resonance with the unit's stray capacity is set up in the frequency band of interest. In this way attenuation of 20 to 30 decibels can be obtained.

(c) For v.h.f. radio communication apparatus in the vehicle

When a very high standard of suppression is necessary, as on vehicles carrying sensitive communication receivers, this may be provided by the vehicle manufacturer. If after leaving the manufacturer a vehicle has to be adapted to carry v.h.f. radio communication apparatus, specialist advice should be sought. The complicated nature of the installation makes it difficult to lay down a simple recommended procedure to meet the suppression requirements for this apparatus and special measures may be necessary.

(d) For radio reception within a vehicle fitted with fibreglass body panels

This is a much more difficult proposition. Much of the screening and bonding afforded by the metal panels in the conventional vehicle is absent and it is therefore often necessary to aim at a much higher standard of interference suppression. As indicated in the earlier paragraph dealing with the reception of v.h.f. signals, practically all items of equipment capable of causing interference will require suppression. In some instances, partial screening of certain components may also be necessary.

FITTING SUPPRESSION EQUIPMENT

From the previous description of the various methods of achieving suppression of radio interference, it will be evident that the source of the interference on a vehicle in service may be difficult to trace. Also when this has been located, care and attention to detail such as cleanliness and firmness of connections are essential in fitting the requisite suppression equipment. The following items in addition to the ignition equipment already referred to require particular attention:

1. *Screenwipers*
   (a) Wound field type. For LW and MW, earthing straps are required on motor body and gearbox.
   (b) Permanent magnet field type. For v.h.f., special screenwiper motors with suppression are made.
2. *Electric clocks (Continuously-driven types).* Fit close to clock body. Capacitors are fitted in shunt, chokes in series with supply line to clock.
3. *Screenwashers (Electric types).* Fit chokes close to motor terminals.
4. *Heater (Fan motors 3 A and 4.5 A).* Mount as close as possible to the motor.
5. *Instrument voltage stabiliser (10-V).* Capacitor must be fitted between B terminal and earth. Chokes in series with B and I leads.
6. *Oil pressure transducer.* In some cases, a 0.2 μF capacitor is also required in parallel with the transducer for LW and MW suppression.
7. *Petrol pumps (electric).*
   (a) Contact types. Capacitor should be fitted as close to pump body as possible.
   (b) Immersed motor driven types. VHF choke included in each pump supply lead.
8. *Radio (12- or 6-V supply line).* Mount suppressor on the radio receiver case.

CHAPTER 17

ELECTRO-MAGNETIC EDDY CURRENT BRAKES

At present braking systems independent of the road wheel brakes are not used to any appreciable extent in the UK but in other European countries and the USA they are employed quite extensively to provide additional safety on long descents in mountainous districts. In France it has been compulsory since 1954 for coaches of gross weight of 8 tons or more traversing certain mountain routes to have an auxiliary braking system or retarder of approved design, independent of the conventional braking system. Such retarders may be of the engine exhaust, electrical, or hydraulic types. In this book we are concerned only with the electrical type, which is generally of the eddy current form.

Nevertheless, it should be mentioned that the engine exhaust type in which a butterfly valve closes the exhaust outlet for braking purposes, is a relatively simple, light and effective retarder. It does, however, have certain disadvantages and, so far as heavier vehicles are concerned, the limitation of the maximum braking force to that obtainable from the engine operating as a compressor is a major one. No such limitation applies in the case of the electrical and hydraulic types, though the size and weight of the fitting influence the maximum braking force obtainable. For heavy passenger and goods vehicles, a higher maximum braking force than that obtainable from the engine exhaust type is generally necessary if the auxiliary system is to be entirely independent of conventional brakes down to speeds of 10 m.p.h. or lower. Both the electrical and hydraulic types are necessarily heavier and more expensive than the exhaust type.

The use of auxiliary retarders is by no means limited to mountainous country. They can advantageously be employed on public service vehicles on city routes with frequent stops. By using an auxiliary retarder very smooth retardation is assured and the likelihood of skidding on slippery road surfaces is minimised. The smooth braking action cuts down tyre wear and since the conventional brakes are relieved of heavy duty, being required only to bring the vehicle to rest from 5 to 10 m.p.h., brake wear is very greatly reduced. The financial saving in servicing brakes can be considerable and warrant the extra cost

439

of the retarder and any higher fuel cost associated with the additional weight of the retarder. For large public service vehicles, the extra weight carried could be in the region of 400 lb (180 kg).

PRINCIPLE OF OPERATION

For the electric type of vehicle brake, the eddy current design is generally adopted because of its constructional simplicity and relatively low cost and without friction or wear causing 'fade'. Bearing in mind the basic principle of braking is the conversion of the kinetic energy due to the forward motion of the vehicle into heat. In this design of electric retarder, its function is based on the creation of electric currents

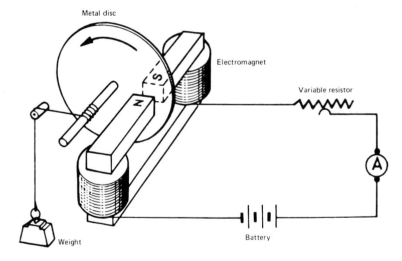

Figure 17.1   Principle of an electric brake using 'Foucault' currents

generated in a metal conductive mass when placed in a variable magnetic field. These currents which revolve around the lines of magnetic flux, called 'Foucalt' or 'Eddy' currents are created within a metal disc rotating between two electro-magnets which set up a force opposing the rotation of the discs. Figure 17.1 illustrates the functioning principle of such an arrangement. If the electro-magnet is not energised, the disc rotates freely and accelerates uniformly under the action of the weight to which its shaft is connected. When the electro-magnet is energised, the rotation of the disc is retarded and the energy absorbed appears as heating of the disc.

If the current exciting the electro-magnet is varied by a rheostat, the breaking torque varies in direct proportion to the value of the current. This principle is used in numerous practical applications, a familiar one being a house electricity meter.

DESIGN CONSIDERATIONS

In road vehicle brakes of the eddy current type, it is customary for one or both faces of an iron disc to rotate in close proximity to stationary magnetic poles arranged near the outside of the disc. These are induced magnetic poles, energised by a winding or windings supplied with

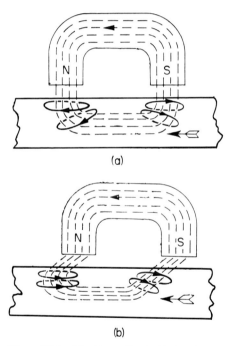

(a)

(b)

Figure 17.2   The 'drag effect' of induced eddy currents

current from the battery. They are provided with means of controlling the excitation current to several set values in order to regulate the magnetic flux and consequently the magnitude of the braking force.

Figure 17.2 illustrates the two poles of an electromagnet located near to the edge of a rotatable iron disc. Magnetic lines of force, induced by current flowing in the magnet winding would take a path through a stationary disc, as shown in diagram (a). When the disc is stationary, no current is induced in the disc, since there is no change in the number of magnetic lines linked with the particular section of the disc under each pole. When, however, the disc is rotated, flux changes occur in the sections of the disc passing the poles, the flux increasing as each section approaches a pole and diminishing as it leaves the pole. These flux changes give rise to circulatory or eddy currents in the disc, depicted approximately by the rings shown around the magnetic lines of force.

The actual pattern of the eddy currents in the mass of the disc in the vicinity of the poles, would, however, be much more complex than that shown. The combined effect of the eddy currents would be to induce N and S poles at the surface of the disc, displaced by half a pole pitch from the electromagnet poles. The eddy current induced poles would be such as to stretch out the air gap flux in the direction of rotation. In consequence, the disc would experience a 'drag' or braking effect, opposing the rotation of the disc amd the flux changes giving rise to the eddy currents.

For a simple eddy current brake employing a thin non-magnetic disc, such as copper, the drag on the disc is given by

$$P = \frac{4H^2AVt}{\rho} \times 10^{-7} \text{ gm}$$

where H = air gap magnetic field strength in oersteds.
         A = pole face area in cm^2.
         V = velocity in cm/sec of the mean radius of the disc under the poles.
         t = disc thickness in cm.
         ρ = specific resistance of the disc material at its operating temperature in microhms per centimetre cube.

Expressing the drag in pounds we have

$$P = \frac{0.88H^2AVt}{\rho} \times 10^{-9} \text{ lb}$$

In applying this formula to an arrangement such as Figure 17.1, where the disc thickness is relatively large, experience has shown that the drag is much less than the value given by the formula. Introducing an empirical constant, k, derived from test data for a particular design, approximate drag values can be obtained from

$$P = \frac{k\,0.88H^2AVt}{\rho} \times 10^{-9} \text{ lb}$$

A numerical example will serve to indicate the order of magnitude of the drag experienced by an iron disc from a pair of poles as illustrated. For this, let is be assumed that the values for the factors involved are

H = B_g = 10 000 oersteds.
A = 5 cm².
V = 4.7 × 10³ cm/sec for a disc speed of 3000 rev/min.
t = 1.0 cm.
ρ = 20 microhms per centimetre cube at an assumed iron disc temperature of 60° C.
k = 0.3.

The drag on the disc in these circumstances will be 32 lb the torque (*Pr*) assuming a mean pole radius of 6 in would be 16 ft lb and the power absorbed 9 hp at a disc speed of 3000 rev/min. With four pairs of poles uniformly spaced around the disc, 36 hp would be absorbed.

The constant k would vary for discs of different thicknesses up to about 1 cm, but not appreciably for thicker discs. The reason for this is

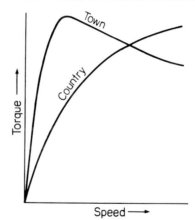

Figure 17.3 Eddy current brake characteristics

that in large circular conductors, for example, the self-induction is greatest at the centre and this results in an uneven distribution of alternating currents over the cross-section of the conductor. In effect, it is equivalent to a reduction of the cross-section or to an increase of resistance. This is known as the 'skin effect', and results in a higher current density near the surface of the conductor. It is an effect which increases rapidly with frequency. In the case of a copper conductor of diameter ¾ in, the resistance to a 100-Hz alternating current is 20% greater than that for a direct current. For a 25-Hz current, the corresponding increase in resistance is only 1%. The skin effect is much more pronounced in magnetic materials, and in the circuit under consideration, the depth of penetration of the eddy currents in the disc would not normally be more than 1 cm and may be much less at the highest operating speed. From the standpoint of eddy currents, the disc thickness need not be more than 0.5 in. Flux density considerations may, however, require a greater thickness.

The foregoing indicates that for a specific design of eddy current brake, there will be an optimum number of poles for certain power absorption characteristics. A unit with poles designed to give a high torque low down in the operating speed range may provide an appreciably lower torque at higher speeds. Whilst these characteristics may not be suitable for long descents in open country, they would be suitable for vehicles on town routes, operating wholly in speed restricted areas. Characteristic torque-speed curves for town and country service are shown in Figure 17.3.

The static flux in the magnetic circuit of an eddy current brake can be calculated with reasonable accuracy. The actual flux values under operating conditions from which braking performance may be predicted are, however, extremely difficult to assess for an initial design. It is in accurately calculating the magnitude, distribution, and overall pattern of the induced eddy currents, together with the interaction of the eddy current and fields that difficulty arises.

Considerable experience has been gained over many years in the design of eddy current couplings for marine applications as recorded in the published work of Dr. S. W. Gibbs who quotes the flux formula as follows:

$$\phi = (157\,l/\epsilon)\,(LNr\rho/n^2)^{1/3}$$

Symbol		Imbricated and salient pole	Homopolar
ϕ	= total operative air gap flux		
l		Pole length in cm	½ total primary pole length in cm is slightly less than 1.0
ϵ	= flux factor	2	
L	= load factor of secondary or disc in watts/cm²	$746\ W/\pi Dl$	$746\ W/2\pi Dl$
N		pairs of poles	number of slots
D	= dia. of air gap in cm		
r	= resistance ratio		$1 + f\lambda/l$
λ	= pole pitch in cm	$\pi D/2N$	$\pi D/N$
f	= resistivity factor depending on ϵ	0.2 approx	$0.2-0.7$
ρ	= specific resistance of secondary in microhms per cm cube		$20-27$ for steel
n	= rotor speed in rev/min		
W	= power absorption in hp at speed n		

* See 'The theory & design of eddy current slp couplings', by W. J. Gibbs, M.Sc., *B.E.A.M.A. Journals,* 53 106–108

The total excitation ampere-turns is the sum of the magnetising ampere-turns necessary to maintain the required flux in the magnetic circuit and those needed to balance the secondary load or eddy current ampere-turns. The former is obtained by the well-known procedure, and for the latter Dr. Gibbs gives the following formula

$$AT_s = 900 \, \lambda/L/\epsilon n\phi$$

EDDY CURRENT BRAKE REQUIREMENTS

An eddy current brake is an energy converter and its function is to convert the kinetic energy of a vehicle into heat and dissipate it at such a rate as to maintain the temperature of the unit within reasonable limits under maximum and prolonged braking conditions.

To meet the needs of mountainous districts, the brake should generally be capable of holding a fully laden vehicle down to a speed of about 25 mph on a steady 8% gradient, irrespective of the gradient length. This means maintaining the braking force for periods of up to half an hour without overheating. For retardation on substantially level roads, a reduction of vehicle speed from 40 to 10 mph at a maximum retardation rate of 0.125 g or slightly more is required. At this retardation rate the slowing down would be effected in just under eleven seconds and in a distance of 400 ft.

The energy absorbed by the brake is transformed into heat by the eddy currents induced in the rotor, and this heat is mainly dissipated in surrounding air through the medium of suitably designed fins on the rotating member.

The vehicle will have an inherent retardation or rolling resistance due to the combined mechanical, tyre, and air drags. Air drag is proportional to V^2 and varies for different vehicles, depending upon the frontal area and design. At the average speed given in the example, the inherent rolling resistance could account for a supplementary retardation of $0.015 \, g$. Since the inherent braking horsepower is proportional to V^3, the retardation due to rolling resistance at 40 mph would be $0.06 \, g$ approximately.

When descending hills the duty of the braking system is to counteract gravity in maintaining a constant vehicle speed. A simple formula for obtaining the power developed by a moving vehicle on a descent is

$$\text{hp} = 6 \, WV \sin \alpha$$

where $W = G.V.W.$ in tons
$V =$ speed in m.p.h.
$\alpha =$ gradient angle to the horizontal.

PRACTICAL APPLICATIONS

The Telma retarder

A very practical design of the eddy current retarder is exemplified by the Telma retarder which has been manufactured for many years and is in service throughout the world.

The first vehicle application of a retarder using the eddy current principle was made by a Frenchman Raoul Serazin produced in 1936*. This was a single-disc type rotating between two groups of coils arranged on each side of the disc. A design incorporating two discs covered by world patents was introduced in 1955. Extensive research and development work was done to determine the best form of cooling fins to provide the most efficient dissipation of the heat energy generated during braking. The concept of radial fins on the face of the disc has been developed to a logical conclusion in the form of curved fins, as in a centrifugal fan. This design is maintained in current Telma models (see Figure 17.4).

The working principle of the Telma retarder †

This retarder is a relatively primitive mechanism yet it employs complex electro-magnetic and thermal phenomena. As a result the calculation theory is mainly empirical.

In order to explain the magnetic function of a retarder, the Maxwell principles may be applied to the following physical arrangement: a ferro-magnetic disc having a permeability, μ and an electric conductivity, ρ, rotates at the face of a ring of magnetic poles of alternate polarity. Each pole produces a magnetic excitation flux, ϕ_0, which is proportional to the excitation current within the coil as long as the core is not saturated. The lines of magnetic flux, ϕ pass through the disc across the small air gap which is arranged between the discs and the poles. When the disc rotates, as a first approximation, this flux varies sinusoidally in a function of time at a given point within the disc according to the following expression:

$$\phi \doteq \phi_o \sin \frac{pN}{60} t$$

where: p = number of pairs of poles.
N = revolutions per minute of the disc.
t = time variable in seconds.

---

* Paper on 'Electro-Magnetic Eddy current Retarders' by T. E. Scharff B.S.c.
† Papers by J. M. Jollois, P. Breton & Ph. Bertrand. Translated from Revue Ingenieurs de l'Automobile, January 1975. Report on Conference on Retarders for commercial vehicles Automobile Division of the I.Mech.E. London (1974)

Figure 17.4   The Telma electro retarder mounted on a Fuller gearbox

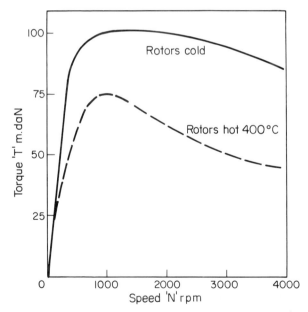

*Figure 17.5   Characteristics of torque T in relation to rev/
min, N of the Telma retarder*

Alternating eddy currents are created within the disc with a strength proportional to the flux, ϕ. The density of these currents, J, is greater in the skin of the disc than its core due to the 'skin effect'. This density in the skin of the disc also varies in proportion with speed, N. The conductivity of, ρ, of the disc material cause these eddy currents to produce heat within the disc.

In contrast to most electrical rotating machines where the torque is a function of the flux and the current, with an electric retarder the retarding torque T is a function of the flux ϕ and the eddy current density, J. The retarding torque expressed as a function of the equivalent electrical power W absorbed by the disc is $T = W/N$.

The variation in torque in relation to the speed of rotation of a conventional retarder with two discs and eight poles gives a saturation curve as shown by the solid line in Figure 17.5.

Whilst steady rotation and excitation of the retarder is maintained, the discs heat up and the torque decreases gradually reaching a stable value which is about half the torque developed when cold, as shown by dotted curve in Figure 17.5. This results from a reduction in the conductivity, ρ and permeability, μ of the discs and also from the reduction in the excitation flux since the temperature of the coils rise, to which must be included the effect of heat expansion of the air gap.

Thermal stability is achieved by means of the convection and radiation of the heat energy at high temperatures. The major part of the heat energy is imparted to the ventilating air which is circulating vigorously through the fan of the heated disc. The value of the energy, Q dissipated by the fan can be calculated by the following expression:

$$Q = MC_p \Delta_\theta$$

where: M = Mass of air circulated.
C_p = Calorific value of air.
Δ_θ = Difference in temperature between the air entering and leaving the fan.

This shows that the heat dissipation efficiency of a fan of this type will be excellent owing to the high temperature of the surface of the disc which is being cooled (in practice in the region of $400°$ C) and also because the mass flow of air through the centrifugal fan is very great. The Curie temperature of the disc material being in the region of $700°$ C, is thus never reached. Due to the intense turbulence of the air current produced by the fan, any increase in the ambient temperature around the retarder becomes negligible and has no adverse effect on the environment.

It might be expected that the mounting of the retarder in a moving vehicle would cause some effect on the ventilating air flow through the retarder. In fact the practical location of the retarder within the vehicle prevents the direct impingement of air on the retarder caused by the motion of the vehicle. Any air flow movement within the chassis of the vehicle is found to have a relatively insignificant effect on the air flow and hence the temperature of both front and rear discs will be practically identical.

The Focal type retarder

The Telma Focal retarder is similar to the CA type retarder described above, except that it has no transmission shaft and bearing assembly. The cross-section of this retarder is shown in Figure 17.6.

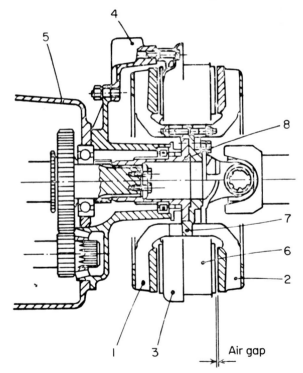

Figure 17.6   A gearbox mounted focal-type retarder

1. *Front rotor*	5. *Axle housing*
2. *Rear rotor*	6. *Coils*
3. *Stator housing*	7. *Adapter*
4. *Stator support*	8. *Pinion flange*

The stator 3 attached rigidly to the axle housing or to the gearbox housing 5 through an intermediate support 4 holds eight induction coils 6 energised separately in groups of two. These coils are made up of varnished copper wire moulded in epoxy resin.

The rotor is made up of two discs, 1 and 2 which differ from those of the CA type of retarder by the form of their radial arms, which are brought back to the centre of the assembly on to a central coupling ring. The adaptor 7, coupled directly to the pinion driving flange of the axle or to the output driving flange of the gearbox 8, supports the rotor and provides a connection to the propeller shaft of the vehicle.

The air gaps are adjusted by shims on the stator mounting and on the rotor hub.

The retarder is energised by a hand control mounted with finger tip access to the driver of the vehicle. This control has five positions, the first is 'off' and the four remaining positions increase the braking power in sequence. This hand-control system can be replaced by an automatic type which can operate either mechanically through the brake pedal or in the braking mechanism or alternatively by means of a pneumatic control from the air-brake pressure system. In these cases the contacts are switched on successively over the initial motion of the brake pedal before engagement of the foundation brakes.

The use of an automatic control must be coupled with a cut-off system operating at very low vehicle speed. This is needed to prevent unnecessary battery drainage when the vehicle is at rest.

Both the manual control and the automatic control activate four solenoid contactors in the relay box which in turn close the four groups of coil circuits within the retarder at either 24 V or 12 V as appropriate.

The maximum electric power demanded by the four groups of circuits within the retarder is approximately 2.5 kW (i.e. 100 A at 24 V). However, the average power usage is, in practice, considerably less than this. Independent tests carried out by the UTAC Organisation in France found that the average current consumption of a retarder lies between 10 and 16 Ah depending on the vehicle weight and the route profile.

In the Telma 'Focal' design, often known as the 'Space Saver', the electro-magnetic unit has been combined with the gearbox or rear axle, resulting in considerable savings in weight on vehicles with very short transmission shafts and rear engined buses. This arrangement has been achieved by close co-operation with vehicle designers.

APPENDIX

METRICATION

SI (Systeme International des Unites) Equivalents for British Units

PRIMARY AND SECONDARY QUANTITIES

The International System of Units (SI) is based initially on six primary quantities, i.e.

Quantity	S.I. unit	Abbreviation
Length	metre	m
Mass	kilogramme	kg
Time	second	s
Electric current	ampere	A
Temperature	degree Kelvin	K
Luminous intensity	candela	cd

In addition, the following secondary and derived units are employed in the system as shown below.

Plane angle	radian	rad
Area	square metre	m^2
Volume	cubic metre	m^3
Frequency	hertz	Hz
Density	kilogram/cubic metre	kg/m^3
Speed	metre/second	m/s
Force	newton	N
Pressure	newton/square metre	N/m^2
Work, energy, or quantity of heat	joule	J
Power	watt	W

451

CONVERSION FACTORS FOR STANDARD UNITS

British unit		SI equivalent
Length		
1 mile	=	1.609 34 km
1 yd	=	0.9144 m
1 ft	=	0.3048 m
1 inch	=	2.54 cm
1 thou	=	2.54 μm
1 μin	=	0.0254 μm
Area		
1 ft²	=	0.092 903 m²
1 in²	=	6.4516 cm²
Volume		
1 ft³	=	28.3168 dm³
1 in³	=	16.3871 cm³
Modules of sections (2nd moment of area)		
1 ft⁴	=	86.3097 dm⁴
1 in⁴	=	41.6231 dm⁴
Velocity		
1 mph	=	1.609 34 km/h
1 ft/s	=	0.3048 m/s
1 in/s	=	2.54 cm/s
Acceleration		
1 ft/s²	=	0.348 m/s²
Mass		
1 ton	=	1016.05 kg
1 lb	=	0.453 592 37 kg
1 oz	=	28.3495 g
Mass/unit area		
1 lb/ft²	=	4.882 43 kg/m²
1 lb/in²	=	70.307 g/m²
Pressure		
1 ton f/in²	=	15.4443 MN/m²
1 lb f/in²	=	6894.76 N/M²
Density		
1 lb/ft³	=	16.0185 kg/m³
1 lb/in³	=	27.6799 g/cm³
Moment of inertia		
1 lb ft²	=	0.042 140 1 kgm²
1 lb in²	=	2.926 40 kg cm²
Momentum		
1 lb ft/s	=	0.138 255 kg m/s
Force		
1 ton f	=	9964.02 N
1 lb f	=	4.448 22 N
Torque		
1 lbf ft	=	1.355 82 Nm
1 lbf in	=	0.112 985 Nm
Energy		
1 Therm	=	105.506 MJ
1 hph	=	2.584 52 MJ
1 Btu	=	1.055 06 kJ
1 ft lbf	=	1.355 82 J
1 ft pdl	=	0.042 140 1 J

Power
1 hp = 745.7 W (J/s)
Heat flow rate
1 Btu/h = 0.293 071 J/s or W
Work (energy); Power (see also table on next page)

Conversion of SI units to British units		
Length		
1 km	=	0.621 371 mile
1 m	=	1.093 61 yard
1 dm	=	0.328 084 ft
1 cm	=	0.393 701 in
1 mm	=	0.039 370 1 in
1 μm	=	39.3701 μin
Area		
1 m²	=	1.195 99 yd²
1 cm²	=	0.155 000 in²
Volume		
1 m³	=	1.307 95 yd³
1 dm³	=	0.035 314 7 ft³
1 cm³	=	0.061 023 7 in³
1 litre	=	0.220 gallon
Velocity		
1 km/h	=	0.621 371 mph
1 m/s	=	3.280 84 ft/s
Acceleration		
1 m/s²	=	3.280 84 ft/s²
Mass		
1 kg	=	2.204 62 lb
1 g	=	0.035 274 oz
Density		
1 kg/m³	=	0.062 428 0 lb/ft³
1 g/cm³	=	0.036 127 3 lb/in³
Force		
1 N	=	0.224 809 lbf
	=	7.233 01 pdl
Torque		
1 N	=	0.737 562 lbf ft
Pressure		
1 N/m²	=	0.000 145 038 lbf/in²
1 kN/m²	=	20.8854 lbf/ft²
	=	0.295 300 in Hg
(see also Table on next page)		
Energy		
1 J	=	0.737 562 ft lbf
1 kJ	=	0.277 778 Wh
(see also table on next page)		

CONVERSION FACTORS (*Continued*)

Work or energy	J	kJ	kWh	Nm	HPh
1 J(Ws)	1	0.001	$2.78 \cdot 10^{-7}$	1	$3.77 \cdot 10^{-7}$
1 kJ	1000	1	$2.78 \cdot 10^{-4}$	1000	$3.77 \cdot 10^{-4}$
1 kWh	$3.6 \cdot 10^6$	$3.6 \cdot 10^3$	1	$3.6 \cdot 10^6$	1.36
1 Nm	1	0.001	$2.78 \cdot 10^{-7}$	1	$3.77 \cdot 10^{-7}$
1 HPh	$2.65 \cdot 10^6$	2650	0.736	$2.65 \cdot 10^6$	1

Power	W	kW	HP	kcal/h
1 W (J/s, Nm/s)	1	0.001	$1.36 \cdot 10^{-3}$	0.860
1 kW	1000	1	1.36	860
	736	0.736	1	632
1 HP	1.16	$1.16 \cdot 10^{-3}$	$1.58 \cdot 10^{-3}$	1
1 kcal/h		(= 0.00116)	(= 0.00458)	

or 1 MJ = 10^3 kJ = 10^6 J.

or 1 MJ = 1000 kJ = 1 000 000 J

Pressure

1 bar = 10^5 Nm^{-2} = 1.019 at = 14.50 psi;

1 at = 1 kp/cm^{-2} = 1 kgf·cm^{-2} = 0.980665 bar;

1 psi = 0.06984 bar

Statutory Regulations

There are two important statutory instruments concerned with automobile electrical equipment which must be complied with for vehicles to be used on public roads in the United Kingdom. These are:

Road Traffic S.I. 1971 No. 694. The Road Vehicles Lighting Regulations 1971. (As amended).
Road Traffic S.I. 1973 No. 24. The Motor Vehicles (Construction and Use) Regulations 1973. (As amended).

The design, dimensions and performance of many electrical components are specified in British Standards.

European regulations

There are a number of European regulations affecting automobile electrical equipment, both in being and also in the process of preparation. These are published as directives in the official journal of the European Communities after consultation with the UN Economic Commission for Europe and the International Organisation for Standardization.

Directive 78/316/EEC dated 21st December 1977 is of particular interest and relates to the interior fittings of motor vehicles (identification of controls, tell-tales and indicators).

INDEX

A.C. ignition, motor cycles, 139–141
Actuating mechanisms, remote control, 339
Actuator, solenoid operated, 319–322
Aerials, 317–319
Air conditioning, 46, 367–374
  automatic, 369–374
Air-cored instruments, 72–75
Air flow, 378
Air-fuel ratio, 405–408
Alloys, 15, 16
Alternator, 89–114
  claw pole, 90–93
  construction of, 93–96
  installation of, 99
  location of, 99
  motor cycles, 132–141
  mounting of, 99–101
  noise, 102
  single pole, 103, 104
  stress on, 101, 102
Amplifier module, 279
Ancillary equipment, 233
Anti-dazzle devices, 219
Anti-theft devices, 327–331
Armature brake, starter motor, 167, 168, 180
Armature current, 151
Automatic cut-out, 122, 123
Automatic dipping, headlamps, 228
'Autosense' test system, 325, 326
Auxiliary lamps, 228, 229

Band theory of solids, 29, 30
Batteries, 183–205
  cable connections, 205

Batteries, (cont.)
  cadmium test, 189, 190
  capacity of, 187, 188
  case, 193
  cell connectors, 195
  charging, 202, 203
  coil ignition, 41
  condition indicator, 72, 73
  construction of, 190–192
  efficiency of, 188
  electrolytes, 196, 197
  installation of, 203
  lead-acid, 183–198
  location of, 204
  motor cycle, 205
  mounting, 205
  principles of, 184–186
  service of, 197
  steel-alkaline, 198–200
  testing, 188
  zinc-air, 201, 202
Beacons, rotating, 248
Beam intensity, 213, 214
Bifocal bulbs, 220
Bimetal switches, 345–352
Booster gap, 305
Boot light, 231
Bosch fuel injection system, 379–382, 401–404
Bosch sparking plugs, 304
Brakes, 439–450
Brightness, 212
British Standards, lighting, 248
Brushes, commutator, 118, 119
Bulb failure, direction indicators, 235–248

Cables, 360–362
Cadmium test, battery, 189, 190
Capacitor discharge ignition system
    (CDI), 143, 145, 267–269,
    283–286
Capacity (batteries), 187, 188
Cell connectors, 195
Centrifugal advance mechanism, 270
Ceramic insulation, 28
Ceramic permanent magnets, 16, 17
Charge indicator light, battery, 96–98
Charging, battery, 202, 203
Circuit breakers, 348–351
Circuit protection, 360
Claw pole alternator, 90–93
Clocks,
    electric, 79, 80
    quartz, 80–82
Coercive force (magnets), 14
Coil ignition system, 41, 253–273
Combustion requirements, 392
Commutator, 117
Conductor resistance, 4, 5
Conductors (electricity) 2, 8
Connectors, 360
Contact-breaker, 50, 265
    operation, 254–260
Coolant temperature indicators, 69–71
Cooling fans, 52
Corrosion of plugs, 297
Cruise control equipment, 306–311
    control switches, 310
    electronic controller, 309
    operation, 310
Current leakage, 297

Darlington pair, 108–109
Dazzle, headlights, 217–221
Demeshing, 166, 172, 179
Depression limiting valve, 400
Diagnostic equipment, 324–327
Diodes, 105, 106
Dipping, automatic, 228
Direction indicators, 235–248
    bulb failure, 242
    operation, 242
Distributors, 277, 278
Door gear, 86, 87
Drive engaging mechanism, 161
Driving media (alternators), 101
Dual level monitoring, 76, 77
Dwell angle, 266
Dynastart, 182

Eddy current brakes, 439–451
Efficiency, 153, 188
Electrical insulation, 23
Electric clocks, 79, 80
Electric fuel pumps, 48–51, 383
Electricity, nature of, 1
Electrodes, 297, 298
Electroluminescent panel, 78
Electrolysis, 186, 187
Electrolytes, 196
Electromagnetic eddy current brakes,
    439–450
    applications, 446
    operation, 440
    requirements, 445
Electromagnetic force production, law
    of, 22
Electromagnetic induction, 20, 89, 90
Electromagnetism, 18–20
Electronic hour meter, 82
Electronic ignition, 41, 42, 274–280
    installation, 281
    operation, 280
    timing, 282
Electronic regulators, 104
Emergency ignition (motor cycle), 138
European headlamp, 222, 223
Exhaust emission control, 400
Exhaust gas circulation system, 400

Flasher units, 235–248
    installation, 245
    Simms, 245
Fluorescent lighting, 232, 233
Flywheel magneto, 290, 291
Focal type retarder, 449–450
Focusing, 217
Four-lamp system, headlamps, 227
Fuel injection equipment, 379–408
    Bosch system, 379–382
    cold start injector, 397
    combustion requirements, 392
    control circuit, 391
    electric fuel pump, 383
    electronic control, 389–395
    extra air valve, 398
    fuel circuit, 391
    start valve, 380
    thermo-time switch, 387–398
    throttle switch, 399
    warm-up regulator, 384
    wiring circuit, 382
Fuel level indicators, 65–68

456

Fuel level transmitter, 66
Fuel pumps, 48–51

Generators, 39, 115–129
    automatic cut-out, 122,123
    output regulation, 121
    simple, 115, 116
    sine-wave induction, 116
    wound field, 119, 120

Headlamps, 44, 206–228
    automatic dipping, 228
    beam intensity, 214
    construction, 221
    dazzle, 217, 218
    European, 222
    four-lamp system, 227
    gas-filled, 207
    safety, change-over circuit, 351
    setting, 220
    tungsten halogen, 207
    two-bulb construction, 225
    wipers/washers, 85
Heating, 46, 365–369
    rear window, 365, 368
Heavy conductors, 148
Horns, 54
    air, 56
    windtone, 55

Ignition, process of, 252
Ignition coil, 260–263
Ignition distributor, 263–265, 395
Ignition systems, 39, 40, 251–287
    battery coil, 41, 253–273
    electronic, 41, 274–287
    magneto, 40, 41, 288–291
Illumination intensity, 211
Impulse tachometer, 58
Indicators, 71–77, 98
Inductive ignition system, 253
Inertia drives, 161
'Inertial Alarm' system, 328
Inertia switches, 343, 344
Injection valves, 400–402
Instrumentation, 56–58
Insulation coverings, 23, 24, 28
    moulded, 26
    sheet,26
Interior lighting, 231
Intermediate transmission, 181, 182
Ionisation, 42, 43

Lambda probe, 407, 408
Laycock de Normanville overdrive,
    409–411
Lead-acid battery, 183–185
Legislation on radio interference, 282,
    428–438
Leyland G2 transmission system,
    413–424
Light formation, 216
Lighting, 44, 206–234
    auxiliary lamps, 228
    brightness, 212
    focusing, 217
    headlamps, 44, 206–228
    headlamp dazzle, 217
    intensity of, 209, 210
    interior, 231
    rear lamps, 230
    reflector theory, 215
    side lamps, 229
Loom construction, wiring, 356
Lucas 'Opus' electronic ignition
    system, 283
Luminous flux, 211

Machine sensing regulators, 108–110
Magnetic lines of force, 10, 11
Magnetism, 9, 10
Magneto ignition, 40, 41, 288–291
    development of, 288
    flywheel, 290, 291
Magnets
    permanent, 12
    sintered, 16
    steels for, 14
Maintenance of alternators and genera-
    tors, 102
Meshing, starter motor pinion, 165,
    166, 169–172, 177, 178
Misfiring, 298
Motor cycles, 129–143
    AC ignition, 140, 141
    alternator for, 132–135
    battery, 205
    emergency ignition, 138
    rectifier, 131
    single-cylinder, 135
    stator, 130
    timing considerations, 141–143
    twin-cylinder, 135–137
Moving coil milliammeter, 62, 63
Motor-in-wheel device, 331
Moulded insulation covering, 26
Multi-disc clutch, 173, 174

Noise, alternator, 102

Odometer, 60, 65
Ohm's Law, 2
Opto-electronic ignition, 287
Overdrive, laycock de Normanville,
    409–411
Overrunning clutch, 166–167

Permanent magnets, 12, 13
    ceramic, 16
    developement of, 13
Pick-up module (transistorised igni-
    tion), 279
Pinion, 161
Planar processes, 109
Polarity, 102
Polytetrafluoroethylene, 27
Pre-ignition, 297
Pressure indicating systems, 71
Pressure sensor, 396
Pressure switches, 340
Printed circuits, 356, 357

Quartz clock, 80–82

Radio aerials, 317–319
Radio interference suppression, 282,
    428–438
    control by legislation, 430, 431
    fitting suppression equipment, 438
    method of, 432
    propagation of, 428
    sources of, 429
Radio receivers, 317
Rear lamps, 230, 249
Rear window heating, 365, 366
Rectifier diodes, 31, 32
Rectifiers, motor cycles, 131
Reflector theory, 215, 216
Refrigeration system, 374–376
Regulators, 104–114, 249
    machine sensing, 108–110
Remanence (magnets), 14
Renault transmission system, 412
Resistances, 4, 5, 6
    parallel, 8
    series, 8
Retarders, 446–450

Reverse battery, voltage, protection
    against, 322
Reversing lamps, 249, 250
Roller clutch model, 160

Selmar alarm unit, 327
Semi-conductor devices, 31, 109, 110
Separators (batteries),194
Series motor, 152
Sheet insulation covering, 26
Short circuiting sparking plugs, pre-
    vention of, 298
Side-lamps, 229
Signalling, 234–250
    direction indicators, 234
    flasher units, 235–248
Silver-zinc accumulator, 201
Simms flasher unit, 245
Sine-wave induction, 116, 117
Single-cylinder motor cycle, 135
Single-pole alternator, 103, 104
Sintering process, 16
Solenoid, 160, 341, 342
Solenoid-operated actuator, 319–322
Solenoid-operated injection valves, 402
Solid state devices, 29
Solid state instrumentation, 77, 78
Spark advance mechanisms, 269
Sparking plugs, 42, 43, 292–306
    auxiliary gap, 300,301
    booster gap, 305
    construction, 302
    durability, 296
    design, 43, 292–295, 299
    gas-tightness, 296
    requirements, 295
    sparking voltage, 301
    tapered seat, 303, 304
Speedometers, 56, 57
    electronic, 60, 64, 65
Starting motor, 44, 148–182
    construction, 157, 158
    Dynastart, 182
    efficiency, 153
    heavy conductors for, 148, 149
    inertia drives, 161
    intermediate transmission, 181, 182
    load, 151
    meshing, 165–166, 169–172, 177,
        178
    motor cycles, 130
    performance, 157, 158

Starting motor, (cont.)
  pre-engaged drives, 160–165
  size of, 154
  sliding armature, 168
  sliding gear, 175
  speed, 149–152
  types of, 155–157
Steel-alkaline battery, 198–200
Suppression, radio interference, 282,
  428–438
Switches, 332–352
  actuating mechanisms, 336
  bi-metal, 345–352
  design of, 333–336
  electromagnetic devices, 341–343
  gears, 332–352
  inertia, 343
  multi-function column, 337
  pressure, 340
  relays, 341
  remote mechanisms, 339
  thermostats, 345–348
Synthetic compound materials, 27

Tachographs, 311–317
Tachometers, 58, 59, 283
Tape recorders, 317
Tapered seat sparking plug, 303, 304
Telma retarder, 446–450
Temperature
  coefficients, 5, 6
  rises, 8
  sensing, 113, 114, 351, 396
Terminal posts, battery, 195
Testing and diagnostic equipment,
  324–327
Testmeters, 324
Thermoswitches, 180, 181
Throttle switch, 399
Timing, electronic ignition, 282
Timing rotor, 277
Torque, 149–152
Trailer lighting, 231

Transient voltage, 322
Transistorised coil ignition system, 277
Transistors, 33–35, 104–106
Transmission systems, automatic,
  409–427
  CAV type 488, 424–427
  Laycock de Normanville overdrive,
  409–411
  Leyland G2, 413–424
  Renault, 412
Tungsten halogen lamps, 207, 208
Tungsten-wire lamps, 207
Twin-cylinder motor cycle, 135–137

Vacuum ignition timing control,
  271–273
Vehicle speed, 62, 63
Ventilating, 365–369
Voltage fluctuation, 117, 118
Voltage regulation, 106–108

Warm-up regulator, 384–386
Warning devices, 54, 55
  horn, 54
  lights, 77
Windings,8
Windscreen washers, 85, 86
Windscreen wipers, 82–85
Wintone horn, 55
Wipers/washers, headlamp, 85
Wiring colour code, 362
Wiring harnesses, 353–364
  cables, 360–362
  circuit protection, 360
  loom construction, 356
  practical applications, 354–356
  printed circuits, 356, 357

Zener diodes, 32, 33, 105, 106, 131,
  143–145
Zinc-air battery, 201